Environmental Coastal Regions

SECOND INTERNATIONAL CONFERENCE ON ENVIRONMENTAL COASTAL REGIONS

CONFERENCE DIRECTOR

C.A. Brebbia
Wessex Institute of Technology

INTERNATIONAL SCIENTIFIC ADVISORY COMMITTEE

R. Bornstein	P. Latinopoulos
C. Borrego	J. Longhurst
A.H-D. Cheng	K. Mizimura
N.F. Ebecken	N. Moussiopoulos
R.A. Falconer	G. Pinder
P. Holz	R. Rajar
G. Ibarra	T. Tirabassi
J.G. Kretzschmar	A.T. Williams

Organised by:
Wessex Institute of Technology, UK

Thanks are due to K. Larm and B. Edge for the use of figure 2 on page 421 which appears on the front cover of this book

Environmental Coastal Regions

EDITOR:

C.A. Brebbia
Wessex Institute of Technology

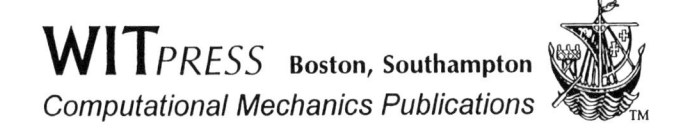

WITPRESS Boston, Southampton
Computational Mechanics Publications

C.A. Brebbia
Wessex Institute of Technology
Ashurst Lodge
Ashurst
Southampton SO40 7AA, UK
Email: carlos@wessex.ac.uk

Published by

Computational Mechanics Publications
Ashurst Lodge, Ashurst, Southampton, SO40 7AA, UK
Tel: 44(0)1703 293223; Fax: 44(0)1703 292853
E-Mail: cmp@cmp.co.uk
WWW: http://www.cmp.co.uk

For USA, Canada and Mexico

Computational Mechanics Inc
25 Bridge Street, Billerica, MA 01821, USA
Tel: 978 667 5841; Fax: 978 667 7582
E-Mail: cmina@ix.netcom.com

British Library Cataloguing-in-Publication Data

A Catalogue record for this book is available from the British Library

ISBN 1 85312 527 X Computational Mechanics Publications/WIT Press, Southampton
ISSN 1462-6098

The texts of the various papers in this volume were set individually by the authors or under their supervision. Only minor corrections to the text may have been carried out by the publisher.

No responsibility is assumed by the Publisher, the Editors and Authors for any injury and/or damage to persons or property as a matter of products liability, negligence or otherwise, or from any use or operation of any methods, products, instructions or ideas contained in the material herein.

© Computational Mechanics Publications/WIT Press 1998

Printed and bound in Great Britain by Antony Rowe Ltd, Wiltshire

All rights reserved. No part of this publication may be reproduced, stored in a retrieval system, or transmitted in any form or by any means, electronic, mechanical, photocopying, recording, or otherwise, without the prior written permission of the Publisher.

Preface

The idea of organising the first conference on Coastal Environment originated from the concern over the State of Rio de Janeiro to protect the environmental quality of its coastal region. The region was under the threat of rapid development, particularly its coastline, and an integrated way of looking at the problem was required. This resulted in the organisation of the first Coastal Environmental Meeting in 1996. The Conference provided a forum for the discussion of a range of air, sea and soil contamination problems and a better understanding of the type of remedial action required. The meeting focussed on the development of computer models for coastal regions and in particular those that could reproduce not only normal behaviour but also extreme conditions. Computer models are a very efficient tool to simulate the behaviour of the air and water masses and this leads to the possibility of using them as tools for a rapid response in the case of an environmental disaster.

The success of the first Conference jointly organised with the Federal University of Rio de Janeiro led to the decision to reconvene the meeting, resulting in this book which contains some of the papers presented at the 2nd International Conference on Environmental Problems in Coastal Regions, held in Cancun, Mexico. The meeting attracted a substantial number of delegates from many different countries. Their papers have been classified into ten different sections:

* Environmental Impact Assessment
* Coastal Management
* Remote Sensing
* Atmospheric Pollution
* Water Pollution
* Pollution Transport and Dispersion
* Hydrodynamic Pollutant Transport Modelling
* Coastal Structures
* Groundwater Studies
* Siltation and Dredging

The high quality of the work published in this book is a reflection of the success of the Conference. This important function deserves to be widely recognised by the scientific community. The Editor is very grateful in this regard to the members of the International Scientific Organising Committee who participated in the review of abstracts and papers.

The Editor
July 1998

CONTENTS

Section 1: Environmental Impact Assessment

Rare and endangered vascular plants of North Carolina's Outer Banks, Emerald Island to Virginia : a case for conservation and preservation
R. Stalter & E. Lamont — 3

The impact of mercury mining on the Gulf of Trieste
M. Horvat, S. Covelli, J. Faganeli, M. Logar, V. Mandić, R. Planinc, R. Rajar, A. Širca & D. Žagar — 11

Environmental impact assessments and legislation - Jamaica's way to protect its coastal zone
A. Haiduk — 21

A new EIA method applied on coastal reclamation projects
K. Jensen, H.K. Bach & C.M.R. Pastakia — 33

Using ecological models to plan and monitor coastal reclamation projects
K. Jensen, H.K. Bach & A. Jensen — 45

An expert system for marine environmental monitoring in the Florida Keys National Marine Sanctuary and Florida Bay
J.C. Hendee — 57

Risk assessment due to water rise of the Caspian Sea coastal area
A. Ardeshir, A. Taher shamsi & M. Fathi — 67

A two-case study on the environmentally-induced damage to materials in marine environments - part 1: Metallic materials
A.M.G. Pacheco & A.M. Maurício — 77

A two-case study on the environmentally-induced damage to materials in marine environments - part 2: Geomaterials
A. Maurício & A.M.G. Pacheco — 87

Section 2: Coastal Management

Shoreline erosion management program for Rosarito Beach, Baja California, Mexico
C.M. Appendini, R. Lizárraga-Arciniega & D.W. Fischer — 99

A digital elevation model of sections of the UK intertidal zone from the integration of satellite imagery and ocean height modelling
I.J. Davenport, D.C. Mason, G.J. Robinson, R.A. Flather, M. Amin, J.A. Smith & C. Gurney — 109

Section 3: Remote Sensing

Monitoring of coastal waters of the Baltic Sea by airborne imaging spectrometer AISA
T. Kutser, K. Eloheimo, T. Hannonen, P. Härmä, T. Kirkkala, S. Koponen, J. Pullianen & T. Pyhälahti — 123

Study of the land-sea interface in the Barcelona area with lidar data and meteorological models
C. Soriano & J.M. Baldasano — 135

Fluorescence bands from georeferenced LIDAR spectra of marine water mapped as RGB images
P.C. Barbosa, R.A. Nunes, R.H. Tabares & S. Paciornik — 143

Section 4: Atmospheric Pollution

Fraction of the overall ozone levels explained by meteorological variables in the Bilbao area
G. Ibarra, I. Madariaga, A. Elías, J. Caamaño, M. Albizu, E. Agirre & J. Uria — 155

Fog formation prediction in coastal regions using data mining techniques
N.F.F. Ebecken — 165

Estimation of acidifying desposition in Europe due to international shipping emissions in the North Sea and the North East Atlantic Ocean
S.G. Tsyro & E. Berge
175

Alternative models for the disposal of flue gases in deep sea
P.N. Tandon & P. Ramalingam
185

Section 5: Water Pollution

Prevention of hydrocarbons sea pollution: sensitivity index maps for the Venice Lagoon as integral component of oil-spill contingency planning and response
F. Cinquepalmi, D. Schiuma, D. Tagliapietra, C. Benedetti & A. Zitelli
197

Evaluation of sewage outfalls by using tracer techniques combined with oceanographics measurements
J. Roldão, J. Pecly, E. Valentini & L. Leal
209

Interactions between dissolved phosphorus and suspended sediments in a tropical estuary
T.E. Payne, R. Szymczak & T.D. Waite
219

Acoustical response of phytoplanktonic volume scatterers at ultrasonic frequencies as an indicator of pollution in sea waters
S. Blanc, P. Mosto, C. Benitez, R. Juárez, M. Milou & G. Lascalea
231

Assessment of natural radioactivity in the marine environment in Croatia
G. Marovic, Z. Franic & J. Sencar
241

Eutrophication modelling of a tidally influenced mangrove area in Bali subject to major dredging and reclamation activities
H.K. Bach, E. Kock Rasmussen & T. Foster
251

Section 6: Pollutant Transport and Dispersion

Climate change and coastal zone: the importance of atmospheric pollutant transport
C. Borrego & M. Lopes 265

A high-accuracy approach for modeling flow-dominated transport
J.L. Lee 277

Section 7: Hydrodynamic Pollutant Transport Modelling

Two- and three-dimensional modelling of mercury transport in the Gulf of Trieste
R. Rajar, D. Žagar, A. Širca & M. Horvat 289

Nonlinear model of random wave load on pipelines
Ž. Vuković & N. Kuspilić 301

Optimal estimation of concentrations in particle-type transport models
J.W. Stijnen, A.W. Heemink, H.X. Lin & H.F.P. van den Boogaard 311

Physical modelling of the hydrodynamics around a group of offshore structures
S. Pan, N.J. MacDonald, B.A. O'Connor & J. Nicholson 321

A Lagrangian 3D numerical model of pollutant dispersion in coastal waters
J.P. Sierra, M. Mestres, A. Rodríguez & A.S. Arcilla 331

Analysis of the hydrodynamical and physico-chemical behaviour of a natural wetland in the Lagoon of Venice
A. Zitelli, A. Bergamasco, L. Zampato, G. Umgiesser & A. Bergamasco 341

Activity distribution of fission products in the surface water of the Adriatic Sea 351
Z. Franić & G. Marović

Section 8: Coastal Structures

Investigation of influences of very large floating structures on exchange of sea water 363
S. Tabeta

A study of anoxic water structures generated in deep bay enclosed by tsunami breakwaters 373
H. Tsuruya & T. Hibino

Section 9: Groundwater Studies

Preliminary signals of groundwater contamination of stressed coastal aquifers: The case of the Eastern Mediterranean groundwater basins 385
A. Melloul & M. Collin

Numerical modelling and management of saltwater seepage from coastal brackish canals in south-east Florida 395
M. Koch & G. Zhang

Cyclic movement of water particles in a coastal aquifer 405
H. Sun & M. Koch

Section 10: Siltation and Dredging

Sediment resuspension in hurricane conditions 417
K.A. Larm & B.L. Edge

Modelling of the sediment transport at the Belgian coast 427
D. Van den Eynde

Index of authors 437

Section 1: Environmental Impact Assessment

Rare and endangered vascular plants of North Carolina's Outer Banks, Emerald Island to Virginia; a case for conservation and preservation

R. Stalter and E. Lamont
Department of Biological Sciences, St. John's University
8000 Utopia Parkway, Jamaica, New York 11439
Email: biosju@stjohns.edu

Abstract

The vascular flora of the Outer Banks, Emerald Island to Virginia, consists of 806 species within the 422 genera and 140 families. Eighteen species occurring on the Outer Banks are listed as rare or endangered in North Carolina. *Amaranthus pumilus* is listed as federally threatened. This number of rare plant species is one of the highest for any area of equal size in North Carolina. Species of foreign origin, though numerous in the flora, are nonetheless only a minor component of the natural vegetation.

1 Introduction

The Outer Banks consist of a chain of Atlantic Ocean barrier islands, extending from Virginia southward along the northern half of the North Carolina coast. The width of the islands ranges from 700 feet (213 m) at Sandy Bay to over 2 miles (3.2 km) in the vicinity of Buxton. The topography is relatively flat along most of its length, with an elevation of less than 5 feet (1.5 m) above mean sea level; however, beach dunes range from approximately 10 feet (3 m) to 30 feet (9 m) in height. Undulating dune fields are found elsewhere on the islands and the highest topography reaches 115 feet (35 m) at Jockey's Ridge.

The 160 mile (258 km) long series of islands surveyed in this floristic study include Bodie, Pea, Hatteras, and Ocracoke Islands, located in Currituck and Dare counties (35 30'N Lat., 75 30'W Long.) and Portsmouth Island, Core Banks, Harker's Island, and Emerald Island in Carteret County. The earliest floristic inventory of the area was conducted by Kearney[6], who reported 136 species of vascular plants from Ocracoke Island. A lapse in floristic collecting ensued during the next 50 years. A second floristic inventory of Ocracoke Island was conducted by Rondthaler[8], who reported 106 species from the island. Brown[2,3] reported 313 species of vascular plants from the Outer Banks, but did not collect voucher specimens.

Burk[4] conducted a comprehensive ecological study of the Outer Banks and presented a voucher-based checklist of 565 vascular plant species. The most comprehensive floristic inventory of the Outer Banks was that of Stalter and Lamont[9,10] who presented an annotated list of 806 species.

Four naturally-occurring plant communities have been described from the Outer Banks by Stalter and Lamont[9]: 1) dune grass community, dominated by *Uniola paniculata*, *Spartina patens*, *Ammophila breviligulata*, *Panicum amarulum*, and *P. amarum*, occurring on sites which are exposed to salt spray but not inundated; 2) shrub community, dominated by *Juniperus virginiana*, *Quercus virginiana*, *Myrica cerifera*, and *Ilex vomitoria*, occurring on sites which are less exposed to salt spray than the previous one; 3) pine-oak-hickory community, dominated by *Pinus taeda* in association with several species of *Carya*, *Quercus falcata*, *Q. nigra*, and *Liquidambar styraciflua*, occurring on sites exposed to little or no salt spray or inundation; and 4) tidal-marsh community, dominated by *Spartina alterniflora*, *S. patens*, and *Juncus roemerianus*, occurring on sites exposed to periodic tidal inundation. The sounds, freshwater pools and ponds, roadsides, and abandoned fields in the vicinities of habitations support less extensive though different plant communities.

2 Methods

Collecting trips were made to the study area approximately once a month during the growing seasons from June 1989 through October 1996. Objectives for each trip included the collection of voucher specimens and accumulation of information on abundance and apparent habitat preference for each species. Rare or endangered plants were determined by consulting Amoroso[1].

More than 2,400 specimens form the basis for this study. Taxonomically problematic specimens were sent to various experts for annotation; see Stalter and Lamont[9,10] for a list of those who assisted in plant identification. A complete set of voucher specimens has been deposited at the herbarium at Cape Hatteras National Seashore, Manteo, North Carolina and Cape Lookout National Seashore, Harker's Island, North Carolina. Partial duplicate sets have been deposited in the herbaria of Brooklyn Botanic Garden (BKL), University of Michigan (MICH), Missouri Botanical Garden (MO), New York Botanical Garden (NY), New York State Museum (NYS), and Academy of Natural Sciences of Philadelphia (PH).

Published reports by earlier investigators of the few species we were unable to collect have been included in the checklist with the statement "not recently observed", accompanied by a literature citation; in these cases we have not examined the original voucher specimens.

Nomenclature primarily follows Radford et al.[7] but recent taxonomic revisions were consulted to update the nomenclature in this flora. In most cases nomenclature presented in this flora agrees with Kartesz[5] but when differences occur, the name as presented in Kartesz[5] is listed as a synonym and enclosed in brackets.

3 Discussion

The vascular flora of the Outer Banks, from Emerald Island to Virginia, consists of 806 species within 422 genera and 140 families. More than six-hundred forty species, or greater than 80% of the flora, are native to the region. A statistical summary is given in Table 1. The major families include the Poaceae (123 species), Asteraceae (88), Cyperaceae (71), and Fabaceae (38); over 25% of the species comprising the total flora are contained in the Poaceae and Asteraceae. Other large families are the Apiaceae (18 species), Caryophyllaceae (17), Polygonaceae (17), and Brassicaceae (15). The largest genus is *Panicum* s. lat. (20 species), followed by *Cyperus* (18), *Juncus* (14), *Carex* (13), and *Eleocharis* (11). Six hundred and sixty-six species are non-woody (82% of the flora), which underscores the generally graminoid and herbaceous appearance of the flora on the Outer Banks.

6 Environmental Coastal Regions

Table 1. Statistical summary of the vascular flora of the Outer Banks from Emerald Island to Virginia. Data from Stalter and Lamont[9,10].

	Fern Allies	Ferns	Conifers	Dicots	Monocots	Total
Families	1	8	9	101	27	140
Genera	1	8	3	290	120	422
Species	1	10	6	511	278	806

Thirty-seven species of southern affinities reach their northern range limit on the Outer Banks (Table 2), whereas only four northern species (*Ammophila breviligulata, Hudsonia tomentosa, Myrica pennsylvanica*, and *Polygonum prolificum*) reach their southern limit on the Banks[7]. *Polygonum prolificum* (determination verified by R.S. Mitchell [NYS]) is a new state record for North Carolina.

Eighteen species occurring on the Outer Banks are listed as rare or endangered in North Carolina[1], while *Amaranthus pumilus* is listed as federally threatened. Plants are rare, threatened or endangered for a multiplicity of reasons. Some plants are rare because they exist at the edge of their normal range; *Ipomoea stolonifera* and *Hudsonia tomentosa* are two examples. Other plants are rare because they have narrow habitat requirements that are met in only a few areas of the state or world (*Amaranthus pumilus* and *Polygonum glaucum*). Natural biological events (plant succession) climatic events (drought, hurricanes, severe cold) and geological events (earthquakes, volcanoes) can reduce the number of species, especially plants in restricted habitats. Additional factors that contribute to the rarity of species are disturbance or loss of habitat from human-induced activities such as development, road building, introduction of exotic (non-native) plants and pests, over collection of useful or attractive plants (*Malaxis spicata, Uniola paniculata*) and pollution.

Occasionally, common plants are protected by state or federal law. One such species, *Uniola paniculata*, sea oats, is the most common plant on the primary dune system, the dune system closest to the ocean. Sea oats can tolerate high concentrations of salt spray that is carried to dunes from the ocean by prevailing winds. Sea oats, at times, may comprise almost 100 percent of the vegetation on the ocean-facing side of the primary dune. In spite of its abundance, *Uniola paniculata* is a threatened species protected by North Carolina law because this rhizomatous perennial grass is an excellent dune stabilizer and dune builder. Before sea oats was protected, tourists would harvest the attractive inflorescences which were taken home to be used in floral displays. Picking the inflorescences reduced the amount of seed for regeneration and the pulling and trampling associated with this unnatural harvesting killed or weakened the plant. Once the plant was killed, the fragile dune system was ravaged by erosion. In the United States, sea oats is protected in coastal environments from Texas to Virginia.

To protect rare or "threatened" plants, comprehensive field surveys, intensive statewide biological inventories, natural area protection and stewardship programs will provide a comprehensive plan for conservation of rare species and their present or potential habitat. The establishment of state and national plant lists including a species threatened with extinction throughout all or a significant portion of their range or "threatened" (a species likely to be endangered within the foreseeable future throughout all or a significant portions of its range) will enable scientists to monitor the rare, endangered or threatened species and protect them. Knowledge of their existence and inventory of their distribution will assist in maintaining their populations.

Eighteen species occurring on the Outer Banks are listed as rare or endangered in North Carolina[1]: *Amaranthus pumilus* (listed as Federally threatened), *Eleocharis cellulosa, E. halophila, E. montevidensis, E. rostellata, Helianthemum corymbosum, Hottonia inflata, Hudsonia tomentosa, Ipomoea stolonifera, Leptochloa fascicularis var. maritima, Lilaeopsis caroliniensis, Ludwiga alata, L. lanceolota, Malaxis spicata, Polygonum glaucum, Quercus minima, Rhynchospora odorata* and *Spiranthes laciniata*. This number of rare plant species is one of the highest for any area of equal size in North Carolina.

Species of foreign origin, though numerous in the flora, are nonetheless only a minor component of the natural vegetation and occur principally in ruderal sites, lawns, and along the edges of trails and roads. The introduction of numerous species of grasses (Poaceae) is especially noteworthy in the flora. Undoubtedly, exotic species will continue to invade and become established on the Outer Banks as the human population grows and traverses the islands, scattering seeds and creating new open habitats of disturbance. The introduction of exotic plants may contribute to the loss of rare and endangered plants in the Outer Banks of North Carolina.

Table 2. Species of southern affinities reaching their northern range limit on the Outer Banks, North Carolina. Data from Stalter and Lamont[9,10].

Aeschynomene indica	Lactuca graminifolia
Apium leptophyllum	Ludwigia maritima
Bulbostylis barbata	Ludwigia microcarpa
Chloris petraea	Ludwigia repens
Corydalis micrantha	Oplismenus setarius subsp. australis
Croton punctatus	Opuntia drummondii
Cynanchum angustifolium	Pariataria floridana
Cyperus sesquiflorus	Paspalum notatum
Daubentonia punicea	Portulaca pilosa
Dichromena latifolia	Prunus caroliniana
Eleocharis cellulosa	Rhynchospora baldwinii
Eleocharis montevidensis	Rhynchospora odorata
Gaura angustifolia	Sabal minor
Habenaria repens	Sisyrinchium rosulatum
Helianthemum corymbosum	Smilax auriculata
Hydrocotyle bonariensis	Spermolepis divaricata
Ilex cassine	Stenotaphrum secundatum
Ipomoea sagittata	Yucca aloifolia
Ipomoea stolonifera	

References

1. Amoroso, J. L. 1997. *Natural Heritage Program list of the rare plant species of North Carolina Natural Heritage Program*, Division Parks and Recreation, Dept. Environment. Health & Natural Resources, Raleigh. 88 p.

2. Brown, C. A. 1957. *Botanical reconnaissance of the Outer Banks of North Carolina*. Technical report No. 8, Part C, Coastal Studies Institute, Louisiana State University, Baton Rouge, LA. 179 p.

3. Brown, C.A. 1959. *Vegetation of the Outer Banks of North Carolina*. Coastal Studies Series No. 4, Louisiana State University, Baton Rouge, LA. 179 p.

4. Burk, C.J. 1961. *A floristic study of the Outer Banks of North Carolina*. Unpublished Ph.D. Dissertation, University of North Carolina, Chapel Hill, NC. 123 p.

5. Kartesz, J.T. 1994. *A synonymized checklist of the vascular flora of the United States, Canada, and Greenland*. 2nd ed. Volume 1 - Checklist. Timber Press, Inc. Portland, OR

6. Kearney, T.H. 1900. *The plant covering of Ocracoke Island: A study in the ecology of the North Carolina strand vegetation*. Contributions to the U.S. National Herbarium

7. Radford, A.E., H.E. Ahles, and C.R. Bell. 1968. *Manual of the vascular flora of the Carolinas*. Univ. North Carolina Press, Chapel Hill, NC. 1183 p.

8. Rondthaler, A.K. 1952. *Flowering plants of Ocracoke Island, North Carolina*. Unpublished B.A. Thesis, Reed College, Portland, OR. 39 p.

9. Stalter, R. and E.E. Lamont. 1997. *Flora of North Carolina's Outer Banks, Ocracoke Island to Virginia*. J. Torrey Bot. Soc. 124:71-88.

10. Stalter, R. and E. E. Lamont. 1997. *Flora of Cape Lookout National Seashore and Emerald Isle, North Carolina*. Bartonia (in press).

The impact of mercury mining on the Gulf of Trieste

M. Horvat[1], S. Covelli[2], J. Faganeli[3], M. Logar[1], V. Mandić[1], R. Planinc[3], R. Rajar[4], A. Širca[4], D. Žagar[4]

[1]*Department of Environmental Sciences, Jožef Stefan Institute, Jamova 39 Ljubljana, Slovenia*
[2]*Department of Geological Environmental and Marine Sciences, University of Trieste, Italy*
[3]*Institute of Biology, Marine Biological Station, Piran, Slovenia*
[4]*University of Ljubljana, Faculty of Civil and Geodetic Engineering, Hajdrihova 28, Ljubljana, Slovenia*
EMail: milena.horvat@ijs.si

Abstract

The Idrija mine, Slovenia has severely enhanced the mobilisation of Hg by mining activities, and Hg-laden material remains in the region. The tailings and contaminated soils are continuously eroded and serve as a continuous source for the river, the flood plains and the Gulf of Trieste. The paper presents data of the recent study which aims to assess the extent of contamination of Gulf of Trieste after the closure of the Hg mine. Mercury and methylmercury were measured in various environmental compartments (estuarine and marine waters, sediments, and organisms) during 1995-97 period. Data obtained show that even 10 years after closure of the Hg mine, Hg concentrations in river sediments and water are still very high and did not show the expected decrease of Hg in the Gulf of Trieste. A provisional annual mercury mass balance was established for the Gulf of Trieste showing that the major source of inorganic mercury is still the River Soča while the major source of methylmercury is the bottom sediment of the Gulf.

1 Introduction

The biogeochemistry of mercury (Hg) has received considerable attention recently because of the toxicity of methylmercury (CH_3Hg^+, abbreviated as MMHg), the accumulation of Hg in biota, and its biomagnification in aquatic food chains. Hg attacks the central nervous system, and concerns about Hg are based on its effects both ecosystems and human health. The principal pathway for human exposure is the consumption of contaminated fish. Numerous recent studies have concluded that the majority, if not all, of the Hg that is bioaccumualted through the food chain is as MMHg [1,2]. Therefore, knowledge of the concentration, transport, and dynamics of MMHg in aquatic ecosystems is needed to predict the potential impact on humans, as well as on aquatic life. Recently, the U.S. Environmental Protection Agency (EPA) set a new guideline for methylmercury in the diet: 0.1 microgram of mercury per kilogram of body weight per day (0.1 µg/kg/day). This is 4.7 times as strict as the World Health Organization's (WHO's) standard of 0.47 µg/kg/day. An average concentration of Hg in fresh and marine fish is about 0.2 mg/kg. In practical terms this means that and average person weighing 60 kg can only consume about 30 g of fish per day [3]. Recent study on mercury levels in humans has shown that inhabitants near the Lagoon of Grado and Marano in the Northern Adriatic, consuming a diet rich in local seafood are exposed to a higher risk of exceeding the recommended weekly intake [4]. In the Gulf of Trieste mercury concentrations frequently exceed 0.5 mg/kg [5], which in practical terms means that local population should limit its consumption of fish to only one meal per week.

2 Study area

The Idrija mercury mine is situated 50 km west of Ljubljana, Slovenia, and it is the site of the second largest Hg mine in the world which was in operation continually for 500 years until about 10 years ago. Over five million metric tonnes of Hg ore was mined, and much of the residues were spread around the town and its vicinity. It has been estimated that 73% of the Hg mined was recovered with the remainder dissipated into the environment. The Idrija mine has severely enhanced the mobilisation of Hg by mining activities, and Hg-laden material remains in the region. Most importantly, the smelting processing of Hg ore over the centuries and the venting of the mine shafts which release native Hg (Hg^0), caused extremely elevated levels in airborne Hg. Concentrations of Hg in air exceed 2.5 µg/m³ during active mining periods, and even today, airborne Hg levels near the abandoned smelter and around the mine shafts are very high at over 300 ng/m³. Hg levels in sediments and flood plain deposits in the area are very elevated as well. Soils in the Idrija valley are also naturally rich in Hg. Hence, although the ultimate source of Hg in the Idrija region is from base deposits, the majority of material that resides in surficial materials, including deep sediments and along the banks of the river are derived primarily from Hg re-mobilized by mining activities, mostly by smelting [6-9]. Some recent studies have shown that the area even today, continues to supply high quantities of Hg into the river systems of the Idrijca and Soča and reaches the Gulf of Trieste some hundred km downstream where river system empties into the NE Adriatic Sea [10-12]. Recent estimates of Hg balance in the Gulf have shown that the annual input through river Soča discharges is about one ton and a half [11].

Figure 1. Location of Idrija, the rivers Idrijca and Soča (the Soča becomes the Isonzo in Italy), and sampling points in the Gulf of Trieste.

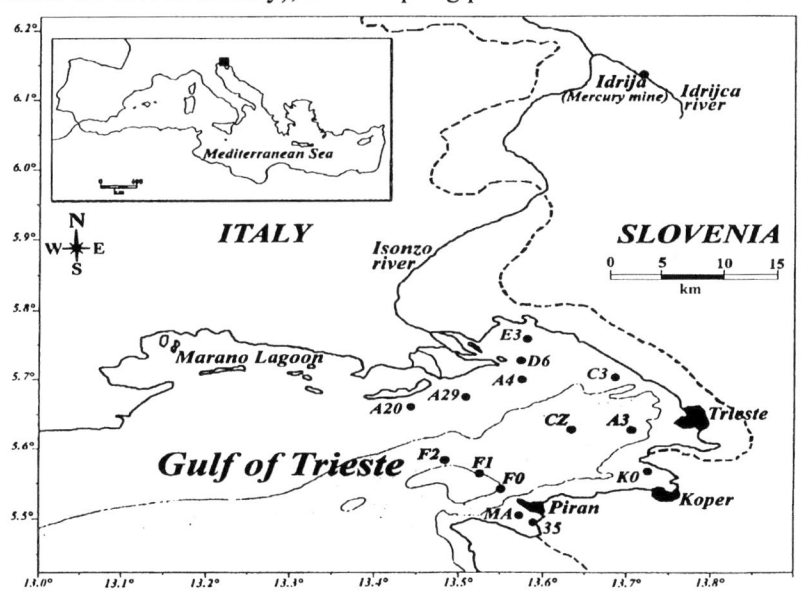

Furthermore, it is well known that the Northern Adriatic is a subject to serious pollution problems, accompanied with eutrofication, anoxic conditions at the bottom, and winter and summer temperature stratification [13, 14]. All these favour transformation of inorganic mercury to more toxic MMHg, which is responsible for the elevated Hg values in marine organisms which frequently exceed the value of 0.5

mg/kg, which is set as the maximum permissible level according to WHO and Slovenian legislation. Moreover, due to deteriorated water quality in the Gulf of Trieste, Hg tends to accumulate in some marine food from mariculture areas, which represents a social and economic problem to the local population [5].

3 Sampling and chemical analysis

Sampling in the marine environment (Figure 1) was performed on three occasions during 1995/96:
- June 1995: high river discharge and biological productivity
- September 1995: low river discharge, stratified conditions in the Gulf, and occasional anoxic conditions at the bottom
- March 1996: low biological productivity period.

Fresh water samples in the river Soča from the bridge near Monfalcone, Italy were sampled twice in November 1997, before and after heavy raining period.

Water samples were collected at the surface (0.5 m depth) and on the bottom (1m above the sediment) by a scuba diver in one litter acid-cleaned Teflon or Pyrex bottles. Samples were stored at 4°C until further processing in the laboratory. Water samples were filtered (0.45 um Nucleopore filters) in order to determine dissolved and particulate bound total Hg and MMHg. Sediment samples were collected using metal free corers (l =30 cm, ϕ = 5 cm) and sectioned at 2 cm intervals. Wet samples were analysed. Results are expressed on a dry weight basis.

Zooplankton was samples using 220 um nets, stored in pre-cleaned Teflon containers and freeze dried. Fish was caught in June 1996 by a local fisherman. Fish muscles were taken for analysis.

Reactive Hg in water samples was determined immediately after sampling using $SnCl_2$ reduction of easy available Hg^{2+} in non-acidified water samples. Total Hg in water samples was analysed by CV AFS or CV AAS after BrCl/UV oxidation [15, 16]. Total Hg in sediment and biological samples was determined by CV AAS after acid digestion[15]. MMHg in water, sediments and suspended matter was determined by solvent extraction, aqueous phase ethylation, gas chromatographic separation, pyrolysis and CV AFD detection [17,18]. MMHg in some of the biological samples was determined by a simplified procedure using HCl leaching, ion-exchange separation, UV oxidation and determination by CV AAS [19]. All analytical methods were regularly validated and performed under good quality control system.

4 Results and discussion

Levels of total, dissolved, and reactive Hg are shown in Table 1. In the vicinity of the estuarine plume, dissolved Hg concentrations were relatively low showing significantly higher concentration in the surface layer than in the bottom. The difference in the central part of the Gulf was relatively small. In contrast, MMHg concentrations were about 10 times higher in the bottom layer (up to 60pg/l) than at the surface (below 5 pg/l) indicating the importance of sediment as the secondary source of this toxic Hg compound. The proportion of reactive Hg was variable ranging from 3-22% of total hg in surface waters close to the river mouth to a max. of 60% in bottom waters. No significant differences were observed in dissolved and reactive Hg during the two sampling periods. The same applies to total Hg in surface waters. However, a significant increase of total Hg in bottom waters has been observed for the September sampling period. The distribution coefficient (K_d) of total Hg between solid and dissolved phase (l/kg) in the estuarine plume was 10^7, while in the coastal waters it varied from 10^5 to 10^6, showing strong association with particulate matter. K_d for MMHg in river water was around 10^4, indicating that MMHg is less strongly associated with particulates than inorganic Hg.

Table 1. Mercury in water samples of the Gulf of Trieste. Results are given in ng/l.

Sampling Station/period	Dissolved Hg Surface	Dissolved Hg Bottom	Total Hg Surface	Total Hg Bottom	Reactive Hg Surface	Reactive Hg Bottom
D6 June-95	4.90	1.31	12.6	4.51	1.20	1.40
Sept-95	4.73	0.89	11.5	12.7	5.1	0.98
A4 June-95	3.47	1.08	10.1	2.40	3.92	0.96
Sept-95	3.74	1.08	11.8	9.73	3.92	0.96
A29 June-95	1.53	1.28	4.10	3.45	1.02	0.75
Sept-95	-	-	-	-	-	-
A20 June-95	2.31	1.34	9.42	2.73	2.00	0.84
Sept-95	2.65	1.28	7.99	4.93	3.72	0.86
CZ June-95	0.97	1.18	1.25	1.31	0.97	1.18
Sept-95	0.52	0.83	1.32	2.37	0.78	0.66
F2 June-95	0.95	1.23	2.47	1.66	0.40	0.64
Sept-95	0.54	0.18	1.71	0.77	0.28	0.33
F0 June-95	0.68	1.09	1.93	2.38	0.64	0.85
Sept-95	0.18	0.78	0.51	1.35	0.29	0.47

Results for total Hg in surface sediments and short cores (0-10 cm) (Figure 2) show that Hg is homogeneously distributed throughout the cores and does not show the expected decrease of Hg in the Gulf of Trieste in recent years. The ratio of MMHg to total Hg in sediments increases with the distance from the river mouth and it is in positive correlation with the percentage of clayey fraction (Figure 3). Benthic fluxes for total and MMHg were also measured in the central part of the gulf and it was estimated that about 74% of total is buried in sediment, while 26% of total Hg, of which 25% is in the methylated form, is annually recycled and released from the water-sediment interface [20].

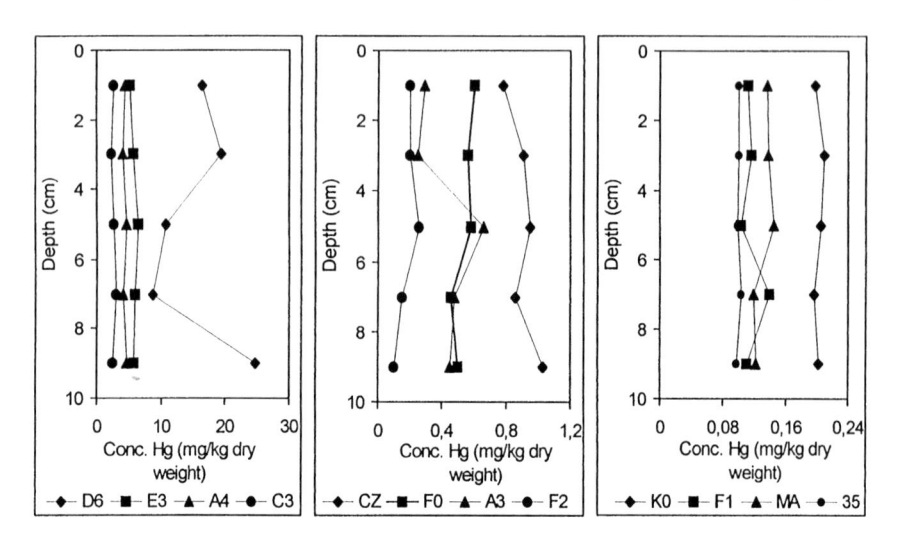

Figure 2. Vertical distribution of total Hg in sediments (1995)

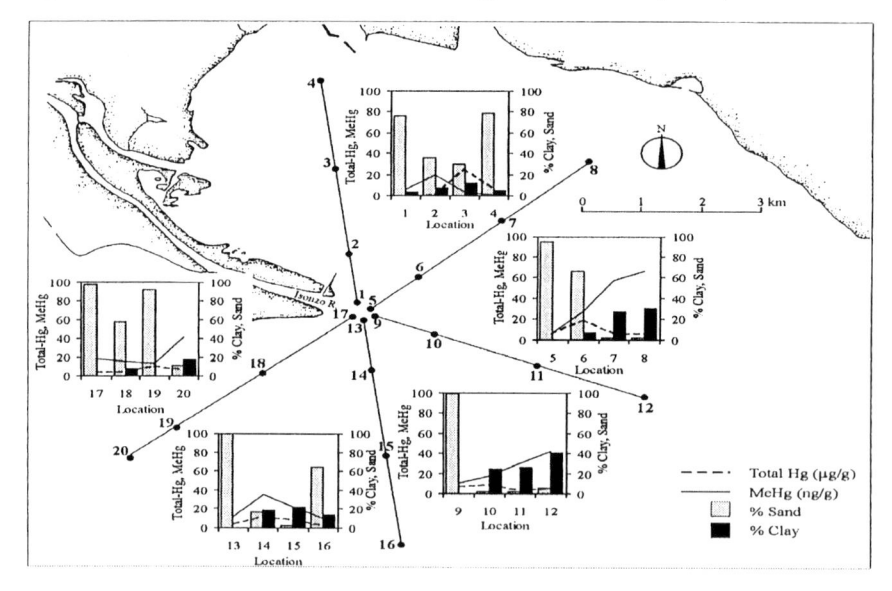

Figure 3. Total and MMHg in sediments near the River Soča mouth

Environmental Coastal Regions   17

The relationship between the size (age) and Hg and MMHg content in fish muscle was studied in two non-migratory fish species, that were caught in the same area (Stations F1 and F0) at the same time (June 1996). Total Hg was measured in fish muscle and the data are presented in Figure 4. The Grey mullet (*Mugil cephalus*) is herbivorous fish and contains lower concentrations of total Hg than Common pandora (*Pagellus erythrinus*) which is a benthic carnivorous fish. The Hg levels are comparable to data obtained for the wider Adriatic[21]. Interestingly, no significant correlation between Hg and MMHg levels and weight (age) was observed. In the case of Common pandora this may be due to the fact that 34 (out of 36) fish were only 2 years old and so had not yet developed the predicted relationship. The Grey mullet shows a better relationship between age and Hg concentrations probably due to better representativenes of the age of the fish. It is interesting to note that the % of MMHg in fish muscle was in most cases much lower than 100%. This is in contrasts with most of the studies which show that with age the % of MMHg should increase. Further investigations are in progress that will explain this unusual behavior of Hg speciation in studied fish species.

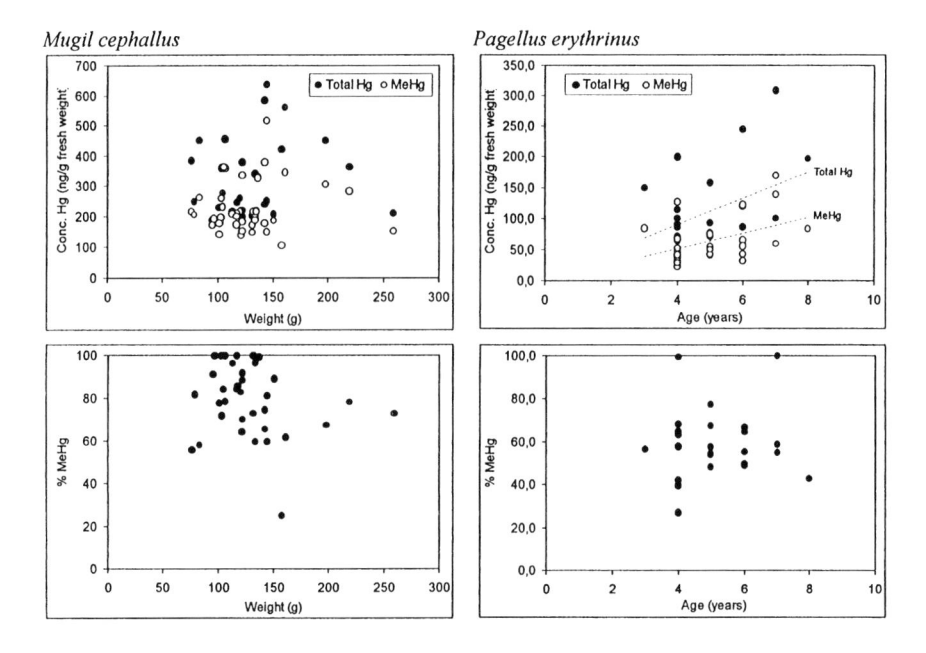

Figure 4. Correlation of total Hg and MMHg concentrations with age (or weight) in two fish species

The results for mixed zooplankton samples are shown in Table 2. Evidently, the concentrations of total Hg and MMHg were 2 to 3 times

18 Environmental Coastal Regions

higher in June compared to September 1995 and about 10 times higher than in March 1996. The percentage of MMHg was also quite variable being lower in June and September and higher in March. The variability of these data can be related to seasonal biological variations in the Gulf, such as phytoplantonic bloom in May/June compared to low productive periods in March. Most probably the this variability is also related to different planktonic species composition to facilitate the interpretation of the results and for the evaluation of the dynamics of the accumulation and release of total Hg and MMHg in stage of the marine food chain. In general, the concentrations levels observed for both analyses do not significantly differ from other studies in the Mediterranean and similar variability of data have also been reported by other researchers [20].

Table 2. Mercury and methylmercury in zooplankton

Sampling Station/period	Total Hg (ng/g, dry weight)	MMHg (ng/g, dry weight)	% MMHg
D6 June-95	-	-	-
Sept-95	200	19.3	9.65
March-96	30.8	9.08	29.48
A4 June-95	959	65.7	6.85
Sept-95	273	10.1	3.7
March-96	12.8	2.08	16.25
A29 June-95	449	46	10.24
Sept-95	-	-	-
March-96	-	-	-
A20 June-95	501	50.5	10.08
Sept-95	193	8.15	4.22
March-96	17.8	2.13	11.97
CZ June-95	282	23.1	8.19
Sept-95	249	12.2	4.9
March-96	14.2	3.68	25.92
F2 June-95	377	52.3	13.87
Sept-95	56.6	5.31	9.38
March-96	24.2	8.82	10.89
F0 June-95	607	59	9.72
Sept-95	242	8.77	3.62
March-96	24.2	2.1	8.68

5 Mass Balance

In 1996, a provisional annual mercury mass balance was established for the Gulf of Trieste [10, 11]. The sources were 1780 kg of mercury carried to the Gulf by the river Soča, 5 kg coming from the atmosphere and 98 kg arriving by currents from the open sea of the Northern Adriatic. With an annual deposition of 1665 kg of mercury, sedimentation was found to be the major sink while a minor one was represented by an outflow of 215 kg to the Northern Adriatic. Numerous recent measurements of mercury concentrations at the bottom of the Gulf, in the river Soča (Tables 1 and 3) enabled us to revise and supplement the mass balance, especially for MMHg data.

Using a benthic chamber experiment it was estimated that only 1308 kg (74 %) of the deposited Hg is buried in the sediment annually, while 26% of the total Hg is recycled. Approximately 25% of the recycled Hg is in MMHg form, whereas in the burial part only 1.8 % of total Hg is methylated [20].

Table 3. Concentration of total and particulate Hg and MMHg in the River Soča in Monfalcone (before discharging into the sea)

Sapling period	Total Hg	MMHg
21.Oct.1997 low water discharge; samples were not filtered (less than 2mg/l of the particulate matter, n=2	Reactive: 0.65; 0,48 ng/l Total Hg (dissolved+ particulate): 1,87; 1,08 ng/l	0.072;0.070 ng/l
8.Nov. 1997 high water discharge; samples were filtered; n=4	Reactive: 1.10±0.15 ng/l Dissolved: 2.98±0.59 ng/l Particulate: 49.6±20.2 mg/kg	0.044±0.010 ng/l 0.092±0.03 ng/g

Measurements of particulate and dissolved mercury during a flood wave with the peak discharge of the last 5-year period in November 1997 (Table 3) confirmed the importance of episodic events on mercury fluxes from contaminated areas to the Gulf of Trieste. With an average of 49.6 µg/g of total Hg content in the suspended matter, an average of 365 g/m^3 of suspended matter and a water volume of 470 million m^3 discharged in 3 days, the total flush of mercury to the Gulf was approximately 8500 kg, which is far above the estimated average annual discharge calculated earlier. The average dissolved MMHg concentrations were around 50 pg/l and particulate bound MMHg around 0.64 ng/g. This means that the total amounts of MMHg in the flood wave was only 0.14 kg . This value is negligible compared to the benthic flux of MMHg into the water column of the Gulf of Trieste, confirming that the main source of MMHg in the Gulf originate from sediment [20].

6 Conclusions

It can be concluded that even though the Hg mine in Idrija was closed completely in 1991, Hg concentrations in sediments and water are still very high and did not show the expected decrease of Hg in the Gulf of Trieste. While some questions have already been answered regarding the main transport mechanisms in the river system and the Gulf of Trieste alone, the main question concerning Hg transformation in the river systems and Idrija remains unanswered. Further studies are in progress in order to understand basic processes of Hg cycling and to develop models for simulation of future scenarios and plan remediation actions, if necessary.

References

[1] WHO-IPCS, Environmental Health Criteria 101 – Methylmercury, WHO, Geneva, 1990.
[2] WHO-IPCS, Environmental Health Criteria 118 – Inorganic mercury, WHO, Geneva, 1991.
[3] US EPA (1996) Mahafey, R.K., Rice, G.E., Schoeny R.: Mercury Study Report to Congress Volume IV: Characterisation of Human health and Wildlife Risk from mercury exposure in the United States (EPA -452/R-97-009), Washington, D.C., December 1997.
[4] Ingrao G., Belloni P., Santorini, J.P.: Mercury levels in defined population groups., presented at Fourth Research co-ordination Meeting, Honolulu, December 1995 organised by the IAEA. Contract No. 6334/CF.
[5] Brambati, A.: Metalli pesanti nelle lagune di Marano e Grado, Piano di studi finalizzato all'accertamento delle presenza di eventuali sostanze tossiche persistenti nel bacino lagunare di marano e Grado ed al suo risanamento, Regione Autnoma Friuli- Venezia Giulia, Trieste, Dicembre 1996
[6] Palinkaš L.A., Pirc S., Miko S.F., Durn G., Namjesnik K., Kapelj S.: The Idrija Mercury Mine, Slovenia. A semi-millenium of continuos operation: an ecological impact. *In: Environmental Toxicology Assessment* /ed. M. Richardson, Taylor&Francis Ltd. 1995. pp. 317-339, 1995.
[7] Gosar, M., Pirc S., Bidovec M.: Mercury in the Idrijca River sediment. *In: Book of Abstracts: Idrija as Natural and Antropogenic Laboratory*, pp.22-26, Idrija, May 1996.
[8] Gnamuš, A., Horvat M., Stegnar, P.: Der Quecksilbergehalt von Rehwild und geästem Laub für die Bewertung der Umweltbelastung im Bergbaugebiet von Idrija-Eine Fallstudie aus Slowenien, *Z. Jagdwiss.*, 41, pp. 206-216, 1995.
[9] Gosar, M., Pirc S., Sajn R., Bidovec M., Mashyanov N.R., Sholupov S.E.: Mercury in the air over Idrija – Report of measurement on 24th September 1994. In: *Book of Abstracts: Idrija as Natural and Antropogenic Laboratory*, pp. 27-32., Idrija, May 1996.
[10] Rajar, R., Četina, M., Širca, A.,.: Hydrodynamic and Water Quality Modelling: Case Studies. *Ecological Modelling*, 101, pp. 209-228., 1997.
[11] Širca, A., Rajar, R.: Calibration of a 2D mercury transport and fate model of the Gulf of Trieste. *Proc. of the 4th Int. Conf. Water Pollution 97*, Eds. Rajar, R. and Brebbia, M., Computational Mechanics Publication, Southampton. pp. 503-512. 1997.
[12] Hess, A.: *Geologija*, 33, pp. 479-486, 1991.
[13] Faganeli, J., Planinc R., Pezdič J., Smodiš B., Stegnar P., Ogorevc B.: Marine geology of the Gulf of Trieste (Northern Adriatic): Geochemical aspects, *Mar. Geol.* 99, pp. 93-108, 1991.
[14] Hines M.E., Faganeli J., Planinc R.: Sedimentary anaerobic microbial biogeochemistry in the Gulf of Trieste, Northern Adriatic: Influence of bottom water oxygen depletion. Submitted to *Geochemistry*, 1998, submitted.
[15] Bloom, N.S., Crecelius, E.A.: *Mar. Chem.*, 14, pp.49-59, 1983.
[16] Gnamuš, A., Horvat M., Stegnar P.: Der Quecksilbergehalt von Rehwild und geästem Laub für die Bewertung der Umweltbelastung im Bergbaugebiet von Idrija-Eine Fallstudie aus Slowenien, *Z. Jagdwiss.*, 41, pp. 206-216, 1995.
[17] Horvat, M., Liang, L., Bloom, N.S.: Comparison of distillation with other current isolation methods for the determination of methyl mercury compounds in low level environmental samples, Part 1: Sediments, *Anal. Chim. Acta*, 281(1), pp.135-152, 1993.
[18] Horvat, M., Liang, L., Bloom, N.S.: Comparison of distillation with other current isolation methods for the determination of methyl mercury compounds in low level environmental samples, Part 2: Water, *Anal. Chim. Acta*, 282(1), pp. 153-168, 1993.
[19] May,K. Stoeppler, M., Reisinger K. (1987) Toxicol. Environ., 13, 153-163
[20] Covelli, S., Faganeli, J., Horvat, M., Brambati, A.: Benthic fluxes of mercury and methylmercury in the Gulf of Trieste. *Geochemistry*, 1998, submitted.
[21] Bernhard,M.: Mercury in the Mediterranean, *UNEP Regional Seas Reports and Studies No. 98*, 1988, UNEP, Nairobi.

Environmental impact assessments and legislation - Jamaica's way to protect its coastal zone

Andreas Haiduk
*Water Resources Authority, Hope Gardens,
PO BOX 91 Kingston 7, Jamaica / WI
Email: ahaiduk@colis.com*

Abstract

Jamaica's coasts are an important economic factor. They attract thousands of tourists every year. But the coastal environment is also under permanent stress as a result of the discharge of millions of litres of primary treated or untreated wastewater. Several legislative tools are implemented to protect the coastal zone such as the Natural Resources Conservation Authority Act (NRCA Act) of 1991 which empowers the Natural Resources Conservation Authority (NRCA) among others to request Environmental Impact Assessments (EIAs) and issue licenses for the discharge of effluents, giving notices to abstain from agricultural practices where the pollution of a receiving waterbody is likely, and carry out operations to prevent any pollution or to remedy or mitigate any pollution.

EIAs for new projects have to be carried out for certain categories of enterprises, constructions and developments. In 1996 twenty three EIAs for new projects were carried out for input into the Permit and Licence System, the final tool prior to the implementation of any development, to control the deleterious effects, direct or indirect on the human health. Trade Effluent Standards as well as Sewage Effluent Standards have been established and Ambient Water Quality Standards and Air Quality Standards are to be finalized. These tools are also used to control any activity with possible negative impacts on the environment.

22   Environmental Coastal Regions

1 Introduction

Within the Caribbean region, Jamaica possesses a central location with respect both to the Caribbean islands and the mainlands of North and South America. Jamaica is situated between latitude 17½° and 18½° north and longitude 76½° and 78½° west, is part of the Greater Antilles, whose other major islands are Cuba, Hispaniola and Puerto Rico. With about 10,976 km^2 of land, Jamaica is the third largest of the Caribbean islands and the largest West Indian Island in the British Commonwealth. The country is 235 km long, has a maximum width of 82 km and possesses about 885 km of coastline. (Town Planning Department [1])

The population of Jamaica in 1997 was estimated to be 2,553,400 (STATIN[2]). Figure 1 shows the settlement strategy. The island's principal commercial and population centres are located on the coast.

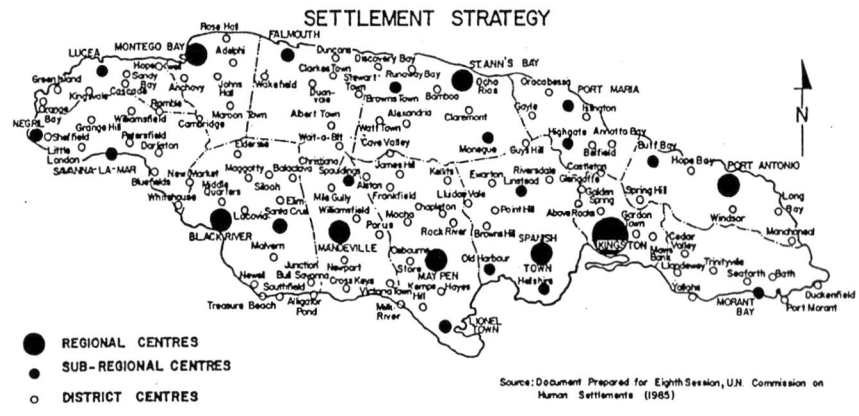

Figure 1: Settlement Strategy (Government of Jamaica [3])

1.1 The Coastal Resources

Jamaica's coast is about 885 km in length, punctuated by numerous inlets and bays. Its varied and irregular coastline gives rise to a unique ecosystem formed by the integration of coastal features that include harbours, bays, beaches, rocky shores, estuaries, mangroves swamps, cays and coral reefs. These natural features provide a coastal resource base that contributes significantly to the economic well-being of the country. Relative to tourism, the worldwide image of Jamaica is based largely on the beauty of the island's beaches, coastal waters and shoreline environment. These coastal resources however, are especially sensitive to the effects of over-use and mismanagement, and many areas exhibit some degree of degradation. Urban development pressures on coastal resources are intense and persistent. (Government of Jamaica [4])

1.2 Importance of Coastal Environment

Apart from the fact that a well functioning coastal environment ie. the existence of reefs has a high ecological value for the marine ecosystem it also plays a vital role for tourism as well as for fisheries. Figure 2 shows the shoreline configuration.Most of the beaches are found at the north coast. However, the south coast also has some beaches which are presently not developed due to lacking access. Between Lucea in the north western side of Jamaica and Ocho Rios in the north a stretch of well developed reefs is found. These reefs are threatened by human activities.

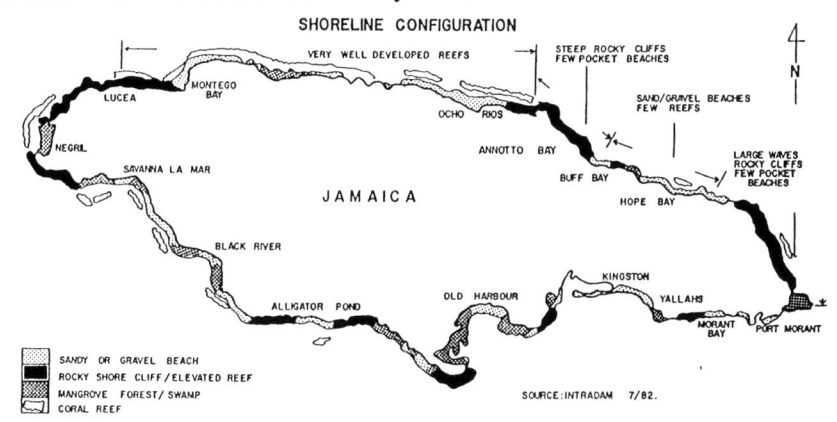

Figure 2: Jamaicas Shoreline Configuration (Government of Jamaica [6])

1.3 Sources of pollution on the marine environment

Industries along the coast have used coastal waters for dilution. Polluting industries are sugar, bauxite/alumina, cement, chemical, distilling and food processing. Kingston and St. Catherine experience all types of pollution. On the Kingston Waterfront, industries such as oil-refinery, fish-processing, detergent manufacturing and the slaughterhouse discharge waste into nearby gullies which enter the harbour. Bauxite activities have polluted coastal environments by spillage of bauxite ore into the water during loading operations which affect corals and sea grass beds.

Thirteen dumps are located close to bodies of water. Highly contaminated surface runoffs result in pollution of nearby streams and gullies which ultimately reach the marine environment. Garbage dumped in gullies and on roadsides reach the sea by heavy rains. No proper disposal system exist for toxic and hazardous wastes. (Yasmin Williams [7])

Sewage is a major cause of pollution. Based on estimates 450,000 m^3/day of sewage are nationally generated.(Michael Betts [8]) Only about 21 % of the total volume is treated in 109 sewage treatment plants, 27 % by septic tanks and soil absorption methods, 40 % by pit latrines and 12 % by no identifiable method. Most of the treatment systems do not function properly. This leads to organic pollution which causes a high oxygen demand, leading to oxygen depletion; eutrophication by an increase in nutrients (nitrates and phosphates) and pathogenic pollution increasing the disease carrying potential of water.

Soil lost from agricultural land-use and deforestation is washed into the marine environment by rivers as silt, covering corals and sea grass beds.

Oil pollution at sea by shipping activities resulting from the discharge of oil contaminated ballast and oil tanker bilge has affected mangroves and cays. Large international ships which do not have on-board systems to compact garbage dispose of their waste at sea.(Yasmin Williams [7])

2 Legislation - an Overview

The following acts deal directly or indirectly with the coastal environment. Harbours Act (1874), Wildlife Protection Act (1945), Beach Control Act (1956), the Litter Act (1986), Watershed Protection Act (1993), the Fishing Industry Act (1976), the Town and Country Planning Act and the NRCA Act (1991). With the NRCA Act of 1991 a more focussed approach was undertaken to tackle the problems of pollution. However the role of the older acts will be explained briefly.

2.1 Harbours Act

The Harbours Act states in section 19 that the deposition of rubbish, earth, mud, stone, sand, ballast or oil into any channel leading into or out of any harbour is prohibited.

2.2 Wildlife Protection Act

The input of any trade effluent to an aquatic area inhabited by fishes, is prohibited. (Section 11) The knowingly killing or injuring of immature fish (Section 9) and the use of dynamite or other noxious materials to kill fish (Section 10) is prohibited.

2.3 Beach Control Act

Section 5 states that it is prohibited to use the foreshore and the floor of the sea without a licence. In section 7 it is mentioned, that the dumping of rubbish or other waste matter, water-skiing, dredging or other disturbance of the floor, using boats other than propelled by wind or oars, destruction or removal of coral, seafans and sedentary marine animals, the searching and removal of any treasure or artefact from the floor of the sea and fishing in protected areas, is prohibited.

2.4 The Fishing Industry Act

It is a ministerial right to declare prescribed areas as fish sanctuaries in which fishing is prohibited. From time to time a closed season can be declared by the Minister.

2.5 The Litter Act

Any person who throws, drops or otherwise deposits and leaves any litter in or into any public place (among others beach or foreshore) is guilty under the Act.

2.6 The Watershed Protection Act

The general duty of the Authority is to promote the conservation of water resources and to institute measures and implement programmes to protect watershed areas. Regulations exist to ensure the proper, efficient and economic utilization of land including the prohibition of felling, barking, destruction of any trees or the clearing of vegetation.

2.7 The Town and Country Planning Act

The general objective of this act is to control developments of the land especially in the view of securing proper sanitary conditions and conveniences and the protection and extension of the reservation of lands for among others the protection of marine life.

2.8 The NRCA Act

This act is the most comprehensive legislative tool and will be discussed on a larger scale. The Natural Resources Conservation Authority is the main regulatory authority on environmental matters. The responsibilities are very wide ranging and include
- pollution control (including satisfactory effluent discharge standards)
- promotion of environmental awareness for the public
- coastal zone management
- designation of national parks and designated areas
- protection of flora and fauna
- Environmental Impact Assessments review and standards
- Jamaican compliance with international environmental treaties (such as the Basel, Montreal and Climate Change Conventions) (Karen Mc Donald [9])

3 Procedures for new developments

Before any development activity can take place the project proponent has to apply for a permit or a licence. The permit is required for any activity within prescribed categories (Table 1) within Jamaican territory. The intension of this permit is to safeguard environmental/natural resources from direct damage due largely to physical development. The licence is required for the intended discharge of any sewage, trade effluent or other pollution matter to air, ground or water.(NRCA[10])

Table 1: Excerpt of the prescribed description of enterprise, construction or development (NRCA [10])
1. Power Generation Plants

2. Electrical transmission lines and substations greater than 69 kV
3. Pipelines and conveyors with diametre of 15 cm and over
4. Port and harbour developments
5. Development projects
   - such as subdivisions of 10 lots and more
6. Ecotourism projects
7. Water treatment facilities (water supply, sewage, desalination plants)
8. Mining and mineral processing
9. Metal processing
10. Industrial projects
11. Construction of new highways and roads
12. River basin development projects
13. Irrigation or water management projects
14. Land reclamation and drainage projects
15. Watershed development and soil conservation projects
16. Modification of wetlands
17. Solid waste treatment and disposal facilities
18. Hazardous waste storage
19. Processing of agricultural waste
20. Cemeteries and crematoriums
21. Introduction of species of flora, fauna
22. Slaughterhouse and abattoir
23. Felling of trees and clearing of land of 10 hectares and over
24. Clear cutting of forested areas of 3 hectares and over

The proponent has to complete a Project Information Form (PIF). This PIF gives information about the project type (see table 1), the site description (slope, soil type, geological formation) and the project description (physical dimension, operational aspects). Based on the PIF an environmental screening is conducted and the decision is made whether an EIA is necessary or not. If no EIA is necessary the permit application is directed to the Technical Committee which makes the final decision.

An EIA is required when an activity is likely to affect significantly the environment. Before the applicant commissions the EIA he has to submit draft Terms of References (TOR). The purpose of the TOR is to set forth the general scope, significant issues to be addressed, methodologies for addressing these issues and the content of the EIA to meet NRCA policies.(NRCA[11]) After approval of the TOR the applicant can commission the EIA. The United Nation Environment Programme has adopted a set of goals and principles of environmental impact assessment. These principles have been taken into consideration for the national guideline. Any EIA should cover the following topics:

Policy, Legal and Administrative Framework
Description of the Environment
Description of the Proposed Project in Detail
Significant Environmental Impacts
Socio-Economic Analysis of Project Impacts
Identification and Analysis of Alternatives
Mitigation Action/Mitigation Management Plan
Environmental Management and Training
Monitoring Programme
Public Involvement (NRCA[11])

The drafted EIA will then be reviewed by the responsible authority and if necessary by other governmental agencies too, such as the Water Resources Authority, the Environmental Control Division and others. The public is also involved by means of distribution of the EIA report to the Parish Library, in which the project is sited and by public presentation. After the presentation the EIA is again submitted to the Technical Committee where the final decision will then be made. If the final recommendation is negative and the permit is denied the applicant can appeal to the Minister. A permit is granted for an indefinite period.

A licence is required for an intended discharge of any sewage, trade effluent or other polluting matter. The process of application is similar to the above mentioned process. A separate EIA is not necessary. A licence is granted for a period of five years.

3.1 Monitoring Programmes

The licence is granted under specific conditions. The applicant is obliged to submit quarterly (or more frequent, if desired) reports about the sampling results for predetermined parameters. The results have to be within the trade effluent standards and the sewage effluent standards.

3.2 Trade Effluent Standards

Fourty two different parameters are mentioned. Starting from general parameters such as Chloride, Nitrates, pH to heavy metals such as Arsenic, Cadmium, Chromium to stream loading parameters like BOD_5, COD and Dissolved Oxygen to bacteriological parameters.

3.3 Sewage Effluent Standards

The standards are as follows

BOD_5	20 mg/L
TSS	20 mg/L
Total Nitrogen	10 mg/L
Phosphates	4 mg/L
COD	100 mg/L
pH	6 - 9
Fecal Coliform	1000 MPN/100mL
Residual Chlorine	1.5 mg/L

4 Process for existing developments

For existing developments which discharge effluent, stack emmissions and so on licences will be phased in on a prescribed schedule. By May 2000 all existing facilities should have applied for licences. (NRCA[10])

5 Fees

The application for a permit/licence costs J$ 1,000 each ($\approx$ 28 US$). The grant of a permit costs J$ 15,000 ($\approx$410m US$) and the grant of a licence costs J$ 5,000 ($\approx$137 US$). The prices are as of January 1, 1997. (NRCA[10])

6 Breaches and consequences

A permit may be suspended or revoked if terms or conditions under which the permit was granted are breached or if the holder fails to submit certain documents. A licence may be revoked if there is a permanent non-compliance of its terms or conditions or if it becomes clear that the licensee has deliberately submitted false information during the application process.(NRCA [10])

7 Practical Experiences with the Granting Process

Since inception of the new Permit and Licence System on January 1, 1997 over 100 applications have been made. Only about 1/3 could be processed so far. As in many other public organisations the lack of staff and inappropriate organisational structures cause the most delays. This is also valid for the review process of EIAs. The selfnamed goal to review and approve EIAs in a period of 90 days was in 1996 only achieved for 30 % of the cases. (Karen Mc Donald [9])

The public is now much more aware of the process. The media (newspaper and television/radio) discusses EIAs in their environmental coloumns and programmes.

8 Conclusions

The multitude of acts and their regulations makes a concerted approach to protect the environment, especially the coastal zone, difficult. With the NRCA Act and the new Permit and Licence System a more focussed approach can be achieved. An adequate number of staff including environmental wardens equipped with the power to enforce the law should improve the situation. Other burning issues like steming the crime wave, alleviating the poverty, improving the education sytem have a high or even higher priority. It remains to hope that with an increased public awareness the degradation of the environment in general can be stopped or at least decreased.

References

[1] Town Planning Department, National Atlas of Jamaica, p. 2-3, 1989
[2] Statistical Institute of Jamaica, 1997 Report
[3] Government of Jamaica, Jamaica-Country Environmental Profile, p.61, September 1987
[4] Government of Jamaica, Jamaica-Country Environmental Profile, p.83, September 1987
[5] Statistical Institute of Jamaica, Population Census 1991, Volume 1 Part 15, p. xiv
[6] Government of Jamaica, Jamaica-Country Environmental Profile, p.86, September 1987
[7] Williams, Y., Overview of Marine Pollution in Jamaica, p. 1-2, December 1995
[8] Betts, M., Waste Water Management in Jamaica, p. 7, October 1995
[9] McDonald, K., Overview of the EIA process in Jamaica, October 1997
[10] Natural Resources Conservation Authority, Permit and Licence System - Guidelines for Project Proponents , January 1997
[11] Natural Resources Conservation Authority, Guidelines for Conducting Environmental Impact Assessment, January 1997

A New EIA Method Applied on Coastal Reclamation Projects

Kurt Jensen, Hanne K. Bach & Christopher M.R. Pastakia
VKI Institute for the Water Environment, 11 Agern Alle, DK-2970 Hørsholm, Denmark
e-mail: kje@vki.dk, hkb@vki.dk & cmp@vki.dk

Abstract

A system called Rapid Impact Assessment Matrix (RIAM) has been developed to organise Environmental Impact Assessment (EIA). RIAM brings together the individual multi-disciplinary parts of an EIA in a transparent and semi-quantitative manner. RIAM allows a holistic and coherent anticipation of problems, and can assist in the repeated assessment of alternatives and project details during the planning phase.

In coastal regions, numerical models are also important tools in the impact assessment process. Some advantages are provided by the combined use of the RIAM and the MIKE21 model system. RIAM defines the important assessment criteria and environmental components as well as a means by which values for each of these criteria can be collated to provide an accurate and independent score for each condition. The MIKE21 model supplies the scientific base for the assessment, evaluating the physical, chemical and ecological impacts of project activities.

RIAM, and MIKE models for hydrography, water quality and ecology, have been used in a dredging and reclamation project in Sabah, Malaysia. The result of the RIAM analysis in combination with MIKE modelling gave a direct quantification of the benefits and disadvantages of different designs of the reclamation options. This combination strongly facilitated the minimisation of environmental impact and the future decision making.

1 Introduction

The completion of an Environmental Impact Assessment (EIA) is a generally accepted precondition for the acceptance and initiation of projects, internationally and nationally[3,4,8]. The term EIA refers to a variety of analyses embracing pure technical and scientific problems as well as social, economic or cultural issues.

EIAs can only be an integral part of the environmental management if they are included in the planning process, and used to chose between different project options to minimise the negative effects (a process which is called environmental optimisation).

EIAs are often accused of being pure approvals of any disturbance of the environment[9], based as they are on a number of subjective, qualitative judgements. The number of options that can be analysed, and the accuracy of the EIA analyses increase significantly if predictive mathematical models can be used. The RIAM concept[1] offers a good tool to include and compare results from numerical modelling as well as pure subjective judgements.

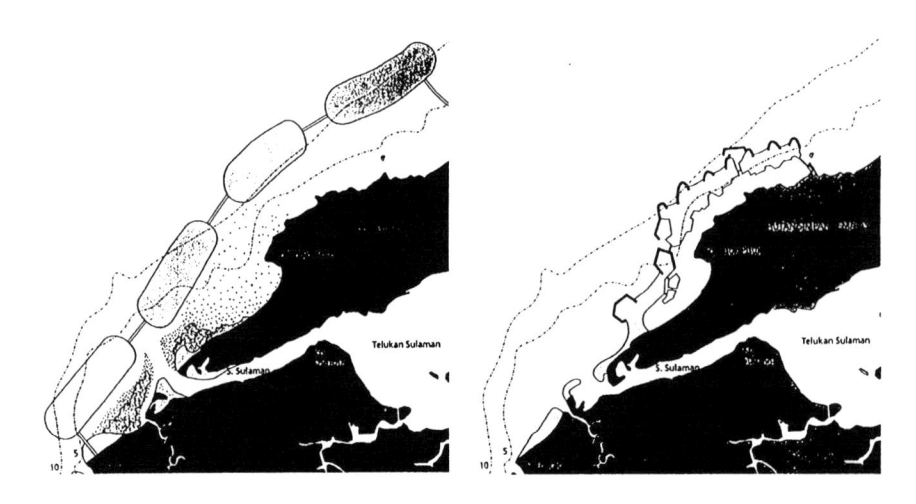

Figure 1: The study project, a tourist island reclamation project along the west coast of Sabah, Malaysia. Left: original design; right: final design.

The paper summarises the RIAM concept, and shows how the RIAM analysis and the EIA process benefit from the predictive data from the

Environmental Coastal Regions   35

mathematical models. Details of the rationale behind the design options or of the choice of components (scoping) are not discussed in this paper.

The study project considers the plans for a land reclamation project along the north-western shores of Sabah, Borneo.

Seaward construction projects of this type that include dredging and reclamation strongly influence the marine environment. In the construction phase sediment spill and discharges may negatively influence the adjacent biotopes. The final construction may change currents and waves, the dilution pattern of present discharges, and may add new sources of discharge. The initial project and the final project emerging from the planning and EIA processes are outlined in figure 1.

2   Methods

2.1   Numerical Models

Numerical models are powerful tools for analysing and forecasting the environmental effects of construction works. In the study of the Sabah reclamation project the numerical hydrodynamic model system MIKE21 has been applied for the advection/dispersion, the water quality and the eutrophication modules[2].

2.2   RIAM

EIA faces the problem of having to compare and present results of analyses ranging from wholly subjective judgements of cultural properties with objective scientific analyses. A conceptual and mathematically simple, sem-qualitative system called the Rapid Impact Assessment Matrix (RIAM)[1] has been developed to organise EIA. The RIAM method brings together the individual multi-disciplinary parts of an EIA. It keeps transparent control of the components in a distinct semi-quantitative manner allowing direct comparison of different problems, and above all it allows a holistic and coherent anticipation of problems.

2.3   Components

The process of defining the issues, or components, which are of importance in evaluating the possible changes due development, is called scoping. In the RIAM these components are considered in a holistic manner and fall into four groups. These groups represent the issues relating to the Physical/Chemical environment (P/C); those relating to

Biological/Ecological (B/E) concerns; human issues defined as Social/Cultural (S/C); and issues dealing with the Economic/Operational (E/O) aspects of development.

2.4 Criteria

In the RIAM analyses, all problems are analysed according to five characteristic criteria. Two criteria relate to properties that are of singular importance to the condition, and three criteria to properties that are of value to the situation.

The first type of criteria is: the importance of the condition, which is assessed against the spatial boundaries or human interests it will affect; and the magnitude, which is defined as a measure of the scale of benefit/dis-benefit of an impact of a condition.

For the importance of condition (I) the scale is defined as:

4 = important to national/international interests
3 = important to regional/national interests
2 = important to areas immediately outside the local condition
1 = important only to the local condition
0 = no importance

For the magnitude of a change or effect (M) the scale is defined as:

+3 = major positive benefit
+2 = significant improvement in status quo
+1 = improvement in status quo
 0 = no change/status quo
-1 = negative change to status quo
-2 = significant negative dis-benefit or change
-3 = major dis-benefit or change

Criteria that are of value to the situation are defined as permanence, reversibility and cumulative properties.

Permanence defines whether a condition is temporary or permanent, e.g. an embankment is a permanent condition even if it may one day be breached or abandoned, whilst a coffer dam is a temporary condition, as it will be removed.

Reversibility defines whether the condition can be changed and is a measure of the control over the effect of the condition. For example an accidental oil spill in the sea has irreversible effects such as death of birds while oil discharge from a refinery is a reversible condition because effect of its effluent can be changed (through effluent control), even though the refinery itself is permanent condition.

Cumulative property is a measure of whether the effect will have a single direct impact or whether there will be an accumulated effect over time, or a synergistic effect with other conditions. For example, the change of certain snails from females to males due to some anti-fouling substances is cumulative in addition to being permanent and irreversible, since it hampers the reproduction of the species and affects future generations of snails.

Table 1: *The scale used for the criteria that are of value to the situation*

Score	Permanent (P)	Reversible (R)	Cumulative (C)
1	no change/not applicable		
2	temporary	reversible	non-cumulative/single
3	permanent	irreversible	cumulative/synergistic

Table 2: *Range Bands used for RIAM*

RIAM Environmental Score (ES)	Range Value (RS) (Alphabetic)	Range Value (RS) (Numeric)	Description of Range Band
108 to 72	E	5	Major positive change/impact
71 to 36	D	4	Significant positive change/impact
35 to 19	C	3	Moderate positive change/impact
10 to 18	B	2	Positive change/impact
1 to 9	A	1	Slight positive change/impact
0	N	0	No change/Status quo/Not applicable
-1 to -9	-A	-1	Slight negative change/impact
-10 to -18	-B	-2	Negative change/impact
-19 to -35	-C	-3	Moderate negative change/impact
-36 to -71	-D	-4	Significant negative change/impact
-72 to -108	-E	-5	Major negative change/impact

2.5 Score and range system

The assessment of the different problems that have been selected for evaluation by the scoping process gives a value ascribed (by the assessor) to each of these criteria. By the use of a simple formula a score (the environmental score) for the individual components can be calculated:

$$ES = I*M*(P+R+C).$$

To use the evaluation system described, a matrix of cells showing the criteria used, set against each defined component, is produced for each project option. See the example for the final chosen option in table 3. Within each cell the individual criterion score is set down. From the formulae given above each ES number is calculated and recorded. To provide a more certain system of assessment, the individual ES scores are banded together into ranges where they can be compared. The ranges cover impacts from a major positive change/impact (+5/E) to similarly negative effect (–5/-E). Conditions that have neither importance nor magnitude will score a zero and be banded together (0/N); and any condition in this band is either of no importance or represents the status quo, or a non-applicable situation.

For the arguments for the score system and its transcription into range bands, see Pastakia.[4]

3 Results

Model calculations of water quality are shown in figures 2, 3 and 4. Primary production is chosen to illustrate model results. Figure 2 shows the situation prior to any development plan for the Sabah coast. There is a pattern of primary production that is mainly determined by the supply of nutrients by the adjacent river Tuaran, and the penetration of light that is impeded by suspended matter from the river. Figure 3 shows the prediction after the construction of the reclamation, where the pattern of primary production is changed due to the reclamation and the predicted future loads from the river. Figure 4 illustrates the difference in primary production between the predicted future load and the present baseline situation. Based on those three model runs the extent of the impact of the proposed construction on primary production can be assessed.

Though model runs may use days of computer time it is comparatively easy recalculate different scenarios in the process of minimising the environmental effects by change of design etc.

Environmental Coastal Regions 39

The results of the model calculations together with many other types of information form the basis for the RIAM matrix of the same development project. It should be emphasised that the zero effect, which forms the comparative baseline for the assessment is the "do nothing" situation. The RAIM matrix is shown in table 3, and the result of the RIAM analysis is summarised in a histogram (Figure 5).

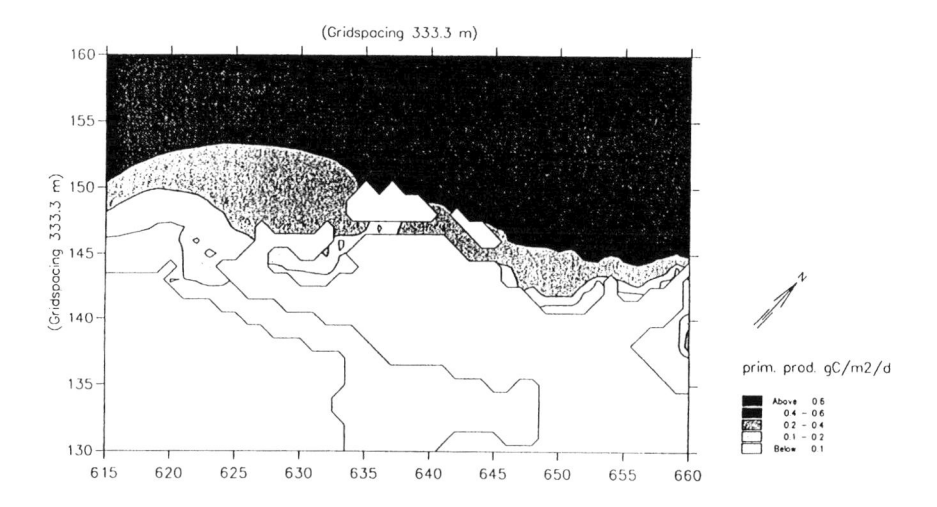

Figure 2: Primary production, modelling of present situation during south-west monsoon in a 333 metre grid (g $C/m^2/d$).

4 Discussion

For an EIA to be of optimal value, the assessment should be undertaken early in the planning stages on relevant aspects of the project. This suggests that the EIA should influence the final design and the implementation process of the project. In order to be able to do so, reassessments must be possible at short intervals and use relatively few resources.

40 Environmental Coastal Regions

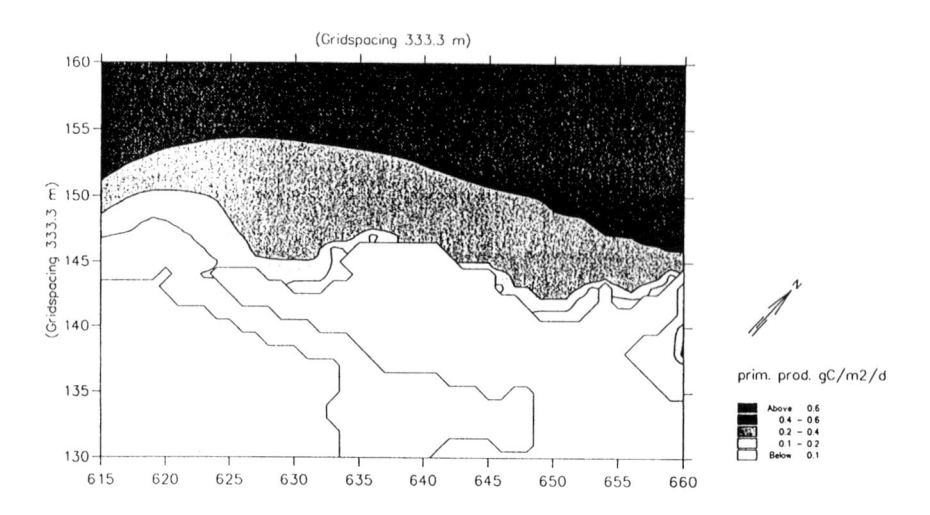

Figure 3: Primary production, modelling of future situation during south-west monsoon in a 125 metre grid (g $C/m^2/d$).

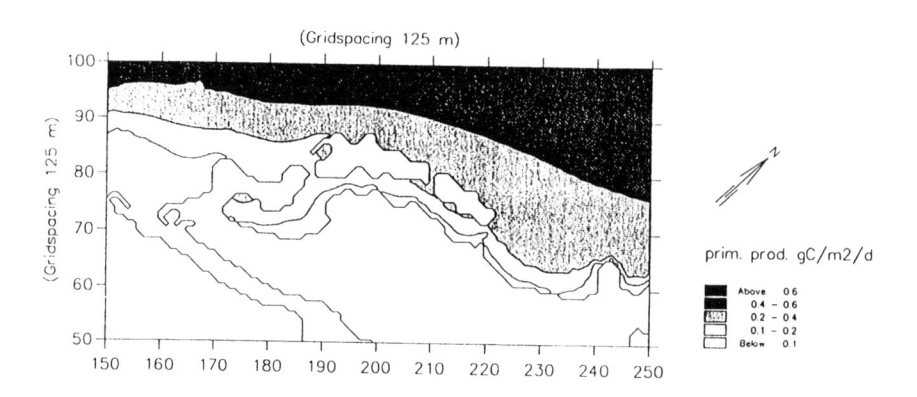

Figure 4: Primary production, modelling of the change between present and future situations caused by discharges outside the reclamation during south-west monsoon in a 333 metre grid (g $C/m^2/d$).

Table 3: The RIAM matrix.

		ES	RS	I	M	P	R	C
PC1	Coastal morphology	-28	-C	2	-2	3	3	1
PC2	Hydraulic conditions	-7	-A	1	-1	3	3	1
PC3	Water quality	-7	-A	1	-1	3	2	2
PC4	Extreme events (natural disasters)	0	N	1	0	3	3	1
PC5	Borrow material	-20	-C	2	-2	2	2	1
PC6	Persistent substances	-14	-B	2	-1	3	2	2
BE1	Coral reefs	-27	-C	3	-1	3	3	3
BE2	Mangroves	0	N	3	0	2	2	2
BE3	Seagrass beds	0	N	2	0	2	2	2
BE4	Endangered species	-24	-C	3	-1	3	2	3
BE5	Eutrophication	0	N	2	0	3	2	2
BE6	Terrestrial ecosystems	-27	-C	3	-1	3	3	3
BE7	Soft bottom macrozoobenthos	-9	-A	1	-1	3	3	3
SC1	Aesthetic and cultural value	-18	-B	1	-2	3	3	3
SC2	Income	14	B	2	1	2	2	3
SC3	Fishery and aquaculture	-8	-A	1	-1	2	3	3
SC4	Housing and squatter development	-14	-B	1	-2	2	2	3
SC5	Cost of living	-36	-D	2	-2	3	3	3
SC6	Recreational value	64	D	4	2	3	2	3
SC7	Employment	28	C	2	2	2	2	3
SC8	Public health and safety	0	N	3	0	1	1	1
EO1	Infrastructure	36	D	2	2	3	3	3
EO2	Maintenance	-14	-B	2	-1	3	3	1
EO3	Utilities	16	B	2	1	3	2	3
EO4	Navigation	7	A	1	1	3	3	1
EO5	Customer base	21	C	3	1	2	2	3
EO6	Regional economy	63	D	3	3	2	2	3
EO7	Quarries	21	C	3	1	2	2	3
EO8	Investment Competition	-36	-D	3	-2	2	2	2

Class	-E	-D	-C	-B	-A	N	A	B	C	D	E
PC	0	0	2	1	2	1	0	0	0	0	0
BE	0	0	3	0	1	3	0	0	0	0	0
SC	0	1	0	2	1	1	0	1	1	1	0
EO	0	1	0	1	0	0	1	1	2	2	0
Total	0	2	5	4	4	5	1	2	3	3	0

42   Environmental Coastal Regions

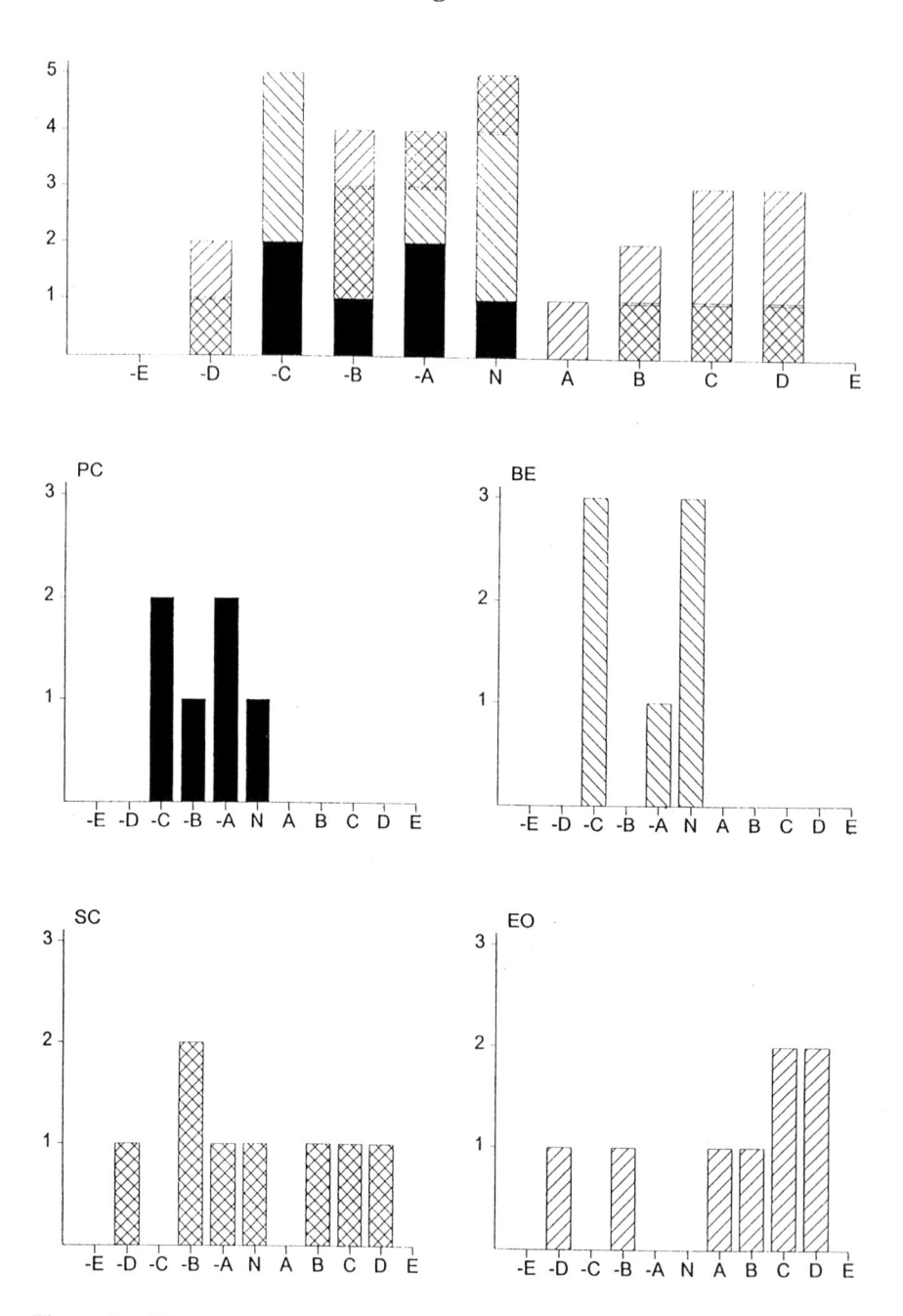

Figure 5: Histogram showing the result of a RIAM analysis. Letters refer to numeric range value (table 2), Y-axes show number of components. PC=physical/chemical, BE=biological/ecological, SC= sociological/cultural, and EO=economic/operational components.

The direct use of EIAs in the planning process of dredging and reclamation projects has been reported[7,10], but these assessments only relate to impacts of exclusive parts of the development. A force of RIAM is that assessments of alternatives and project details can easily be accomplished and repeated during the planning phase allowing an iterative planning process to continue (Jensen[5], Jensen et al[6]). The same applies to the numerical models such as the MIKE21 system. This model system has been used in numerous assessments of the impact of marine construction work[3].

The combined use of numerical MIKE21 models and the RIAM system provides some advantages. RIAM defines the important assessment criteria and environmental components within the project area, as well as a means by which values for each of these criteria can be collated to provide an accurate and independent score for each condition. MIKE modelling supplies the scientific base for parts of the assessment, so that the impacts of project activities can be evaluated against the environmental components, on a without/with project basis, as well as allowing for comparison between different project options.

The assessors are faced with the problem of having to relate very different problems. How does one compare an assessment of substantial impact on employment with significant impact on coral reefs? One cannot do so directly. However, RIAM provides a set of inborn characteristics that relates to most types of impacts or changes. Applying those allows a transparent, visual comparison of the results of the assessment, which more easily aggregate the various negative or positive impacts in one illustration. The human brain is able to keep a tremendously high number of complicated pictures in mind. Similarly RIAM histograms function as a means to contain much information in one file (picture) helping to compare and chose between different solutions.

Acknowledgements
The authors wish to express their thanks to project participants C. Heang Knudsen, Tom M. Foster, Karsten Mangor, Dan B. Hasløv and Sidsel M. Dyekjær for their contribution to provide information needed for this paper, and to Ministry of Tourism and Environment, Sabah, for financing the project.

References

[1] Pastakia, C.M.R., The Rapid Impact Assessment Matrix (RIAM) – A New Tool for Environmental Impact Assessment. *Environmental Impact Assessment Using the Rapid Impact Assessment Matrix*

(RIAM), ed. K. Jensen, Olsen & Olsen, Fredensborg, Denmark, pp. 8-19, 1998.

[2] Warren, I.R. & Bach, H.K., MIKE21: a modelling system for estuaries, coastal waters and seas. *Environmental Software*, 7, pp. 229-240, 1992.

[3] CEU (Council of the European Union), Council Directive on the assessment of the effects of certain public and private projects on the environment. Official Journal of the European Communities, Dir. 97/11/EC, 1997.

[4] World Bank, Environmental Screening, Environmental Assessment Sourcebook Update No. 2, Environmental Department, World Bank, Washington D.C., 1993.

[5] Jensen, A., Environmental Impact Assessment of Halong City Sanitation Project, Vietnam. *Environmental Impact Assessment Using the Rapid Impact Assessment Matrix (RIAM)*, ed. K. Jensen, Olsen & Olsen, Fredensborg, Denmark, pp. 8-19, 1998.

[6] Jensen, K., Bach, H.K., Knudsen, C.H., Foster, T.M., Mangor, K. & Hasløv, D.B., Initial Impact Evaluation of a Tourist Development Project in Malaysia. *Environmental Impact Assessment Using the Rapid Impact Assessment Matrix (RIAM)*, ed. K. Jensen, Olsen & Olsen, Fredensborg, Denmark, pp. 8-19, 1998.

[7] Jensen, K., Møller, J. & Randløv, A., An environmental impact assessment of the construction of bridges and tunnels across the Øresund. *Proc. 12th. BMB Symp. Int. Symp. Ser.*, pp. 87-90, 1992.

[8] Department of the Environment, A Handbook of EIA Guidelines. Ministry of Science, Technology and Environment, Kuala Lumpur, Malaysia. 1995

[9] Ursic, A., Environmental impact assessment: an easy task or a demanding challenge. *Proc. 3rd EIA Conf. Praha,* pp. 64-69, 1996.

[10] Poulsen, K.M., Environmental Optimization of the Great Belt and Øresund Links. *IABSE Journal,* 5, pp. 87-90, 1995.

Using Ecological Models to Plan and Monitor Coastal Reclamation Projects

Kurt Jensen[1], Hanne K. Bach[1] & Anders Jensen[2]

[1]VKI Institute for the Water Environment, 11 Agern Alle, DK-2970 Hørsholm, Denmark
EMail: kje@vki.dk & hkb@vki.dk

[2]Danish Hydraulic Institute, 5 Agern Alle, DK-2970 Hørsholm, Denmark
EMail: anj@dhi.dk

Abstract

An environmental management system called feedback monitoring is being tested on a dredging and reclamation project. The system includes planning, impact assessments and monitoring. It is being applied in connection with a large infrastructure project involving dredging and reclamation of more than 7 million m^2 of seabed.

The system incorporates the use of computerised numerical models of hydrography, sediment transport and sedimentation, water quality and ecological variables. The tool is applied for the planning, environmental impact assessment, development of detailed design and monitoring of the environmental impacts to control the compliance with environmental criteria during the construction.

The use of numerical models has strongly contributed to the fact that the negative impacts of the project have been minimised, and that it has been possible to introduce mitigating actions while the work is still in progress. A two-dimensional hydraulic model (MIKE21) covers the area around the construction work. This forms the basis for models of sediment spreading and sedimentation, and biology of the important eelgrass (*Zostera marina* L.) ecosystem of the area. The system has been developed to include mussel (*Mytilus edulis* L.) bed community in the feedback monitoring system

1 Introduction

Land reclamation projects have become increasingly important to achieve land for development all over the World. The environmental management of dredging and reclamation projects is important to avoid irreversible damage to flora and fauna. The present paper presents an environmental management system that uses computerised numerical models of hydrography, sediment transport and sedimentation, water quality and ecological variables in combination with field monitoring.

A system of land reclamation, an immersed tunnel and a high bridge is being built across the strait between Denmark and Sweden (Figure 1). The project involves the dredging and deposition of approximately 7 million m^3 of dredged material. It consists of limestone varying within few meters from soft, un-lithified to strongly indurated limestone with flint overlaid by a few metres of glacial deposits consisting of tills, and limited deposits of sand and gravel of postglacial age.

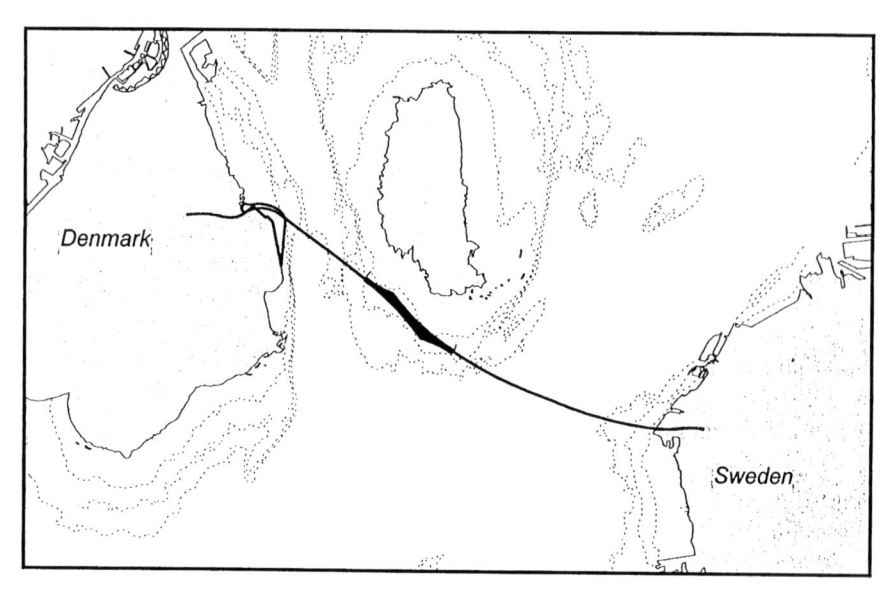

Figure 1:   The link between Denmark and Sweden.

Two biological communities dominate the alignment area: an eelgrass bed community and a mussel bed community (Figure 2). The eelgrass beds are estimated to cover 136 km². Mussel beds with more than 40% coverage of live mussels cover 46 km², and 30% of the populations are developed in shallow water (<6 m). The biomass of mussels in this area is estimated at 92,000 metric tons wet weight including shells.

Sediment will be spilled during the construction work creating plumes of suspended material and the sediment will eventually deposit at the seabed. The sediment plumes may adversely affect the growth and survival of the eelgrass (*Zostera marina*) mainly by the shading effect,[1] and the deposition of sediment at the seabed may affect the mussels (*Mytilus edulis*) by burying or by hampering recruitment.[2,3]

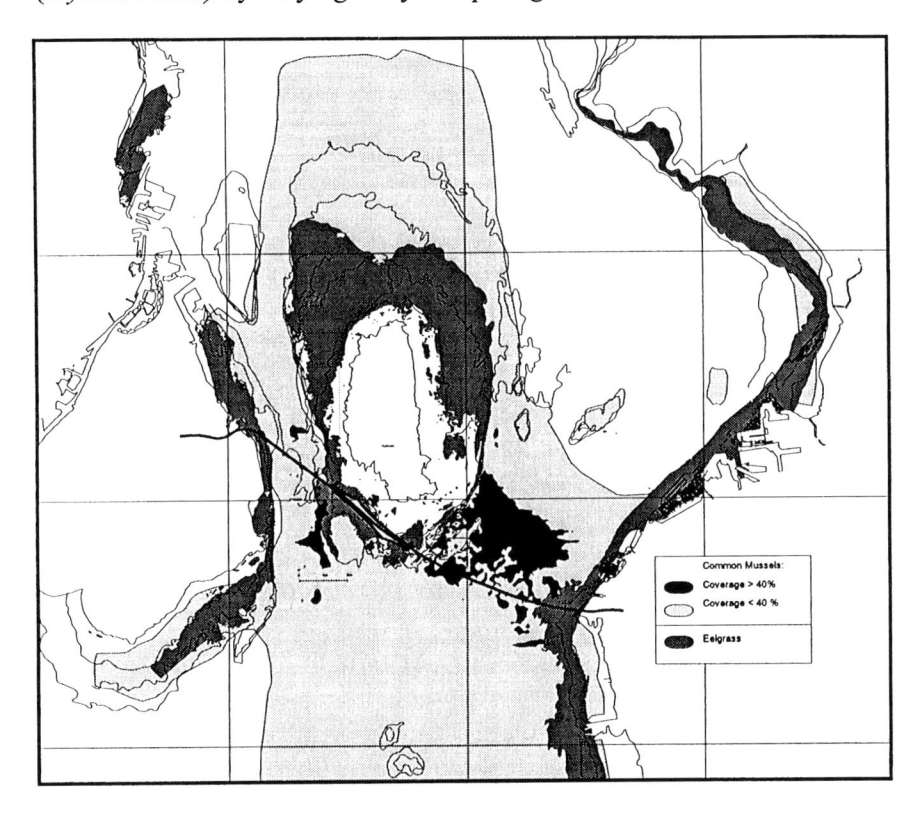

Figure 2: The dominant ecosystems of the study area.

2 Environmental Criteria

The Danish and Swedish authorities have established a set of environmental criteria in order to control the environmental effects. The art of the environmental management of the constructor is to comply with the regulations and to complete the construction work at a minimum of expenses. In practise this means to avoid delays and interruptions.

The criteria of the authorities concerning mussels can be summarised as: 'The total area of common mussel banks with coverage exceeding 40% must not be reduced by more than 25% relative to determined baseline levels as a result of the construction work'. 'The biomass of the mussels inside each area outside the work zones must never in any of the years through the effects of dredging be reduced with more than 25% compared with normal conditions'.

The criteria concerning eelgrass were outlined as: 'The distribution and biomass of eelgrass inside each area outside the work zones must never in any of the years through the effects of dredging be reduced with more than 25% compared with normal conditions'.

3 Work plans and forecast modelling

In order to plan the dredging and reclamation work, detailed work plans were elaborated. The spill rate of sediment was estimated on basis of experience relating to the used equipment and the dredged material. The time, position and magnitude of work were determined. The next step was then to forecast the impacts of these spills on the mussel and eelgrass communities, and assess if a work plan would comply with the environmental criteria.

The environmental criteria of the authorities relating to mussels were transformed into more operational criteria describing sedimentation. This transformation was based on literature and experience from a similar large bridge and tunnel building project across the Great Belt in Denmark. The mussels are able to survive rather substantial sedimentation, and the mussel beds were anticipated to survive as long as sedimentation rate were kept below $15 kg*m^{-2}*y^{-1}$.[3,4]

Mussel larvae are more susceptible to sedimentation as their settling may be hampered at lower concentrations. At concentrations above approximately 60 $g*m^{-2}*d^{-1}$ settling was reduced in the Great Belt.[3] This is of course only of interest in the period of the year where settling actually occurs.

The environmental criteria of the authorities relating to eelgrass were transformed into operational criteria stating that shoot density and biomass should not decrease more than 25% at any sampling point, biomass and distribution should not decrease more than 25% in predefined sub-areas. The amount of soluble carbohydrates in the rhizomes should not fall below 70 $mg*g^{-1}$ DW, based on Bach et al.[2]

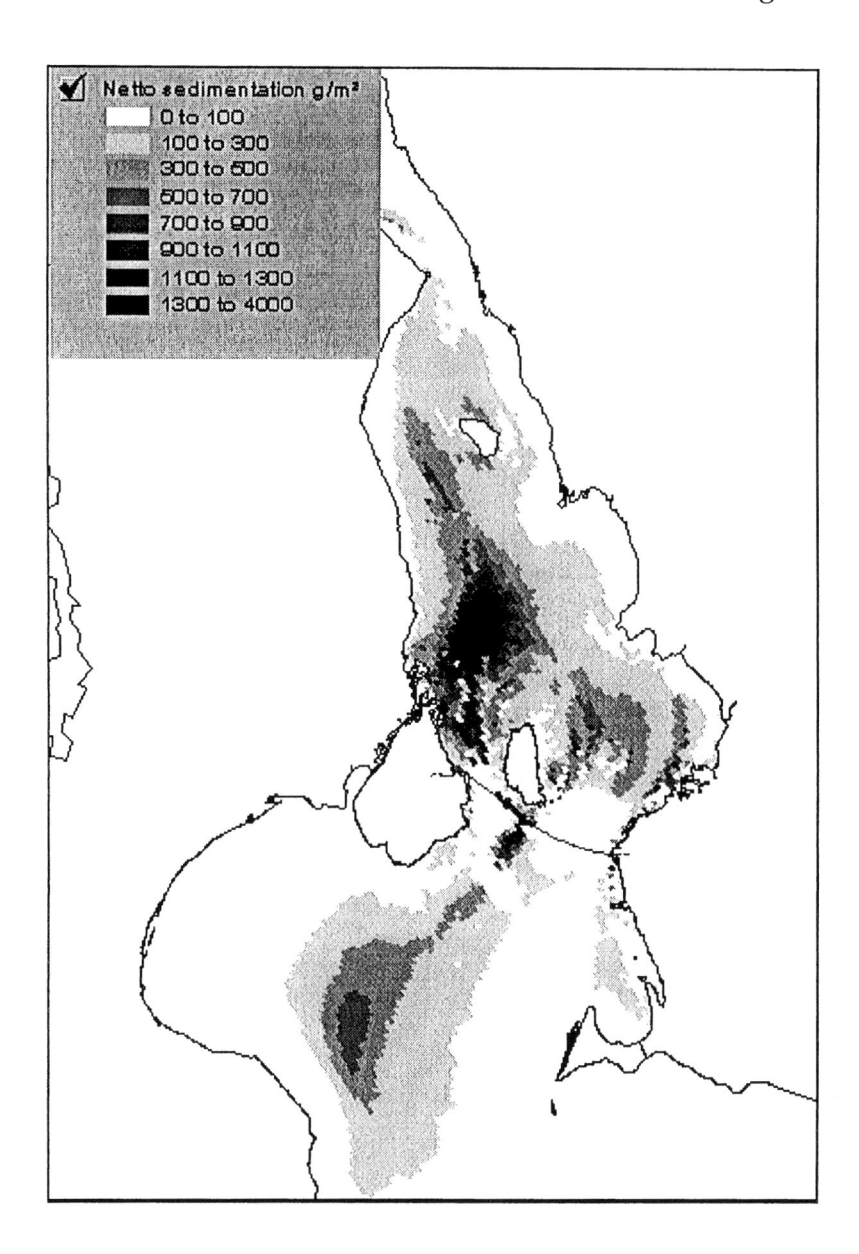

Figure 3: Result of modelling of accumulated sedimentation using MIKE21 PARTICLE.

For the forecasting of the effects on the mussel beds of the dredging and reclamation work were done by forecasting sedimentation using the PARTICLE module of the MIKE21 model system, which has been established covering the relevant marine area. The underlying

50 Environmental Coastal Regions

hydrodynamic MIKE21 model system has been described by Warren & Bach,[5] and the principles of the sediment model PARTICLE by Brøker et al.[6] An example of the forecast modelling of sedimentation is shown in Figure 3.

The forecast modelling of the effects on the eelgrass populations were done using an eelgrass community model. The model applies the previously described sediment model as basis and describes the eelgrass community in 18 state variables.[1,7] An example of the result of a forecast model calculation of the eelgrass biomass is shown in Figure 4.

If a model forecast showed that violation of the environmental criteria would most likely occur during the construction work if a work plan was implemented, the work plan was changed until compliance with the criteria. During the construction the work plans were changed for different reasons. Accordingly, the new work plans were assessed, constantly ensuring compliance with criteria.

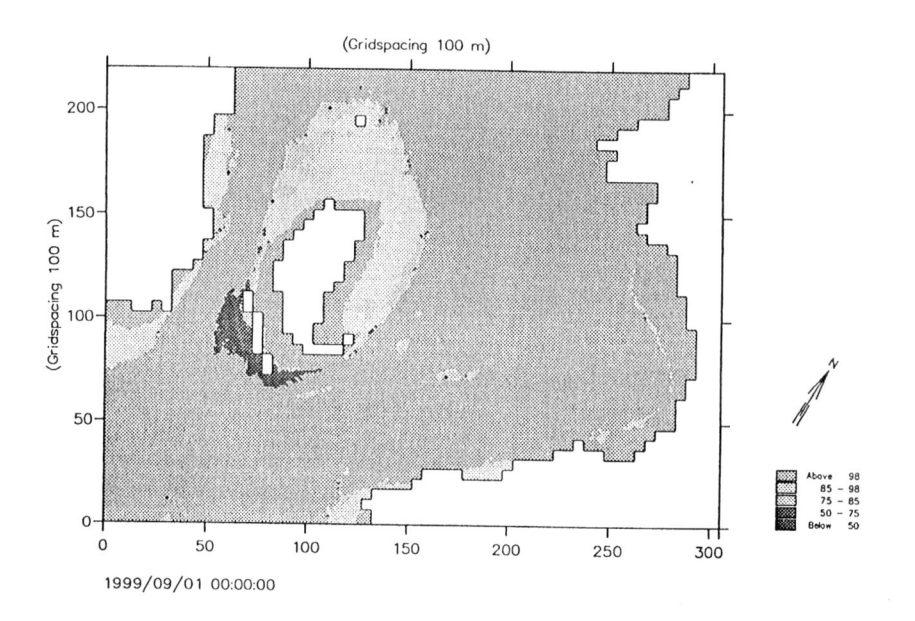

Figure 4: Result of modelling of the biomass of eelgrass in September 1999 using MIKE21 EU (eelgrass) model.

4 Monitoring and hindcast modelling

When the construction work was initiated in accordance with the accepted work plan, a feedback-monitoring programme was continuously running. The programme contained three elements (sediment spill, mussels and eelgrass) that aimed at controlling the forecast model results and the compliance with the environmental criteria.

Sediment spill from the construction activities was measured using acoustic dobbler profiler (ADCP) and optical backscatter sensors mounted on a streamer. Sedimentation was monitored by several different methods, i.e. sediment traps permanently deployed at fixed stations, video inspection, and sidescan sonar.

Mussel distribution was monitored using a non-destructive photo-sampling technique.[4] The mussel biomass was determined as wet weight of samples collected by SCUBA diving.

The eelgrass variables that were monitored were shoot density, biomass and soluble carbohydrates in the rhizomes. Sampling was done by SCUBA diving.

In the feedback-monitoring programme the sampling and analyses were done at weekly to biweekly intervals. Further, the distribution of mussels was controlled annually by acoustic methods and eelgrass distribution by aerial photography.

The forecast modelling was based on anticipated environmental conditions with respect to temperature, wind, prevailing current directions and speed. However, the actual spill may differ from the planned, and occurring environmental conditions may differ significantly from the average situation that has been used in the forecast modelling. Consequently, model calculations were repeated at regular intervals, then based on the actually observed environmental conditions, so-called hindcast modelling.

The results of the hindcast modelling were used as basis for hot-start of forecast modelling resulting in increased level of precision of the assessment of the effects of the proceeding plans.

An example of the results of the monitoring programme compared with the modelling results is shown in figure 5.

Monitoring gives results that record the environmental state in one sampling point at one occasion. The results of the hindcast modelling were used to interpolate in time and space, and produce area-related information.

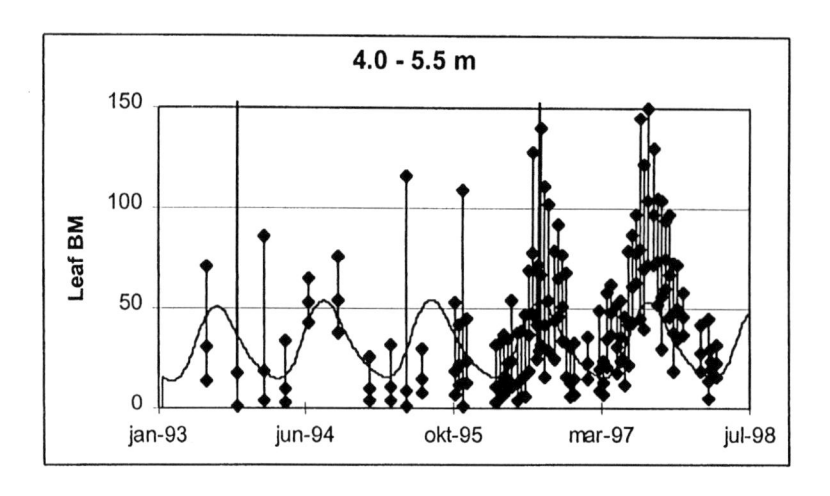

Figure 5: Comparison of monitoring and hindcast modelling results, leaf biomass (gC/m^2). Construction activities started October 1995.

5 Discussion

In the past the major efforts in connection with the environmental management of large construction projects at the sea and at shores have been put into the plain recording of the environmental disturbance, or it has been targeted towards compliance test of environmental impact assessment leaving little space for mitigating actions. This holds for most other monitoring programmes. If the environmental management shall achieve the aim of maximum protection of the environment, such monitoring approach has little chance of fulfilling its aim. Gray & Jensen[8] have suggested another monitoring strategy, so-called feedback-monitoring approach. Those ideas have been the basis for the environmental management of the programme described here. Also the London Convention Dredged Material Assessment Framework suggests feedback of monitoring although in a less detailed form.[9]

Environmental Coastal Regions 53

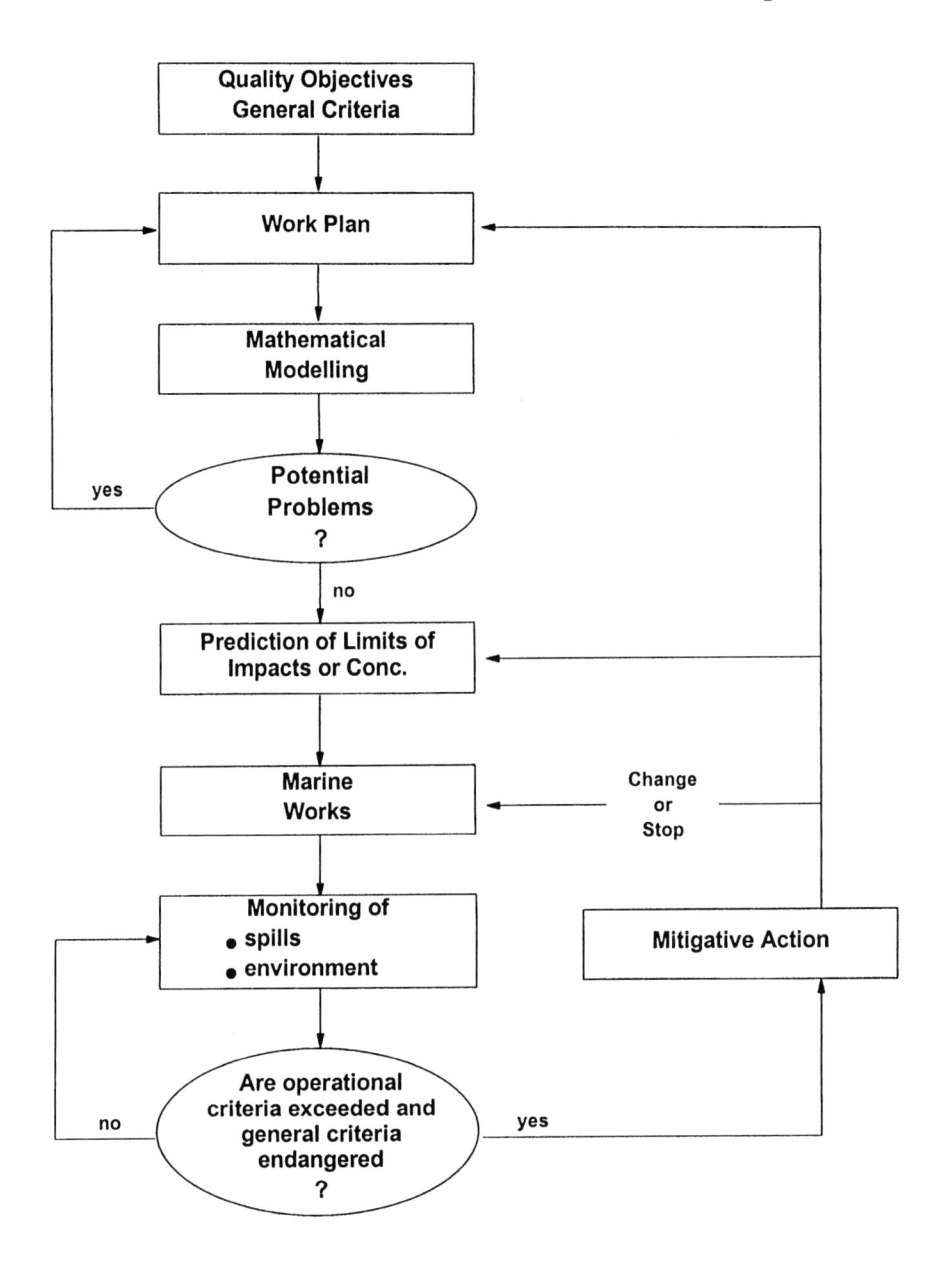

Figure 6: *Principles of feedback monitoring.*

The cautious planning, cf. the testing of the work plans is important for the success of a feedback-monitoring programme. This can only be done if two things are fulfilled, namely knowledge of certain dose-response relationship between the impact and the effect, and a model of this relationship.

The knowledge of the dose-response relationship from the dosage of spilled sediment to response by an eelgrass variable and the modelling of this run through the following steps: spilled sediment ⇒ sediments plumes ⇒ turbidity/light ⇒ eelgrass variable. Information about the relationship between the steps has been obtained for the first step through the MIKE21 PARTICLE model.[5] A series of laboratory experiments have shown the connection between concentration of sediments in water and turbidity[10], and the influence of light and shading on eelgrass have been studied in field experiments.[1] On the modelling side the steps are: sediment spill ⇒ MIKE21 PARTICLE [6] ⇒ MIKE21 EU (eelgrass). [7]

In the case of impact on mussel beds knowledge could be improved on the effects of sedimentation. At present no model is available that allows for a direct modelling beginning with the spilled sediment and ending with a resulting variable of the mussel population. The MIKE21 PARTICLE can bring us so far as to modelling sedimentation. From here a primitive verbal model has to be used such as 'sedimentation above 60 $g*m^{-2}*d^{-1}$ will prevent recruitment of young mussel'. Nevertheless, the transformation of the information into operational criteria, and the ensuring of compliance with those have prevented effects on the mussel beds of sediment spill from the dredging operations.

The present study demonstrated the high degree of environmental protection that can be obtained by the use of a feedback monitoring programme which includes variables complying with the definitions of Gray & Jensen[8] and Randløv & Jensen[11] as shown in figure 6.

References

[1] Bach, H.K., Jensen, K. & Lyngby, J.E., Management of Marine Construction Works Using Ecological Modelling, *Estuarine, Coastal and Shelf Science,* **44** (Supplement A), pp. 3-14, 1997.

[2] Bender, K. & Jensen, K., The effects of the construction works at the Great Belt link on mussel (*Mytilus edulis*) beds. *Proc. of the 12th Baltic Marine Biologists Symposium,* eds. E. Bjørnestad, L. Hagerman & K. Jensen, Olesen & Olsen, Fredensborg, pp. 17-22, 1992.

[3] Jensen, K., Madsen, K.N. & Bach, H.K., Feedback Monitoring of Dredging Operations using Mussel (*Mytilus edulis*) as a Key Organism. *Environmental Monitoring and Assessment* (in press).

[4] Brinke, W.B.M. ten, Augustinus, P.G.E.F. & Berger, G.W., Fine-grained Sediment Deposition on Mussel Beds in the Oosterschelde (The Netherlands), Determined from Echosoundings, Radio-isotopes and Biodeposition Field Experiments. *Estuarine, Coastal and Shelf Science*, **40**, pp 195-217, 1995.

[5] Warren, I.R. & Bach, H.K., MIKE21: a modelling system for estuaries, coastal waters and seas. *Environmental Software*, **7**, pp. 229-240, 1992.

[6] Brøker, I., Johnsen, J., Lintrup, M., Jensen, A. & Møller, J.S., The Spreading of Dredging Spoils During Construction of the Denmark-Sweden Link. Proceedings 24th International Conference Coastal Engineering, Kobe, Japan, 23-28 October 1994, pp. 1-14, 1994.

[7] Bach, H.K., A Dynamic Model Describing the Seasonal Variation in Growth and Distribution of Eelgras (*Zostera marina* L.). I Model Theory. *Ecological Modelling*, **65**, pp. 31-50, 1993.

[8] Gray, J.S. & Jensen, K., Feedback Monitoring: A New Way of Protecting the Environment. *Trends in Ecology and Evolution*, **8**, pp. 267-268, 1993.

[9] Bray, R.N., Bates, A.D. & Land, J.M., Dredging, Arnold, London, 428 pp., 1997.

[10] Øresundskonsortiet, Copenhagen, Report no 93/113/1E and 93/118/1E.

[11] Randløv, A. & Jensen, K., Design and Results of a Biological Feedback Monitoring Programme. *IABSE Rep.*, **63**, pp. 107-114, 1991.

An Expert System for Marine Environmental Monitoring in the Florida Keys National Marine Sanctuary and Florida Bay

James C. Hendee
Atlantic Oceanographic and Meteorological Laboratory
National Oceanic and Atmospheric Administration
4301 Rickenbacker Causeway
Miami, FL 33149-1026
EMail: hendee@aoml.noaa.gov

Abstract

The National Oceanic and Atmospheric Administration's (NOAA, U.S. Department of Commerce) Atlantic Oceanographic and Meteorological Laboratory (AOML) works cooperatively with the Florida Institute of Oceanography (FIO) in the implementation of the SEAKEYS (Sustained Ecological Research Related to Management of the Florida Keys Seascape) network, which is situated along 220 miles of coral reef tract within the Florida Keys National Marine Sanctuary (FKNMS). This network is itself actually an enhanced framework of seven Coastal-Marine Automated Network (C-MAN) stations for long-term monitoring of meteorological parameters (wind speed, wind gusts, air temperature, barometric pressure, relative humidity). To the C-MAN network SEAKEYS adds oceanographic parameters (sea temperature, photosynthetically active radiation, salinity, fluorometry, optical density) to the stations. As a recent enhancement to the SEAKEYS network, an expert system shell is being employed to provide daily interpretations of near real-time acquired data for the benefit of scientists, fishermen and skin divers. These interpretations are designed to be automatically emailed to Sanctuary managers and to the FIO maintainers of the network. The first set of interpretations include those dealing with environmental conditions conducive to **coral bleaching. Other marine environmental interpretations will be forthcoming**

1 Introduction

1.1 Background

The Florida Keys National Marine Sanctuary (FKNMS) was established to provide protection for a unique marine ecosystem in the Florida Keys so that generations of visitors could enjoy the beauty of coral reefs and the enjoyment of water sports such as skin and scuba diving, boating and fishing. "Protection" entails not only the policing of the Sanctuary to prevent plundering of its natural resources by poachers, but also monitoring the status (i.e., "health") of the ecosystem so that appropriate steps may be taken in the event anthropogenic stressors threaten the environment. The FKNMS also seeks to support basic research of the marine ecosystem to further understand its mysteries, and, where applicable, to apply the findings of that research not only to the general fund of knowledge, but toward greater enjoyment for its visitors. Conservation and understanding of the Florida Keys natural marine resources are its prime directives.

At the National Oceanic and Atmospheric Administration's (NOAA) Atlantic Oceanographic and Meteorological Laboratory (AOML), Ocean Chemistry Division (OCD), in Miami, Florida, environmental data such as wind speed, barometric pressure, dew point, sea temperature and salinity are acquired from remote sites on lighthouses and navigational aids situated at reefs along and within the FKNMS via a satellite data archival site at Wallups Island, Virginia. The data are collected at the sites hourly by oceanographic instruments which are maintained by the Florida Institute of Oceanography (FIO) and meteorological instruments which are maintained by the National Data Buoy Center (NDBC) of NOAA. There are currently six sites, which have been termed the SEAKEYS (Sustained Ecological Research Related to Management of the Florida Keys Seascape; Ogden [1]) network: Fowey Rocks (in Biscayne National Park), Molasses Reef (off Key Largo), southern Florida Bay (off Long Key), Sombrero Key (off Marathon), Sand Key (near Key West), Dry Tortugas, and northwestern Florida Bay. Field maintenance for the oceanographic instruments, and data management for these sites, are being funded by NOAA's South Florida Ecosystem Restoration, Prediction and Modeling (SFERPM) program, which seeks to describe, model, predict and, where indicated, restore marine and coastal ecosystem processes in and near Florida Bay. The SFERPM program has extended the existing monitoring capabilities of the SEAKEYS network by funding the addition of instruments for measuring turbidity, fluorometry and water level at three sites critical to the interaction of Florida Bay and the FKNMS.

There are many physical, chemical and biological events of interest and concern to the managers of the FKNMS, marine biologists, oceanographers, fishermen and divers. Some of these events would of course be observable if it were possible to continuously be present at a remote site of interest, or if instrumentation could monitor the remote site and the observer could in turn monitor the output of the instrumentation. Except in very critical cases, however, the truth is that large volumes of data are generated by instruments which may be distributed over many sites (as in the SEAKEYS network), and no one has the time to look at every printout of data from every station, every day, seven days a week. It is highly desirable to have an automated system that can monitor meteorological and oceanographic parameters and produce specialized alerts of specific events. The expert system described here collects data from one station (Sombrero Key) in the SEAKEYS network and produces automated email and World-Wide Web (the "Web") alerts when conditions are conducive to coral bleaching. This represents the first of many applications for this system.

1.2 Expert Systems

Expert systems, or knowledge-based systems, are a branch of artificial intelligence. Artificial intelligence is the capability of a device such as a computer to perform tasks that would be considered intelligent if they were performed by a human (Mockler & Dologite [2]). An expert system is a computer program that attempts to replicate the reasoning processes of experts and can make decisions and recommendations, or perform tasks, based on user input. Knowledge engineers construct expert systems in cooperation with problem domain experts so that the expert's knowledge is available when the expert might not be, and so that the knowledge can be available at all times and in many places, as necessary. Expert systems derive their input for decision making from prompts at the user interface, or from data files stored on the computer. The knowledge base upon which the input is matched is generally represented by a series of IF/THEN statements, called production rules, which are written with the domain expert to approximate the expert's reasoning. The degree of belief the expert has in her conclusion is represented as a condition factor (CF) in the expert system. For instance, the expert may feel that the conclusion based upon the input has a 95% probability of being correct, so the CF would equal 95.

1.3 Expert Systems in Oceanography

There appear to have been relatively few expert systems constructed, or envisioned, for oceanographic and maritime purposes. Ryan and Smith [3] proposed an expert system for fisheries management purposes; however, the senior author developed interests elsewhere and the project idea was dropped (P. Smith, personal communication) . Holland [4] proposed the use of an expert system and other artificial intelligence tools for the extraction of marine environmental information, but apparently was never able to follow up with funding to develop a system. Sigillito et al [5] and Wachter and Sigillito [6] developed a prototype system, XCOR, to aid human analysts in detecting and correcting errors in oceanographic data collected

60 Environmental Coastal Regions

from research and merchant voluntary observing ships throughout the world. Groundwater [7] developed a preliminary "ocean surveillance information fusion expert system" which modelled the thought processes of a surveillance watch analyst who assesses vessels' missions and destinations given their previous record of tracks, history and locations/status of other vessels in a particular domain of interest (e.g., shipping). Lybanon and associates [8, 9, 10, 11] developed a successful expert system for the interpretation of image data for the purpose of predicting movements of the Gulf Stream, as did Thomason [12]. Dantzler and Scheerer [13] and Scheerer and Dantzler 14] apparently spent a considerable amount of time and effort in the design of an expert system for predicting the coastal ocean environment. However, these authors eventually split their efforts along different lines to develop the Tactical Oceanographic Monitoring System [15], which does not appear to use an expert system framework, and explorations into the relationship between upwelling and fog formation using sea surface temperatures and the Regional Atmospheric Modeling System (D. Scheerer, personal communication). Cochrane and Hutchings [16] suggested that an expert system could be developed for forecasting recruitment of anchovies, using observed values of copepod biomass and oocyte development in adult female anchovies, but the system has yet to be developed. Hendee [17] constructed an expert system to determine whether oceanographic data collected at sea represented reasonable values as compared to those collected by other ships at sea visiting the same area.

1.4 Coral Bleaching

Coral bleaching may be described as the general whitening of coral tissues due to the release of symbiotic zooxanthellae and/or reduction in photosynthetic pigment concentrations in the zooxanthellae residing within the tissues of the host coral. Bleaching can be a generalized stress response to harsh environmental conditions such as high sea temperature or abnormal salinity (abiotic-induced bleaching), bacteriological or viral infection (biotic-induced bleaching), or for other unknown reasons (Glynn [18], Kushmaro et al [19]). In most reported incidences of coral bleaching, however, high sea temperature is in evidence and appears to be the chief environmental stressor [Brown & Ogden [21], whether those temperatures may be considered "abnormal" or not (see, for example, Atwood & Hendee [23]). One other event that appears to influence abiotic-induced bleaching is the presence of low wind speed, as this apparently favors localized heating and a greater penetration of solar radiation (Glynn [18]; Causey [22]; Jaap [23]; Lang [24]). The expert system described below screens near real-time incoming wind speed, sea temperature and photosynthetically active radiation (PAR) data to determine if conditions are optimal for abiotic-induced coral bleaching at Sombrero Key. The use of the expert system allows researchers to model the environment to further understand the phenomenon, as the knowledge base can be easily reconfigured. The automated transmittal of the bleaching alerts also gives the FKNMS managers some immediate feedback on one facet of the "health" of the sanctuary. It similarly alerts researchers so that they may travel to the site for further study, and so that they may provide feedback to the knowledge engineer for further fine-tuning of the system (e.g., whether or not bleaching is beginning to occur or not). Finally, development of the framework for monitoring and reporting provides a springboard for other environmental modeling, for example, harmful algal blooms (i.e., "red tide").

2 Methods

2.1 Expert System Development

CLIPS (C Language Integrated Production System) was used for the expert system shell, and Hypertext Markup Language was used for the presentation of data and expert system analyses via the Web. The expert system proceeds in different stages.

2.1.1 Stage 1
Every day at 4:05 am, an automated data acquisition program is initiated. The program uses communications software and a modem to call a location at Wallups Island, Virginia, which holds the archived raw data. The data held are those which were collected five minutes earlier via a satellite from the SEAKEYS stations in the FKNMS, which transmit the data on the hour, every hour. From the archived site, a caller can acquire data for the last 72 hours worth of transmissions from the stations. After the program makes contact with the archival site, using a password, a special prearranged data suite identifier is supplied. The required data are then supplied to the screen, and the program captures the whole session in a file. The data represent values recorded from instruments designed to measure barometric pressure, wind speed, wind gusts, wind direction, air temperature, sea temperature, salinity, photosynthetically active radiation (PAR), and sensor depth, which provides and indirect measurement of the state of the tide. With these data are also sent the station name, date and time (GMT).

The raw data file captured by the data acquisition module is parsed to extract the data of interest. The data are written to an ASCII text file in a format that can be read and interpreted by Stage 2, described below. This file is more easy to work with than the complex raw file. CLIPS is used as the data parser, as programming language classes for each SEAKEYS station are the same, and code changes due to data stream changes at each station are more easy to effect without having to recompile an executable, as in a language like C. Also, program code is easily portable across operating systems when using CLIPS.

2.1.2 Stage 2
ASCII data files produced in Stage 1 are screened against production rules to determine whether the values for the instruments are within realistic ranges, or whether the instrument appears to be malfunctioning or offline (garbled or no data). To aid

in the analysis of data, which may vary widely depending upon the time of day and the season of the year, values are averaged for eight three-hour periods per day, termed midnight (2200 to 0100 hours local time), pre-dawn (0100 to 0400 hours), dawn (0400 to 0700), morning (0700 to 1000 hours), all day (1000 to 1300), pre-sunset (1300 to 1600), and evening (1900 to 2200 hours). These groupings are convenient because meteorological phenomena quite often show predictable fluctuations during these periods of the day, for instance the change of wind direction and wind speed with sunrise and sunset. The averaged values within each of these categories are then determined to fall within one of eleven categories: unbelievably low, drastically low, very low, low, somewhat low, average, somewhat high, high, very high, drastically high, and unbelievably high. These groupings are arbitrary names, of course, and parameters such as wind direction are further translated to different regions of the compass (e.g., NE-ENE). The assignment of values to these categories depends upon what the season of year is (spring, summer, fall or winter), so, for instance, what might be considered "somewhat high" in winter might otherwise be considered to be "average" during summer. Values which are determined to be "unbelievably high" or "unbelievably low" represent values which are considered to be totally unrealistic for the parameter in question. However, should it happen that these values begin to represent real-life values, the ranges may be easily reset to encompass the newer values.

The instruments themselves are visited once a week by field technicians of the FIO so corrective maintenance of the oceanographic sensors can be attended to as necessary. In any case, oceanographic instruments become fouled and must be cleaned periodically. Sea temperature and salinity sensors are "sea-truthed" during the station visits. That is, calibrated instruments are taken into the field and measured at the same time the *in situ* instruments make their automated measurements to see if the *in situ* instrument needs to be replaced, or its data need to be corrected. Meteorological instruments are maintained by NDBC. If any of the instruments are malfunctioning, the expert system code is easily adjusted so that those values are not accounted for in the process. Under Stage 3 (below), production rules requiring measurements from the malfunctioning instrument simply do not fire, but the fact of a malfunctioning instrument is sent via email and saved as a file for that day's worth of data interpretations.

The status of the parameters, that is, where they are on the continuum from unbelievably low to unbelievably high, and when these values occurred (i.e., the period of the day), are saved as "facts" in a text file, which is loaded into the program under Stage 3 processing.

2.1.3 Stage 3
The "facts" loaded from the processing under Stage 2 (which requires about 1800 production rules) represent subjective interpretations of the measured data, and therefore represent *information*, not just columns of numbers. This information

synthesis, proceeding according to how experts view the ranges of data, makes the further processing for the benefit of determining environmental trends and events easier in the context of an expert system. At this point, information represented as occurring through time (since the "facts" show data progress through the previous 72 hours, or longer still, if desired), presents an enormous range of possibilities for further processing, limited only by the skill of the programmer and the time required for coding.

There is an excellent chance that abiotic-induced coral bleaching can be predicted, as this appears to be a function of environmental stress. As mentioned above, high sea temperature alone, or in combination with low wind speeds and/or high solar radiation appear to be the chief stressors involved, whether they are the sole causative agents or not. The following pseudocode represents one of the production rules of the coral bleaching module of this expert system:

> IF sea temperature is drastically high all day,
> AND wind speed is very low all day,
> AND wind gusts are very low all day,
> AND PAR at 1 m is drastically high during daylight hours,
> THEN there is a high probability (CF = 95) that
>   abiotic-induced coral bleaching will occur.

Additional production rules make CF assertions depending upon the subjective values of all the variables. If a parameter is missing because an instrument is malfunctioning, rules fire which do not depend on the parameter in question; thus, the CF would be lower.

As the model proves to be right or wrong, using feedback from sanctuary divers, the knowledge base is adjusted to more accurately predict further bleaching events, and researchers begin to further understand the phenomenon.

3 Results

In a test of the system, data from 1997 during the time of widespread coral bleaching at Sombrero Key, were used as input, however values for PAR came from another station, as the Sombrero Key station did not have operable PAR sensors at that time. Unfortunately, data were not kept during summer 1997 as to the first occurrence of coral bleaching, so more refined testing of the ability of the system to predict bleaching could not be done. However, the real values and fabricated values resulted in the system correctly detecting and reporting conditions that would be expected to be conducive to coral bleaching. The expert system continues to execute every morning after receipt of the daily data, so it is anticipated that the first alerts may come this summer, if the conditions occur again. These alerts will be available via the Web at http://coral.aoml.noaa.gov/sferpm/seakeys/es. NOAA and FIO divers will be alert to the first signs of coral bleaching at Sombrero Key, and will offer feedback on the further development of the system.

At this time, less refined output of the system, available at the same Web address, reflects the status of the last seven days of processed data for all parameters. This portion of the system still requires much work.

4 Discussion

It is anticipated that further development of the expert system described here might provide alerts such as conditions conducive to the onset of harmful algal blooms (e.g., "red tide"), conditions conducive to the arrival of commercially important fisheries stocks (such as pink shrimp, *Penaeus duorarum*), conditions conducive to larval fish (and other animal) survival or death (e.g., extended high or low temperatures), incidences of extended duration of clear or turbid water (via optical density measurements), phytoplankton blooms (measured directly through fluorescence), influx of hypo- or hypersaline waters from Florida Bay through the Florida Straits, excessive dissolved nutrient encroachment (reflected in sustained high levels of fluorescence), and the influx of cool or warm water from Florida Bay through the Florida Straits. Such predictive models, through iterative development with domain experts, will help environmental managers and scientists understand natural cycles in Florida Bay, and serve as a backdrop for feedback on, and development of, environmental regulations.

Acknowledgements

I should very much like to thank Chris Humphrey and Trent Moore of FIO for their intrepid, careful and often thankless work at maintaining the SEAKEYS network.

References

[1] Ogden, J., Porter, J., Smith, N., Szmant, A., Jaap, W. & Forcucci, D. A long-term interdisciplinary study of the Florida Keys seascape. Bulletin of Marine Science, **54(3)**, 1059-1071, 1994.

[2] Mockler, R.J. & Dologite, D.G., *Knowledge-Based Systems. An Introduction to Expert Systems*, Macmillan Publishing, New York, 1992.

[3] Ryan, J.D & Smith, P.E., An "expert system" for fisheries management. *Oceans '85 Proceedings: Ocean Engineering And The Environment*, **2**, pp. 1114-1117, 1985.

[4] Holland, C.R., Artificial intelligence and the extraction of marine environmental information. *Proceedings Marine Data Systems '86: Marine Data Systems International Symposium*, pp. 433-436, 1986.

[5] Sigillito, V., Wachter, R. & Hunt, R., Jr., XCOR—A knowledge-based system for correction of oceanographic reports. *IEEE Symposium: Expert Systems in Government* [Proceedings held in McLean, VA, October 24-25. Available from IEEE Service Center (Cat no. 85CH2225-1), Piscataway, NJ.], pp. 190-195, 1985.

[6] Wachter, R.F. & Sigillito, R.G. Man-machine interface for a knowledge based system for validating oceanographic reports. *Second Conference on Artificial Intelligence Applications: The Engineering of Knowledge-Based Systems*, [Available from IEEE Service Center (Cat no. 85CH2215-2), Piscataway, NJ], pp. 342-346, 1985.

[7] Groundwater, E.H., An expert system for ocean surveillance. *IEEE Symposium: Expert Systems in Government* [Proceedings held in McLean, VA, October 24-25. Available from IEEE Service Center (Cat no. 85CH2225-1), Piscataway, NJ.], pp. 196-199, 1985.

[8] Bridges, S.; Chen, L-C & Lybanon, M., Predicting and explaining the movement of mesoscale oceanographic features using CLIPS. *Proceedings 3rd Conference on CLIPS, September 12-14, 1994. LBJ Space Center*, p. 211-216, 1994.

[9] Romalewski, R.S. & Lybanon, M., Implementation of an oceanographic expert system: Problems, feedback, solutions. *NOARL-PR-89-069-252*, [Available from: Naval Oceanographic and Atmospheric Research Lab.,Stennis Space Center, MS], 6 pp. ,1990.

[10] Lybanon, M. & Romalewski, R., An expert system to aid the oceanographic image analyst. *SPIE 1293, Applications of Artificial Intelligence VIII*, 918-928, 1990.

[11] Lybanon, M., McKendrick, J., Blake, R., Cockett, J. & Thomason, M.G., A prototype knowledge-based system to aid the oceanographic image analyst. *SPIE 635, Applications of Artificial Intelligence III*, 203-205, 1986.

[12] Thomason, M.G., *Oceanographic expert system development: Final report—Year 1*. Knoxville, TN: Perceptronics Corporation. [NTIS Order No.: AD-A189 134/0/GAR. Contract N00014-87-C-6001], 1987.

[13] Dantzler, H.L. & Scheerer, D.J., An expert system for describing and predicting the coastal ocean environment. *Johns Hopkins APL Technical Digest,* **14(2)**, 181-192, 1993.

[14] Scheerer, D.J. & Dantzler, H.L., Jr., Expert system tools for describing and predicting the coastal ocean environment. *IEEE Ocean* **2(II)**, 11-16, 1994.

[15] Dantzler, H.L., Jr., Sides, D.J. & Neal, J.C., An automated tactical oceanographic monitoring system. *Johns Hopkins APL Technical Digest* **14(3)**, 281-295, 1993.

[16] Cochrane, K.L. & Hutchings, L., A structured approach to using biological and environmental parameters to forecast anchovy recruitment. *Fisheries Oceanography* **4(2)**, 102-127, 1995.

[17] Hendee, J., PELAGOS: An expert system for quality control and feature recognition of oceanographic data from the open ocean. *National Oceanic and Atmospheric Administration, Technical Memorandum, Environmental Research Laboratories*, **AOML-87,** 1995.

[18] Glynn, P., Coral reef bleaching: ecological perspectives. *Coral Reefs* **12**, pp 1-17, 1993.

[19] Kushmaro, A., Rosenberg, E., Fine, M. & Loya, Y., Bleaching of the coral *Oculina patagonica* by Vibrio K-1. *Marine Ecology Progress Series* **147(1-3)**, pp. 159-165, 1997.

[20] Brown, B.E & Ogden, J.C., Coral bleaching. *Scientific American*, **268**, pp. 64-70, 1993.

[21] Atwood, D.K. & Hendee, J.C., An assessment of global warming stress on Caribbean coral reef ecosystems. *Bulletin of Marine Science,* **51(1)**, pp. 118-130, 1992.

[22] Causey, B.D., Observations of environmental conditions preceding the coral bleaching event of 1987—Looe Key National Marine Sanctuary. *Proceedings Association of Island Marine Laboratories of the Caribbean,* **21**, pp. 48, 1988.

[23] Jaap, W.C., The 1987 zooxanthellae expulsion event at Florida reefs. *NOAA's Undersea Research Program Research Report* **88(2)**, pp. 24-29, 1988.

[24] Lang, J.C., Apparent differences in bleaching responses by zooxanthellate cnidarians on Colombian and Bahamian reefs. *NOAA's Undersea Research Program Research Report* **88(2)**, pp. 30-32.

Risk assessment due to water rise of the Caspian sea coastal area

A. Ardeshir, A. Taher shamsi, M. Fathi
Civil Eng. Department, Amirkabir University of Technology, Havez Avenue, Tehran, Iran

Abstract

This paper deals with the effect of water rise in the coastal area of the Caspian Sea, its affects on hydraulics and the morphology of the coast and its future serviceability.

In the last 100 years the water level was at -26 meters with very few fluctuations. Therefore, most industrial and urban developments were based on this level.

The case study shows that public and private facilities and urban development will be overtaken by the water rise, which could complicate their usability. It will also cause the nonuniform settlement of buildings, and a reduction in drainage. The intrusion of sea water into drinkable water is another problem. The study's findings indicate that future land use and coastal development should be subject to reevaluation.

1 Introduction

The Caspian Sea is one of the largest natural lakes, with a surface area estimated between 360000 to 400000 square kilometers (km^2), and retains about 80000 cubic kilometers (km^3) of water[3].

The Caspian Sea is a closed system, being separated from the sea and the oceans. This separation has caused the water level of the Caspian Sea to be lower than the zero geographical surface of the oceans. For shipping the surface elevation of the sea, based on the (1940-1945) measurement, is -28 meters below zero ground elevation [5].

The Caspian Sea environment can be categorised into three regions:
1) The northern region has an average depth of 5-6 meters; more than 86 per cent of the sea's water enters from this region [1,2,3]. Because of this, there is a diffusion flux from north to south. In winter the temperature of the water goes down to -10° and the surface area is frozen.
2) The mid region is deeper, with a depth of about 600-700 meters.
3) The southern region is the deepest, having a maximum depth of 1000m. This region has a mild climate, with an average temperature in the winter of about +12°C.

Due to the inflow of the potable water from the Volga river basin from the north to the Caspian sea, the salt concentration varies between 2 percent in the north to 14 percent in the south and southwest [1,2].

2 Studies Carried Out

Among all the existing problems relating the Caspian sea and the coastal area is the fluctuation of the water surface. This has caused a great deal of damages to the bordering countries. Investigation of the causes of the water rise and the economical solution to the problem is of the first priorities for the prosperity of the region.

In this study, we will cite some of the investigations, but we will try to asses the costs of rehabilitations and co-existance with nature.

The historical data indicates the fluctuation of the water surface being 15 meters, (meaning the average water surface was from -20 m. to -35 m.) [5,6].

In the last 100 years the fluctuation of water level was between 4 meters, (-25.3 m. in 1873 to -29 m in 1977) shown in fig.1.

Environmental Coastal Regions 69

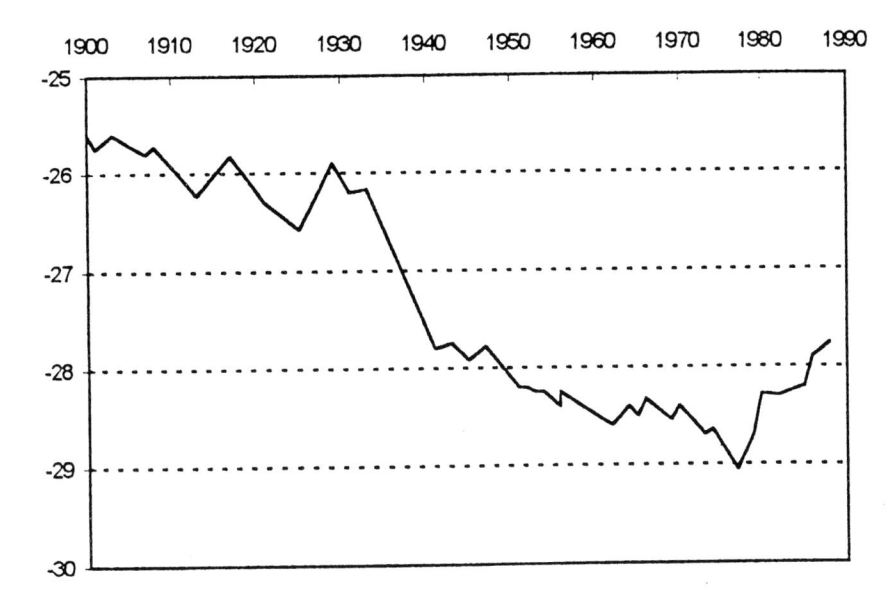

Figure 1. Water level of Caspian sea (1900-1990).

During the observed years 1873 to 1991, the yearly water rise in three years were more than 30 cm (1867, 38 cm ; 1979,32 cm ; 1991, 39 cm). The yearly water fall in two years were more than 30 cm (1851, 32 cm; 1937, 31 cm) [5].
Average yearly water surface fall between 1930-1941 were 16 cm.

2.1 Causes of Water Rise

Due to the global climate changes, the climate of the Caspian sea has also changed. These changes, in general, causes warmer winters and cooler summers [3,4]. Therefore, it has caused more precipitation and less evaporation.
According to some reports[3], the precipitation, in the last decade in comparison with the early ninteen hundred, shows an increase of 5 cm (17.3 cm to 22.2 cm) and evaporation has decreased by 4 cm [A. Mansouri].

More precipitation in the region has given economic opportunities. Economic activities have expanded. Deserted lands have been cultivated. Dams being constructed. Wet land being dried for farming. More water being conveyed for agricultures, industrial needs, and for urban consumption.

The study shows, the rise of water level (1978-1991) and increase of pollution depends on human activities.

Some researchers believed, the cause of water rise is due to the Tectonic movement of the earth crust, Darvishzadeh (1990) [3] states no movement of the plates have been reported in the past 12 years.

Monthly variation of the inflow from Volga river shows that more than 58% of the flow occurs between April and May [1,2]. It causes more influx to the sea and less waste of water from the Volga river [fig.2].

The loss of water such as evaporation from the surface of the river and evapotranspiration has an inverse relation to the water level rise.

Time series analysis carried out by Peyvandi, H.[4] shows the correlation between Volga river basin influx and the Caspian sea between 1900-1990 [fig. 3]. The water surface rise has created a lot of problems in recent years, especially from 1994 to present. It has caused the seepage of domestic and industrial sewage and other pollutants into the river basin and finally to the Caspian sea. This has caused an increase in coliform and bacteria levels on the shore and coastal area. Because of the water rise, from June to August 1991, the odour in the beach area of Anzali was very noticeable.

Researchers believe that due to the aforementioned reasons, the water surface rise will continue until at least the year 2010 (up to +3 meters) [3]. They also predict that the elevation will be at -25m in the next decade and will reach the elevation of -21m by the year 3035.

What should we do to cope with the problems? Our stretegic planning should be socio-economic investigation of the shoreline area. Management and economic aspects of the water level rise and fall must be evaluated. Offshore structures to be constructed based on the assessed risks.

Environmental Coastal Regions 71

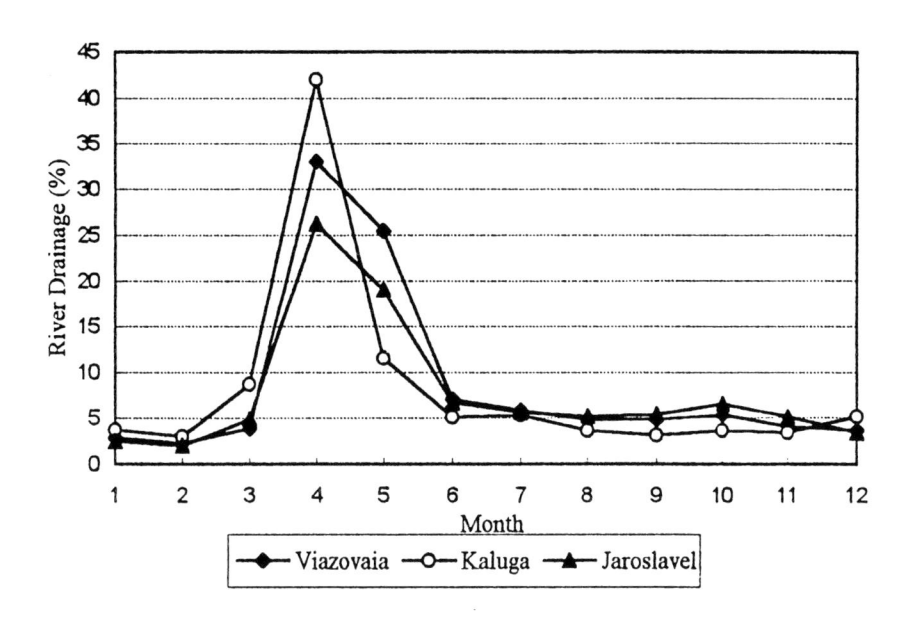

Figure 2. Monthly variation of Volga river basin

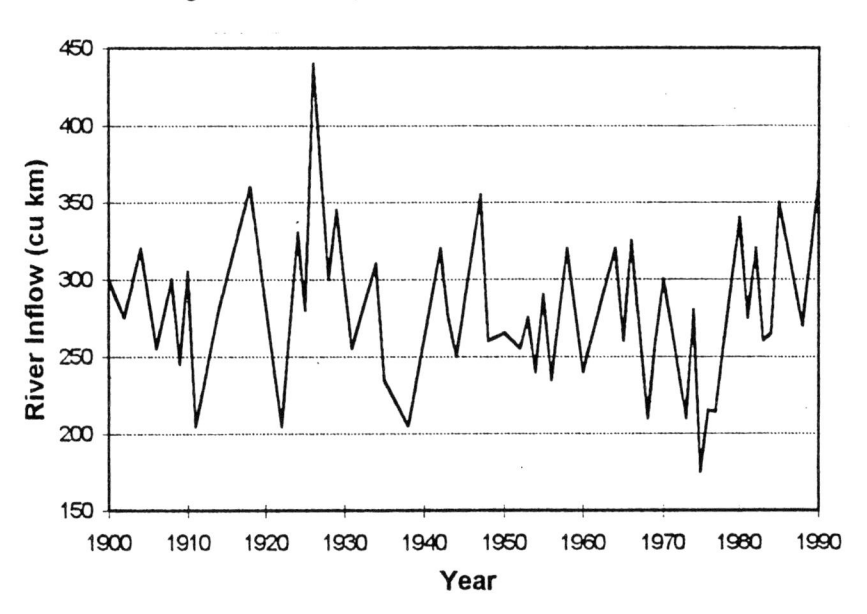

Figure 3. Caspian sea river inflow (1900-1990).

3 Model Studies

The study conducted is based on the Iranian Institute of Urban Planning and Development, which gives permission for land usage, such as urban, agricultural, industrial, commercial, etc. The prices are estimated based on the usage.

The water level rise has created a great deal of damage to the houses and urban development in the shoreline and the coastal area [7]. Based on our studies, the elevation of the water from -25m to-22m for low risk and high risk assessment are chosen respectively.

From the topographic map, the estimated cost of land losses are evaluated.

At present, very few shoreline protecting structures have been constructed in Iran. They are very expensive and the cost/benefit ratio is not reasonable.

Our studies indicate that the priorities of shoreline protection are not very well known. We must prioritize the protection based on the cost/benefit ratio of the region. This has been carried out for a one kilometer long beach area in Bander-e-Anzali. The land usages were estimated with an average market value placed on them (see table 1).

Based on a linear model, the cost of damages before and after the shoreline protection is evaluated.

$$B_{ih}^{J} = (D_{ih}^{J} - P_{ih}^{J}) - C_{ih}^{J} \tag{1}$$

In eq. (1) we assumed the damage is D_{ih}^{J} at any point i based on the protected elevation (h), and P_{ih}^{J} is the predicted damage cost after the protection by method J. C_{ih}^{J} is the cost of method J. The Index J includes protection by gravel and rocks, concrete hangars, reinforced sandbags, concrete retaining walls, circular concrete piles.

The most economical means of protection is reinforced sandbags with gravel and rocks at toes, and the most expensive being concrete retaining walls.

This study shows the delta area near Bandar-e-Anzali in Gylan province has more benefit, than the beach area of Ramsar in Mazandaran province (Table 2),[fig.4].

From row 8 of table-2, we can see that the Cost /Benefit ratio for elevation of -26m to-24m is reasonable. As the water rises to the elevation of -22m the benefit decreases. With elevation reaching -21 to -20m, there is very little gain in construction of shoreline protection.

In this study, the estimated cost of sewage disposal was inconclusive.

Environmental Coastal Regions 73

Table 1. Land area under taken by water rise

Hight(m) area underwater	-26	-25	-24	-23	-22	-21	-20
Urban ($\times 1000\ m^2$)	108	123	143	188	215	242.5	253.5
Industrial ($\times 1000\ m^2$)	22.7	27	28.1	28.2	32.7	35.5	35.6

The estimated damages are based one Kilometer long of beach area in Anzali port.

Urban unit cost = 600,000 Rials/m^2 ; Industerial = 1,100,000 Rials/m^2

Table 2. Cost of shoreline protection with water rise

	Water level Damage*	-26	-25	-24	-23	-22	-21	-20
1	Urban	6.5	7.4	5.6	11	13	15	15
2	Industrial	2.5	3.0	3.1	3.1	3.6	3.9	3.9
3	Facilities	1	1.2	1.3	1.6	1.8	2.0	2.1
4	Total damage cost ($D_{ih} - P_{ih}$)	10	12	13	16	18	20	21
5	Shoreline protection cost (P_{ih})	2.7	3.3	4.1	5.1	6.6	8.0	9.5
6	Damage cost after protection (C_{ih})	4.1	6.3	7.5	9.6	11	12	13
7	Benefit (B_{ih})	5.9	5.7	5.5	6.4	7.0	8.0	8.0
8	Cost-Benefit ratio	2.2	1.7	1.4	1.3	1.2	1.0	0.8

1$=3000 Rials ; * Damage based on 1×10^{10} Rials

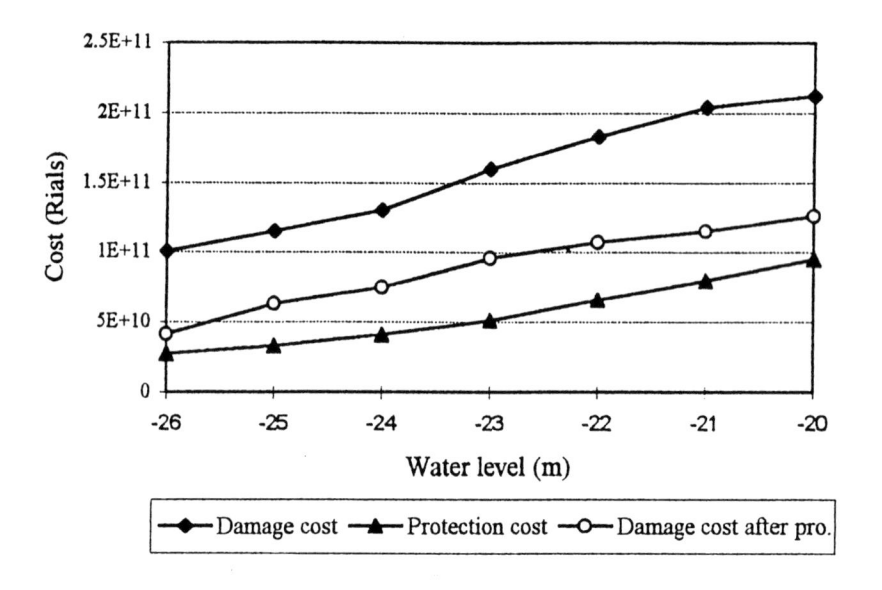

Figure 4. Shoreline protection costs.

4 Conclusion

1. This study shows the probability of water rise in Caspian sea is very high in the next decade [fig. 5].
2. Due to the rise of water surface and the ecological effects of the water level, destruction of the coastal area at Talab-e-Anzali is eminent.
3. Identification of a zero line as a boundary and development in the shoreline based on this criteria.
4. It is necessary to accept, but not to confront, the mechanism of water level changes in the Caspian sea with least damages.
5. Because of the limited resources, the risk assesment based on the economic evaluation and managment must be made to prioritize the protection of the coastal area.

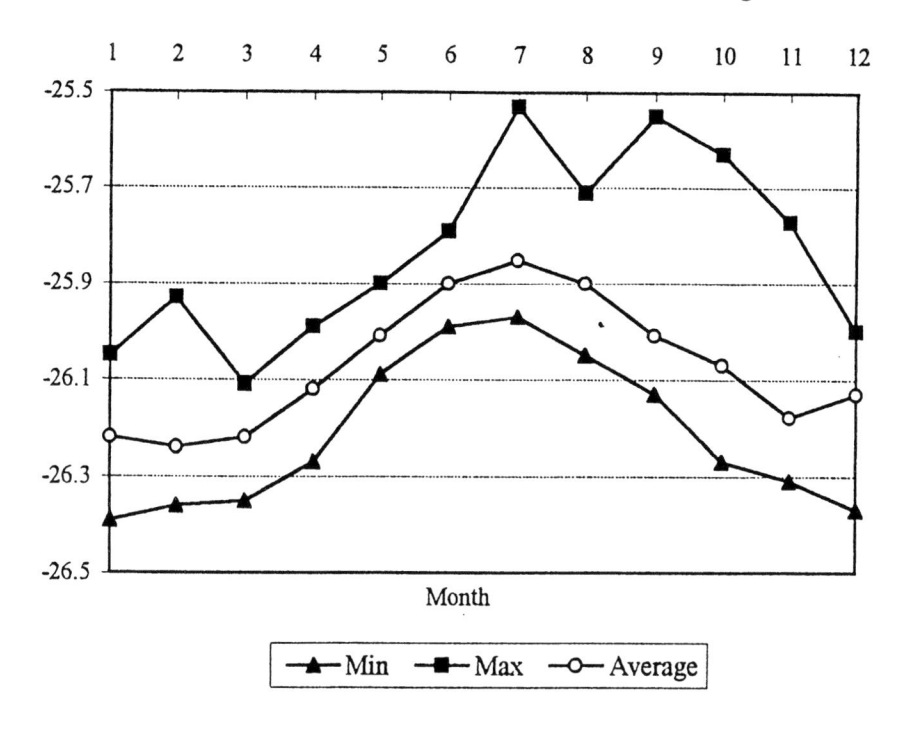

Figure 5. Variation of surface level in 1995 at Anzali station.

References

[1] Avacyan Ab^Kornilov, Ba^Ilitvinov, As^yaskovlev, V.N.;Predicting the Ecological Conditions in the Volga Basin in connection with Interbasin water transfer, Vodn Resur. [Moscow]; Vol.5, pp. 122-127 ; (1986).
[2] Monita, K.M. ; Hydrological Annual, 1973, Vol.4 ; the Caspian sea basin the Oka basin ; 458 p. (1977).
[3] The Caspian sea Reseaches. Iranian Institue of Engineering Reserches. Vol.1 ; 1995.
[4] The Caspian sea Reseaches. Iranian Institue of Engineering Reserches. Vol.2 ; 1996.
[5] Iranian Hydro meterological Institue, Fluctuation of water level, 1989.
[6] Iran power supply Institute, Fluctuation of Caspian sea water level, 1988.
[7] Iranian Remote sensing center, Satilite survay of Caspian sea, 1987.

A two-case study on the environmentally-induced damage to materials in marine environments – Part I: Metallic materials

A.M.G. Pacheco[1] & A. Maurício[2]

[1]*Dept. Engenharia Química and* [2]*Lab. Mineralogia e Petrologia, Instituto Superior Técnico (Technical University of Lisbon), Av. Rovisco Pais 1, 1096 Lisboa Codex, Portugal; EMail: qamgp@alfa.ist.utl.pt*

Abstract

This two-part paper addresses the specific hazards that most materials are faced with in coastal areas, particularly in their atmosphere. Pretty common features like high humidity and airborne salts of marine origin, which are inherent in such an environment, may turn into a nightmare for conservationists, architects and materials scientists, that is for everyone involved with old (historic) or new infrastructure. Two cases are presented and discussed herein. Neither of them was designed or singled out specially for the occasion: both were taken from extended programs of metal-corrosion and stone-decay monitoring in the open. The first case (Part I) deals with the implication of saline contamination for the time of wetness (TOW) of a metallic surface. The results show that standard procedures for assessing TOW from weather data can severely underestimate the duration of surface wetness and, in the final analysis, lead to some misclassification of atmospheric corrosivity. The second case (Part II) follows the evolution of salt efflorescences at an ancient building as a function of local (microclimatic) conditions, in order to get the time probability associated with deliquescence- crystallisation transitions at a given location. By doing this, it was possible to identify more-or-less risky areas in the stone monument, which could then be subjected to differential surveillance and/or care. Both studies seem pertinent to illustrating the need for establishing risk thresholds for materials selection and infrastructure maintenance that can really hold in marine environments.

1 Introduction

In the living world, environmental assets and liabilities are often a question of perspective. Some few examples of that could be easily given. Stratospheric ozone, for instance, is absolutely necessary for shielding the earth from the outer-space radiation and yet, should such a reactive species pervade the lower troposphere, its presence would be a serious concern for the world as we know it. A similar situation may occur when an environmental component is pushed away from its place and/or extent, or as long as a negative synergy arises between components even if they are quite harmless when kept apart from each other and sensitive items.

Probably due to the global concern about the spreading of man-made pollution all over the four ecosystems – atmosphere, hydrosphere, lithosphere and biosphere – comparatively little attention has been paid to the fate and extent of natural inputs, and to their consequences in everyday life. There is no reason for such an intellectual apartheid. Hazardous substances can as well occur naturally and pose some major threats on life-supporting systems and infrastructures. Besides, once an anthropogenic substance is released into the environment, natural forces take over its interphase transfer and intraphase variability. Ultimately, both natural and man-made ("unnatural") substances are moved across phase boundaries by *natural* driving forces and conditioned within phase boundaries by *natural* factors of variability. Significantly enough, in a modern textbook like Thibodeaux's [1], *chemical* is used in a broad sense that goes from water and oxygen to DDT...

A good example of the aforesaid, and most germane to this paper, is airborne salinity. The basic mechanism responsible for the production of spray is the bursting of air bubbles at the sea surface [2-4]. It is generally accepted that some 3 to 4 percent of that surface is covered with bubbles at any moment [5], which means that the yearly production of salt lies between 10^9 and 10^{10} tonnes [5-13]. About 10% of such an amount falls off the ocean [6,12,14-16], turning inland saltfall into an air pollution issue. Needless to say that high dampness as well as airborne salinity are inherent in marine environments and thus essential for the equilibrium of any coastal ecosystem. Notwithstanding, it is by no means less true that they invariably pose a serious threat to infrastructure, cultural artefacts and life-supporting facilities, not to mention vegetation and soil/water resources. Such an issue is definitely not liable to an abatement strategy whatsoever. In the realm of materials, for instance, only careful selection, protective measures and close surveillance can prevent extended damage in the open, in salt-laden environments.

The present paper (Parts I and II) deals with the nasty environmental (electro)chemistry set by common atmospheric moisture and regular airborne salts upon the surface of a plain metal (I) or into the bulk of a composite geomaterial (II). Both studies seem pertinent to illustrating the specific hazards that most materials are faced with in coastal areas and the need for establishing risk thresholds for materials selection and infrastructure maintenance that can really hold in marine environments.

2 Case study

2.1 First case: Salt-induced corrosion of an atmospheric cell

2.1.1 Background

Other than its economic impact, which makes it accountable for more than half the whole corrosion bill [17], atmospheric corrosion bears other two distinctive features: transient nature and thin-layer electrochemistry. The (cumulative) period during which a metallic surface is covered by adsorptive and/or phase films of electrolyte that are capable of causing or sustaining corrosion activity in the open, is the time of wetness (TOW) by ISO 9223 [18]. Along with sulphur dioxide and airborne chloride levels, TOW is a primary variable in categorising atmospheric environments as to their aggressiveness, according to that Standard. The classification of atmospheres provides a basis for the selection and protection of materials, so there should be little room left for questioning its technical relevance.

The wetting of surfaces is quite a complex phenomenon and the TOW itself is an event, location and time-dependent variable [19]. For practical purposes, however, TOW is usually derived from general data and defined as the length of time when both relative humidity (RH) and temperature (T) remain above 80 % and 0 °C, respectively [18]. It has been argued that such an approach should prove accurate enough to be used in corrosivity assessment – see [20], for example. This could be an exception rather than the rule, though, especially in marine atmospheres. That standard procedure is likely to yield very conservative estimates (to say the least...) and lead to a most significant divergence between the wetness actually experienced by a decaying surface and its nominal duration from weather records, as we shall see below.

2.1.2 Experimentation, results and discussion

For several months, an exposure program was carried out in order to evaluate printed-circuit cells (in brief, iron lines on an epoxy substrate) as atmospheric corrosion monitors (ACMs). Details on the cell construction and preparation prior to work as well as on the experimental set-up were given elsewhere [21,22]. The rationale behind the experiments is quite straightforward. Each cell consists of a multilinear array of 40 iron strips (30.00±0.10 mm x 0.50±0.05 mm; line spacing: 0.20 mm) alternately shorted by an external track, resulting in an interdigitated, two-electrode configuration. There is no contact between an iron strip and its immediate neighbours (they are parallel to each other), and every cell is exposed outdoors in a completely dry condition at an angle of 45° to the horizon, roughly facing south, under a shield for rain and direct sunlight.

Once enough moisture builds up to form an electrolyte layer, corrosion starts or resumes at the surface and an electric signal can be recorded from a cell. Should the dampness fade away and the surface get dry, the thin-layer electrochemistry comes to a halt and ceases to drive a current through the data-acquisition system. There is neither an applied electromotive force (emf) nor any other kind of external polarisation: these cells are self-driven devices and their output relates to nothing but what really happens on the corroding surface.

The present results refer to a 96-h term during a mild winter season. Temperature went from a low of 8 °C to a high of 18 °C, with an average of 13 °C, whereas relative humidity stayed between 48 % and 84 % for the whole term (mean RH: 63 %). The galvanic output (short-circuit current, I_{SC}) was measured at zero external impedance from a NaCl-contaminated cell with no applied emf, under a simulated chloride rate of 100 mg m^{-2} day^{-1} (mmd Cl$^-$). This figure falls within the S_2 category of airborne salinity, according to ISO [18]: in practical terms, it is likely to stand for an average saltfall rate at a distance of at least 250 m downwind from the surf [23]. The entire exposure program was factorially designed for assessing the influence of primary atmospheric stimuli (T and RH) plus Cl$^-$ levels on the corrosion of iron ACMs. In the particular series which this experiment was taken from, an average rate of Cl$^-$ presentation was fixed (simulated) in order to ensure that cells were responding to changes in T and RH only, and to prevent any major kinetic ("ohmic") control during the periods of corrosion activity [24].

For zero-resistance ammetry (ZRA), the cell signal was put through a low-noise (0.01 pA Hz$^{-1/2}$), chopper-stabilised operational amplifier, and read as an ohmic drop across virtually-null resistors by a high-impedance (> 10^{12} ohm) voltmeter. Continuous records on cell output as well as on site relative humidity and air temperature were kept throughout the run (onset: 22.00 GMT). An overview of the full test interval in terms of RH and I_{SC} is given in Figure 1; logarithmic (right) scale is just a plotting device to accommodate current data over several orders of magnitude.

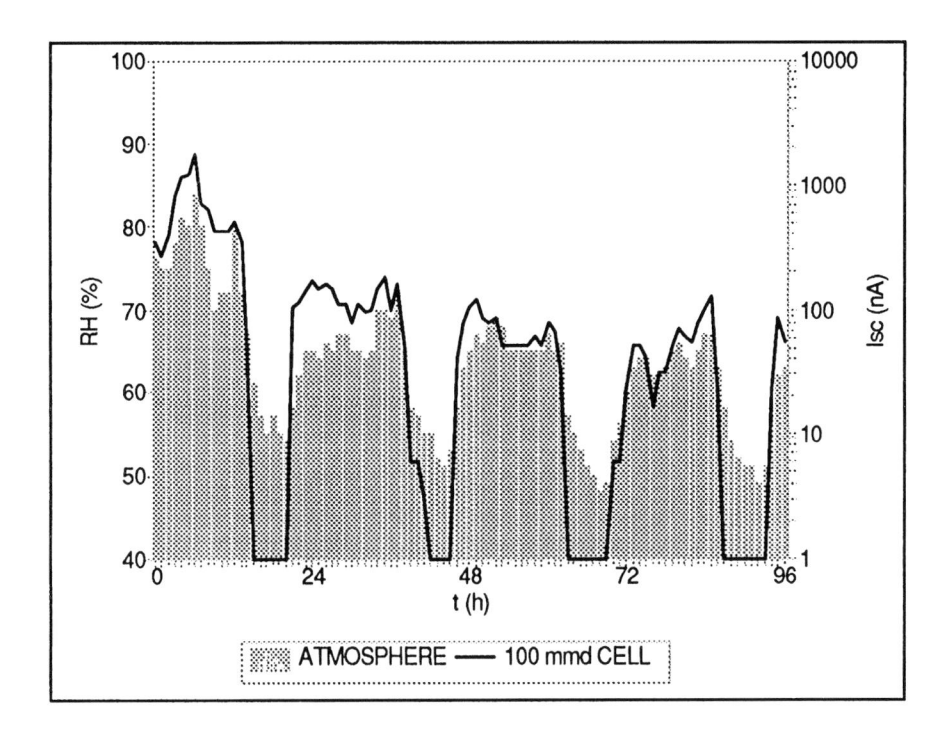

Figure 1. Short-circuit current (I_{SC}) at zero external impedance vs relative humidity (RH), from an iron cell under 100 mg m^{-2} day^{-1} (mmd) of Cl$^-$.

Other than an overall impression of close agreement, the quality of association between corrosion activity on the cell and relative humidity in the air was checked by means of nonparametric measures of correlation. Why nonparametrics? First, rank-order statistics imply fewer and/or less stringent assumptions about data (if at all...) than parametric measures of correlation usually do. Second, even though all observations attain an interval scale of measurement, data structure seems much more important an asset here than numerical magnitude itself, so the assignment of ranks to raw scores does not result in any waste of information [25]. Third, ranks are less sensitive to the experimental error inherent in translating data from analogical records.

The correlation between I_{SC} and RH was found significant at any level by either nonparametric measure – Spearman, Kendall and gamma statistics – which means that the cell acts as a reliable moisture sensor: coefficients are 0.897, 0.763 and 0.812, respectively (0-96 h; raw data ≡ 97 cases). Censoring data for tied-at-zero observations did not lead to an improvement in any correlation. Nonparametrics seem robust enough to account for a few ties in the ranks and the risk of randomness does not exhibit an obvious trend along the exposure. Now, the total time during which an electric signal could be recorded from the cell amounts to 76 h for an evaluation interval of 96 h: in other words, the cell was wet for about 80 % of the experiment. An appraisal of the time-of-wetness based on temperature-humidity data (ISO 9223 [18]) would give just 3 h, a most conservative estimate to say the least. Needless to comment further on this issue: Figure 2 speaks for itself as showing the cumulative build-up of TOW by either approach.

The experiment was designed to exclude visible water inputs, which makes it possible to locate an average threshold for current flow around 58 % RH, for the entire period of evaluation. Variate by variate, corrosion activity can be traced down to 54 % RH. These results are in excellent agreement with early studies on mild-steel surfaces inoculated with NaCl nuclei [26,27], which indicate that corrosion becomes appreciable slightly below 60 % RH, actually at about 58 % RH.

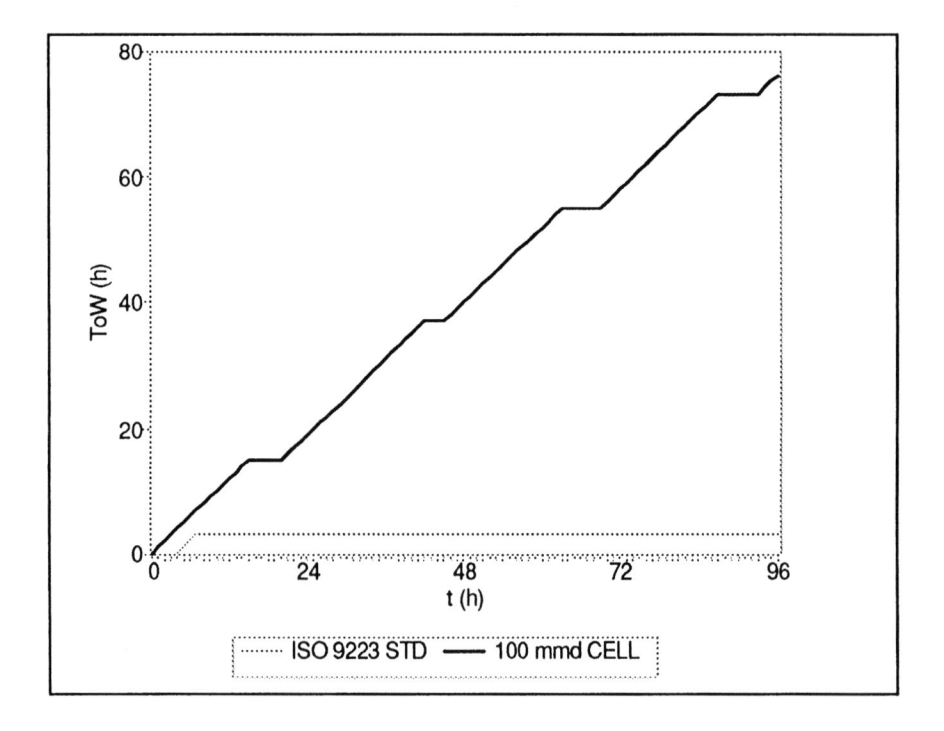

Figure 2. Time-of-wetness conforming to ISO 9223 (calculated TOW) and corresponding to actual corrosion activity (measured TOW).

On the other hand, the same results cast a serious doubt on the validity of TOW estimates, as defined by ISO 9223, at least in what concerns marine atmospheres. It is true that the Standard itself includes several disclaimers and waives exclusiveness in assessing TOW for classification purposes. However, it is by no means less true that most surface-wetness assessment, whether for atmospheric classification or not, still follows the predetermined levels of T and RH in accordance with that Standard, despite the sound evidence coming from a few other sources – see [28] and references therein, for instance.

Finally, it should be emphasized that the case reported herein is just one in many proprietary experiments pointing to an underestimation of TOW to a fairly large extent, whenever the ISO-standard approach is used under saline conditions. This is neither an odd, special case, nor should the simulation be viewed as an accelerated, unrealistic procedure. In fact, the present rate of Cl^- deposition fits into the lower half of both the S_2 category by ISO [18] and the reference range by Mattsson [29], so it hardly qualifies as an extreme-pollution event (S_3 category goes up to 1500 mmd Cl^- and, beyond that, there are still splash and spray episodes). Besides, the situation could only be worse in the presence of real sea-salt, that is in an actual marine environment. The critical humidity for rusting in the presence of sea salt is somewhat lower due to its composite nature. The existence of highly hygroscopic components in sea water, especially $MgCl_2$, $MgSO_4$ and $CaCl_2$, may trigger corrosion at about 40 % RH [30], whereas little corrosion should be expected below about 60 % RH in the presence of pure NaCl, according to Evans and Taylor [31] and our own experience.

3 Conclusions

As far as corrosion resistance is concerned, and apart from user-defined products, most materials for outdoor service are selected and protected under some previous classification of the service location in terms of time of wetness and air contamination, that is on an *ab initio* basis. However, the ability of an environment to induce and/or sustain an atmospheric corrosion process can lie way far from what might be anticipated through standard procedures for corrosivity assessment and classification. This is an almost inevitable issue when dealing with salt-laden environments, though it seems that major Standards – like ISO's, for instance – failed to accommodate such an issue as yet.

The present results show that, even in mild conditions, a significant divergence was found between standardized (estimated) and experimental (measured) time of wetness, an increase by a 25-fold factor to be precise.

The situation could only be worse in some harsher conditions of exposure and/or in an actual marine environment, due to the enhanced hygroscopicity of real sea salt when compared to pure sodium chloride. Conservative estimates of TOW to a fairly large extent are most likely to occur in coastal environments and thus lead to an ill classification of atmospheric aggressiveness for service locations down there. In the final analysis, such an underrating would undermine the basis of selection criteria and protection systems, that is the whole front trench in the fight against corrosion in the open.

This first case-study (Part I) is an example of how some peculiarities of marine environments should be accounted for *before* damage is done. An approach to an *after* situation will be dealt with in the second study (Part II of this paper), which is going to focus on the salt-induced stress to an ancient, stone building.

(to be continued in Part II)

Acknowledgements

Research contracts PBIC/C/QUI/2381/95 and PBICT/C/CTA/2127/95 (JNICT - Portugal) assisted in meeting the production cost of this paper (Parts I and II).

References

[1] Thibodeaux, L. J., *Environmental Chemodynamics – Movement of Chemicals in Air, Water, and Soil*, Wiley-Interscience, New York NY (USA), 1996.

[2] Boyce, S.G., Source of atmospheric salts, *Science*, **113**, pp. 620-621, 1951.

[3] Woodcock, A.H., Salt nuclei in marine air as a function of altitude and wind force, *J. Meteorol.*, **10**, pp. 362-371, 1953.

[4] Blanchard, D.C. & Woodcock, A.H., Bubble formation and modification in the sea and its meteorological significance, *Tellus*, **9**, pp. 145-158, 1957.

[5] MacIntyre, F., The top millimeter of the ocean, *Scientific American*, **230**(5), pp. 62-77, 1974.

[6] Eriksson, E., The yearly circulation of chloride and sulfur in Nature; meteorological, geochemical and pedological implications – Part I, *Tellus*, **11**, pp. 375-403, 1959.

[7] Blanchard, D.C., The electrification of the atmosphere by particles from bubbles in the sea, *Prog. Oceanogr.*, **1**, pp. 71-202, 1963.

[8] Robinson, E. & Robbins, R.C., *Emission, Concentration and Fate of Particulate Atmospheric Pollutants (SRI Project SCC-8507 – Final Report)*, SRI, Menlo Park CA (USA), 1971.

[9] Hsu, S.A. & Whelan III, T., Transport of atmospheric sea salt in coastal zone, *Environ. Sci. Technol.*, **10**, pp. 281-283, 1976.

[10] Petrenchuk, O.P., On the budget of sea salts and sulfur in the atmosphere, *J. Geophys. Res.*, **85**, pp. 7439-7444, 1980.

[11] Várhelyi, G. & Gravenhorst, G., Production rate of airborne sea-salt sulfur deduced from chemical analysis of marine aerosols and precipitation, *J. Geophys. Res.*, **88**, pp. 6737-6751, 1983.

[12] Blanchard, D.C., The oceanic production of atmospheric sea salt, *J. Geophys. Res.*, **90**, pp. 961-963, 1985.

[13] Möller, D., The Na/Cl ratio in rainwater and the seasalt chloride cycle, *Tellus*, **42B**, pp. 254-262, 1990.

[14] Cullis, C.F. & Hirschler, M.M., Atmospheric sulphur: natural and man-made sources, *Atmos. Envir.*, **14**, pp. 1263-1278, 1980.

[15] Möller, D., On the global natural sulphur emission, *Atmos. Envir.*, **18**, pp. 29-39, 1984.

[16] Friend, J.P., Natural chlorine and fluorine in the atmosphere, water and precipitation, *Scientific Assessment of Stratospheric Ozone: 1989 (WMO Report)*, WMO, Washington DC (USA), 1989.

[17] Haagenrud, S.E., Mathematical modelling of atmospheric corrosion and environmental factors, *NATO ARW on Problems in Service Life Prediction of Building and Construction Materials*, NATO, Paris (France), 1984.

[18] ISO Standard 9223, *Corrosion of Metals and Alloys - Classification of Corrosivity of Atmospheres*, ISO, Geneva (Switzerland), 1992.

[19] Hechler, J.J., Wetness monitoring on the exterior of infrastructures, *Corrosion Forms and Control for Infrastructure (ASTM STP 1137)*, ed. V. Chaker, ASTM, Philadelphia PA (USA), pp. 126-139, 1992.

[20] Dean, S.W., Classifying atmospheric corrosivity – A challenge for ISO, *Mater. Performance*, **32**(10), pp. 53-58, 1993.

[21] Pacheco, A.M.G. & Ferreira, M.G.S., An investigation of the dependence of atmospheric corrosion rate on temperature using printed-circuit iron cells, *Corros. Sci.*, **36**, pp. 797-813, 1994.

[22] Pacheco, A.M.G. & Ferreira, M.G.S., The outdoor performance of printed-circuit iron cells in self-driven operation: electrochemical and environmental features, *Corros. Sci.*, **in press**, 1998.

[23] Kain, R.M., Using ASTM Standards to combat corrosion, *Standardization News*, **24**(10), pp. 34-39, 1996.

[24] Askey, A., Lyon, S.B., Thompson, G.E., Johnson, J.B., Wood, G.C., Sage, P. & Cooke, M.J., The effect of fly-ash particulates on the atmospheric corrosion of zinc and mild steel, *Corros. Sci.*, **34**, pp. 1055-1081, 1993.

[25] Townsend, J.T. & Ashby, F.G., Measurement scales and statistics: the misconception misconceived, *Psychological Bull.*, **96**, pp. 394-401, 1984.

[26] Preston, R.St.J. & Sanyal, B., Atmospheric corrosion by nuclei, *J. Appl. Chem.*, **6**, pp. 26-44, 1956.

[27] Ericsson, R., The influence of sodium chloride on the atmospheric corrosion of steel, *Werkst. Korros.*, **29**, pp. 400-403, 1978.

[28] Forslund, M. & Leygraf, C., Humidity sorption due to deposited aerosol particles studied *in situ* outdoors on gold surfaces, *J. Electrochem. Soc.*, **144**, pp. 105-113, 1997.

[29] Mattsson, E., The atmospheric corrosion properties of some common structural metals – A comparative study, *Mater. Performance*, **21**(7), pp. 9-19, 1982.

[30] Chandler, K.A., The influence of salts in rusts on the corrosion of the underlying steel, *Br. Corros. J.*, **1**, pp. 264-266, 1966.

[31] Evans, U.R. & Taylor, C.A.J., Critical humidity for rusting in the presence of sea salt, *Br. Corros. J.*, **9**, pp. 26-28, 1974.

A two-case study on the environmentally-induced damage to materials in marine environments – Part II: Geomaterials

A. Maurício[1] & A.M.G. Pacheco[2]

[1]*Lab. Mineralogia e Petrologia* and [2]*Dept. Engenharia Química, Instituto Superior Técnico (Technical University of Lisbon), Av. Rovisco Pais 1, 1096 Lisboa Codex, Portugal; Email: pcd 2045@alfa.ist.utl.pt*

Abstract

This two-part paper addresses the specific hazards that most materials are faced with in coastal areas, particularly in their atmosphere. Pretty common features like high relative humidity and airborne salts, which are inherent in such an environment, may turn into a nightmare for conservationists, architects and materials scientists, that is for everyone involved with old (historic) or new infrastructure. Two cases are presented and discussed herein. Neither of them was designed or singled out especially for the occasion: both were taken from extended programs of metal-corrosion and stone-decay monitoring in the open. The first case (Part I) deals with the implication of saline contamination for the time of wetness (TOW) of a metallic surface. The results show that standard procedures for assessing TOW from weather data can severely underestimate the duration of surface wetness and, in the final analysis, lead to some misclassification of atmospheric corrosivity. The second case (Part II) follows the evolution of salt efflorescence at an ancient building as a function of local (microclimatic) conditions, in order to get the time probability associated with deliquescence- crystallisation transitions at a given location. By doing this, it was possible to identify more-or-less risky areas in the stone monument, which could then be subjected to differential surveillance and/or care. Both studies seem pertinent to illustrating the need for establishing risk thresholds for materials selection and infrastructure maintenance that can really hold in marine environments.

1 Sta Marija Ta´ Cwerra case study

Evaporite minerals are particularly significant in the Mediterranean basin because most cultural (historic-architectonic) heritage is concentrated in coastal areas where they can be exposed to marine spray or salt-rising damp. Their presence contributes significantly to the weathering of building stones because of their response to cycles of relative humidity. Since the critical reative humidity points of dissolution or change of state of hydration of the minerals are usually within the typical ranges of relative humidity (RH) observed in most temperate climates, they can oscillate frequently between solution and crystal phases. They can also oscillate from a crystalline phase to another [1]. The present case study is aimed at estimating when and where different pure salts may crystallise from solutions, evaluating the probability of the salt system being crystallised or deliquescent. The assessment of the salt weathering potential on the surface of the stone, by means of crystallisation/dissolution and transition probability estimations (TPE) along the year will also be considered, in a given monitored site.

1.1 Diagnosis and data collection

To evaluate some effects of coastal environments on the weathering of historic buildings, an extensive study has been carried out at four pilot monuments along the east-west axis of the Mediterranean basin – Cathedrals of Cadiz (Spain) and Bari (Italy), and Church of Sta Marija Ta'Cwerra (Malta). This was done with a specific interest in the action of marine salts and air pollution. The various locations of the monuments reflect dissimilar conditions of salinity, extent of marine and atmospheric pollution, and topographical aspects of the area, leading to different types and grades of weathering and decay patterns [2].

The church of Sta Marija Ta´ Cwerra is located in the village Siggiwi, in the south west of the Malta island, at a distance 3 km far from the sea. It is a free standing building from the XVII century, less than 10 x 10 m^2 plan view. The church is built entirely of Globigerina limestone. This limestone has a total porosity of 35% with mainly small pores (2-5 µm). The chemical composition of the stone is dominated by calcium carbonate (88 to 97 %). The four external walls show severe deterioration, for about two-third of their height, the lower courses are cemented. The middle courses are deteriorated in the form of alveolar weathering as well as powdering of several areas. Most of the mortar has been lost from the joints in this area. The uppermost courses are better preserved. At the inside of the building, the plaster has fallen away in several areas revealing powdering and flaking stone underneath and even some of the carvings have almost completely disappeared [2]. In the outside walls, granular disintegration and relief by rounding and notching are the prevailing weathering forms. The intensity of salt weathering is basically controlled by stone characteristics, especially porous matrix properties, and degree of salt accumulation [3]. It can be seen that stone samples from the outside show clear enrichments in Na^+ and Cl^- and a bit in SO_4^{2-}. On the inside, efflorescences are enriched in Na^+ and Cl^-, indicating mainly the influence of sea as a cause of chemical deterioration of the stone. Anthropogenic chemical emission impacts on the stone are of negligible importance in this church [2].

The evaporite minerals found on the monument are nitrocalcite nitromagnesite, nitratite, halite, thenardite gypsum, mirabilite and niter [4]. Nitrates are almost always dissolved owing to their low deliquescence humidities when compared to the environmental relative humidity range usually found inside and outside the church. Regarding crystallisation pressures, halite is the most dangerous salt. This can be easily understood comparing crystallisation pressures (atm) of different salts, under thermodynamic conditions found in some environments: 554 for halite; 282 for gypsum; 292 for thenardite; 72 for mirabilite. Considering the molar volume (cm^3/mole): 220 for mirabilite, 28 for halite, 55 for gypsum, 53 for thenardite, then mirabilite can be considered the most dangerous salt. Thenardite and mirabilite are sometimes found together showing that phase transitions between anhydrous and hydrate forms easily occurs in the stone. They produce also hydration pressure in the stone porous matrix that is particularly effective because of the rapidity of the change. The transition of thenardite to mirabilite is more rapid than hydration of other salts, taking about 20 minutes at 39 C [5].

The environmental data were collected on a hourly basis at different sites (one outdoors and four indoor), from April 1994 to June 1995, by means of a monitoring station. The system comprises a network of sensors: contact thermometers attached to the surface of the stone and thermohygrometric sensors located 5cm far from the stone surface at different heights from the ground. There are four indoor contact thermometers and four indoor thermohygrometric sensors. They are positioned in Local 1, Local 2, Local 3 and Local 4 (respectively, in South wall - 3,5 m; North wall - 3,5 m; South wall - 0,5 m; North wall - 0,5 m). There is also one outdoor thermohygrometric sensor facing south/south-west, attached to the dome.

In the search for interactions between atmosphere, salt-induced weathering and stone condition, accurate temperature and relative humidity measurements were carried out in the atmospheric layer close (5 cm) to the

90 Environmental Coastal Regions

surface (Atm condition). Stone surface temperatures measurements (Surf condition) were also monitored.

1.2 Data processing

In order to evaluate the potential damage on porous-stone materials resulting from pure-salt crystallisation, it is essential to become aware of their deliquescence thresholds (boundary conditions), RH_{eq}. The evaluation should be made for a given salt on the range of temperature and relative humidity existing on temperate climates. These functions are very important since they enable to establish phase diagrams for each pure salt, as well as to follow the evolution of deliquescence humidity along time as a function of ambient temperature.

To deal with such an issue, a computer program was conceived, based on a few underlying hypotheses [4] in order:
  i) to estimate atmosphere boundary layer (thermohygrometric) conditions in equilibrium with the stone-surface;
  ii) to estimate the phase diagram corresponding to each pure-salt system;
  iii) to study the expected behaviour of each salt system upon its phase diagram, by means of scatter plots corresponding to the actual atmospheric (Atm) or to the estimated atmosphere boundary layer conditions on stone-surface (Surf);
  iv) to estimate the equilibrium relative humidity (RH_{eq}) of some pure-salt systems as a function of time;
  v) to look into the probable time-course evolution of each system, using estimates of deliquescence humidity (RH_{eq}), monitored temperatures and relative humidities corresponding to Atm conditions, or estimated relative humidities corresponding to Surf conditions;
  vi) to estimate the probability of a given salt system being crystallised ($f(RH < RH_{eq})$) or deliquescent ($f(RH > RH_{eq})$), and the probability of a crystal/solution transition (TPE).

This can be done for any given set of thermohygrometric data during monitoring time. The following definitions apply here: TPE - is the percentage ratio of the number of crystal/solution transitions across RH_{eq} to the total number of data points. Both concepts refer to an interval which T and RH chronograms are available.

The program can process data from different sites in a monument. However, it should be noticed that the relative humidity of the atmosphere close to a stone surface is an estimate. It is based on the assumption of thermal equilibrium between the stone-air boundary layer and the stone itself. An empirical approximation by Tetens to the Clausius-Clapeyron equation enables such computations to be performed [6].

The RH_{eq} phase diagrams of each salt system were estimated through Lagrange interpolation method, from tabulated (experimental) data. Tabulated deliquescence humidities for practically all common salts that could be relevant for building materials are available from the literature.

1.3 Results

1.3.1 Deliquescence humidity estimation

As an example of an output of the computer program for a monitored site, Figures 1-a and 1-b are shown, concerning data processing for Atm monitoring conditions. Halite behaviour can be followed outside the church from April to October 1994. In Figure 1-a, the estimated phase diagram and outdoor-environment data (scatter plot) is derived. These results allow estimating deliquescence humidity along time as a function of varying local ambient temperature as it is shown in Figure 1-b.

In Figures 1-b, chronograms of exterior monitored temperature and relative humidity are shown. Estimations of the equilibrium relative humidity (RH_{eq}) of halite along monitoring time, as a function of instantaneous local temperature can be seen as well.

1.3.2 Transition Probability Estimations

The classification of each monument as to its saline-risk potential can be made on the basis of TPE values. This is because such values depend simultaneously on: i) all salts present, ii) all monitoring sites and iii) all measurement conditions (Surf, Atm) [7-9]. From the set of all possible relationships that can be derived between TPE values, it is shown quantitatively the transition behaviour likely to be expected of some salt systems (ex: nitratite, halite and niter) by means of TPE values. The difference between TPE estimations for Surf and Atm conditions can also be easily evaluated (Figures 2-a, 2-b). It should be noticed that niter is expected to be always crystallised since RH_{eq} is always above environment RH (not presented in this paper). So, there is no need to present graphically its TPE values.

A summary of the computation results obtained for the three salts is presented in Table 1.

92   Environmental Coastal Regions

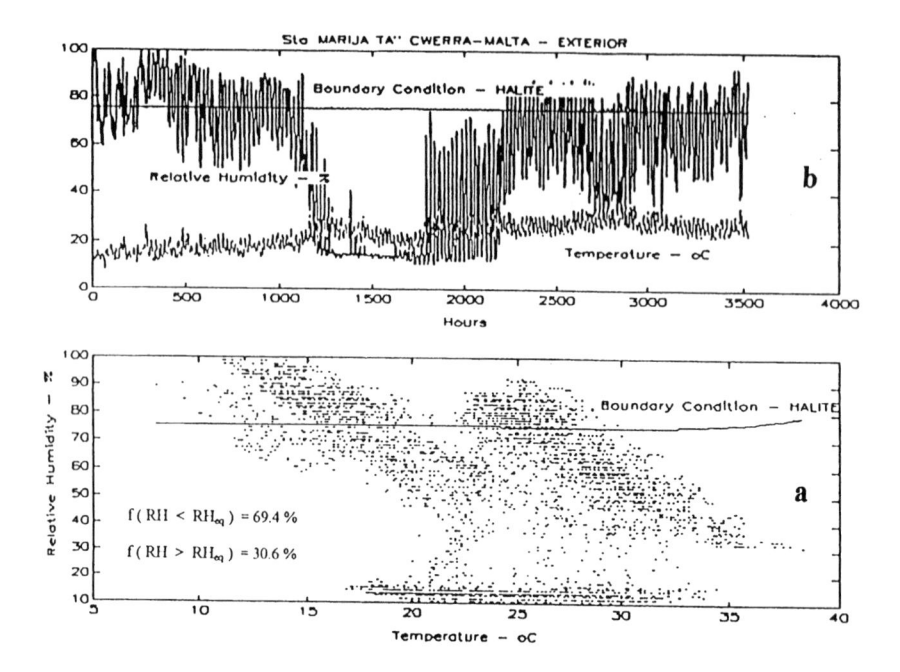

Figure 1. Chronograms and phase diagrams corresponding to sensor location "Exterior". a: Phase diagram of Halite (calculated) and scatter plots of air temperature and relative humidity (monitored data); b: Chronograms of Halite deliquescent conditions (calculated), air temperature and relative humidity (monitored data).

1.4 Discussion of results

A computer model was conceived and presented elsewere [4]. An example was given herein for some salt systems likely to be found in the Church of Sta Marija Ta'Cwerra (Figures 1-a,b) and (Table 1). The behaviour of pure-salts can thus be forecast on the basis of indoor and outdoor varying environment conditions if it is assumed that the kinetics of salt transitions is fast enough. However, it should be emphasised that such thermodynamically based results are merely indicative of what might happen at the stone surface of monuments under surveillance.

Contamination by a single salt is very uncommon, if not at all: a mixture of salts is present in (almost) every situation, owing to air pollution and/or rising damp. Recently, Price and Brimblecombe [10], as well as Steiger [11] dealt with the thermodynamics of the much more complex case of salt mixtures, for temperature conditions of 15, 20 and 25 C. They

Table 1 - A summary of the computation results partially presented on Figures 2-a,b.

Salt \ Local		Nitratite	Halite	Niter
	Atm.	M	M	0
Level 1	Surf.	H	H	0
	Surf-Atm	MD+	MD+	0
	Atm.	V	V	0
Level 2	Surf.	V	V	0
	Surf-Atm	LD+	LD+	0
	Atm.	M	M	0
Level 3	Surf.	H	H	0
	Surf-Atm	VD+	VD+	0
	Atm.	V	V	0
Level 4	Surf.	V	V	0
	Surf-Atm	LD-	HD-	0

In the Table:
  L low values of TPE $0 < L \leq 1\%$;
  M mean values of TPE $1 < M \leq 2\%$;
  H high values of TPE $2 < H \leq 4\%$;
  V very high values of TPE $V > 4\%$;
  D difference;
  +,-,0   positive, negative, or no differences between TPE;
  LD, MD HD, VD low, mean, high, or very high differences between TPE;
  $|LD| \leq 0.5\%$ $0.5 < |MD| \leq 1\%$ $1 < |HD| \leq 1.5\%$.

used the approach made by Pitzer to calculate the relative humidity in equilibrium with any mixed-salt solution. Pitzer approach can in principle be extended to any situations of varying temperature. Unfortunately, the nature of the data available does not enable us to use such an approach in this paper.

An index of the environmental-weathering potential should be the next step beyond, in the near future. Such an index could turn into an important tool for assessing monuments as to stone decay, provided that mineralogical, texture, porosity and interfacial (chemical and physical) characteristics could be considered and quantitatively modelled.

The kinetics of deliquescence/crystallisation and crystallisation/

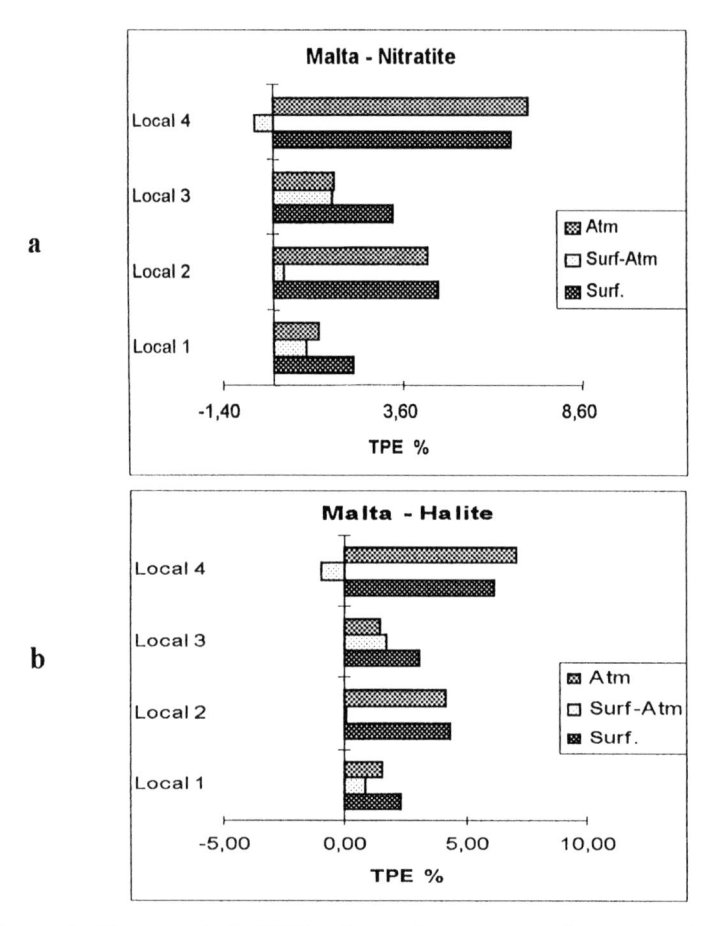

Figure 2. Characteristic TPE values of two pure salt systems inside the church. a: nitratite, b: halite.

hydration transitions as well as the processes of salt-solution transport for mixed-salt systems inside the porous stone should be modelled too. This allows a view to a deeper understanding of their time-course evolution and to a more accurate simulation and forecast of stone-decay patterns. Given this, an optimal choice of the sampling rate to the environment factors acting on a stone monument or any other historic-architectonic artefact, could be envisaged as well [8,9].

2 Conclusions

The results presented above show that the behaviour of the pure-salt systems conditioned to varying thermohygrometric conditions can be significantly different when open-atmosphere or stone-surface conditions are considered. It is not enough to measure thermohygrometric variables some 5cm far from the stone surface and then extrapolates the results of RH_{eq} estimations made thereby directly for that surface. At least, thermohygrometry of the atmosphere nearby the surface as well as surface temperature must be monitored.

Considering transition-probability estimations (TPE), it is possible to ascertain which salts should be considered potentially more dangerous in a given context (monument plus environment). The projection of monitored data on the estimated phase diagram, and the monitored and estimated deliquescence humidity chronogram allows visualising immediately when and where phase transitions are likely to occur in the salt system. The state of the salt system (deliquescent or crystallised) are also easily ascertained. Some qualitative conclusions can also be made on the overall appearance of the chronograms, describing local environment behaviour along time.

Further research should account for a very important aspect of salt-induced damage on historic buildings: actual salts are seldom pure. An effort must be put on adapting the presented methodology for the real situation, that is: the joint occurrence of several saline species, inside a natural, multiphase and heterogeneous porous matrix, whose behaviour is conditioned by a varying environment on every site under monitoring.

To study different monuments simultaneously, it is very important that monitoring and surveillance programs are set up and carried out on a uniform (standard) basis and synchronised, for accurate data and results comparison. This should be done in what concerns either salts (origin, composition, extent) or buildings (sampling sites, material properties, etc.).

Acknowledgements

Research contracts PBICT/C/CTA/2127/95 and PBIC/C/QUI/2381/95 (JNICT-Portugal) assisted in meeting the production costs of the present paper (Parts I and II).

References

[1] Livingstone, R., Influence of evaporite minerals on gypsum crusts and alveolar weathering, *Proc. of the III Int. Symp. on the Conservation of Monuments in the Mediterranean Basin*, eds. V. Fassina, H. Ott and F. Zezza, Venice, pp. 101-107, 1994.

[2] Torfs, K., Van Grieken, R. & Cassar, J., Environmental effects on deterioration of monuments: case study of Sta Marija Ta´Cwerra, Malta, *Proc. Protection and Conservation of the European Cultural Heritage: Research Report N° 4 (European Commission Research Workshop)*, ed. F. Zezza, Bari, pp. 441-451, 1996.

[3] Fitzner, B., Henrichs, K. & Volker, M., Model for salt weathering at maltese globigerina limestones, *Proc. Protection and Conservation of the European Cultural Heritage: Research Report N° 4 (European Commission Research Workshop)*, ed. F. Zezza, Bari, pp. 331-344, 1996.

[4] Aires-Barros, L., & Maurício, A., Chronology, probability estimations and salt efflorescence occurrence forecasts on monument building stone surfaces, *Proc. 8th Int. Cong. on Deterioration and Conservation of Stone*, ed. J. Riederer, Berlin, pp. 497-511, 1996.

[5] Fassina, V., Neoformation decay products on the monument's surface due to marine spray and polluted atmosphere in relation to indoor and outdoor climate, *Proc. Protection and Conservation of the European Cultural Heritage: Research Report N° 4 (European Commission Research Workshop)*, ed. F. Zezza, Bari, pp. 37-53, 1996.

[6] Monteith, J.L. & Unsworth, M.H., *Environmental Physics*, Edward Arnold, London, pp.20-30, 1990.

[7] Aires-Barros, L. & Maurício, A., Transition frequencies of evaporitic minerals on monument stone decay, *Proc. of the 4th Int. Symp. on the Conservation of Monuments in the Mediterranean Basin*, eds. A. Moropoulou, F. Zezza, E. Kollias and I. Papachristodoulou, Rhodes, Vol. 1, pp. 33-51, 1997.

[8] Maurício, A. & Aires-Barros, L., Salt systems and monument stone decay in coastal marine environment, *Chemistry, Energy and the Environment*, Royal Society of Chemistry, Cambridge (in the press).

[9] Maurício, A., Aires-Barros & Pacheco, A,M,G., Forecast of salt occurrences on monument stone surfaces, *Chemistry, Energy and the Environment*, Royal Society of Chemistry, Cambridge (in the press).

[10] Price, C., & Brimblecombe, P., Preventing salt damage in porous materials, *Preventing Conservation: Practice, Theory and Research*, International Institute for Conservation, London, pp. 90-93, 1994.

[11] Steiger, M., Crystallisation properties of mixed salt systems containing chloride and nitrate, *Proceedings of the European Commission Research Workshop on the Conservation of Brick Masonry Monuments*, Leuven (Belgium), pp. 1-9, 1994.

Section 2: Coastal Erosion

Shoreline erosion management program for Rosarito Beach, Baja California, Mexico

C.M. Appendini, R. Lizárraga-Arciniega and D.W. Fischer
Institute of Oceanographic Research and Faculty of Marine Sciences, Universidad Autónoma de Baja California, Carr. Tij-Ens Km 103, Ensenada, B.C. Mexico
Email: christ@cicese.mx   roman@bahia.ens.uabc.mx

Abstract

Rosarito Beach has a population of nearly 50,000 and was recently created as a municipality where tourism is its most important economic activity. The town area includes 11 km of sandy beaches which constitutes the most important natural asset for the region's economy and provides protection from marine forces. Recently the beach width along the sandy stretch has decreased due to shoreline erosion, resulting in damages to recreational and urban infrastructure. Although beach erosion is considered a critical problem by the municipality, there is no coastal policy to address this problem nor a plan for beach preservation. A shoreline erosion program was elaborated by the local university to promote beach preservation. As a first step, a sediment budget for the area was generated and the shoreline was characterised as a function of coastal morphology, land use, population density and shore protection structures. From this information an erosion hazard analysis was conducted to define erosion vulnerability of the entire beach. A set of strategies and policies for erosion management was established for discussion with key actors in the community of Rosarito Beach in order to develop a plan accordingly to their interests.

1 Introduction

Rosarito Beach is a recently formed municipality on the Northwestern coast of Baja California, Mexico, located 20 km south of the Mexico-US international border (Fig. 1). The most important economic activity of the municipality is tourism, employing 65% of the economically active population[3], and it is the economic impulse for the region. The 11 km of sandy beaches represent a natural asset which generates the tourism vocation of the municipality, having an important role in the regional economy and providing protection to the infrastructure and developments adjacent to the beach areas. Despite the importance of the sandy beaches for the municipality, there is a lack of coastal management and accelerated development is threatening this important asset[1]. Beach erosion is now considered a critical problem by the local government since severe damages have occurred along the shore. The local residents have responded to erosion on an individual basis, but the problem has not been addressed in an integrated manner, considering the whole stretch of sandy beach.

2 Area description

The coast of Rosarito Beach is characterised by the presence of cliffs and low lying beaches. Cliffs are mostly igneous in the north zone of Rosarito Beach, although there is an important component of sedimentary cliffs north of Rosarito. The low lying beach is present in the rest of the area and was once backed by extensive dune fields (as described by local residents), but today development encroaches on the beach (Fig. 1).

The littoral cell for Rosarito Beach has not been established, but this study considers only the sandy beaches of the town of Rosarito Beach, although the littoral cell most likely begins at a headland some kilometres north of Rosarito Beach and ends several kilometres to the south. It is considered that in the study area there is a sandy sediment input and output through the area boundaries.

The primary land uses adjacent to the beach of the study area are residential in the north zone, industrial (represented by the Mexican oil company PEMEX and a thermal electric power plant of the electricity company CFE) and local housing in the central zone, and residential, mass tourism and local housing in the south zone[1].

Environmental Coastal Regions 101

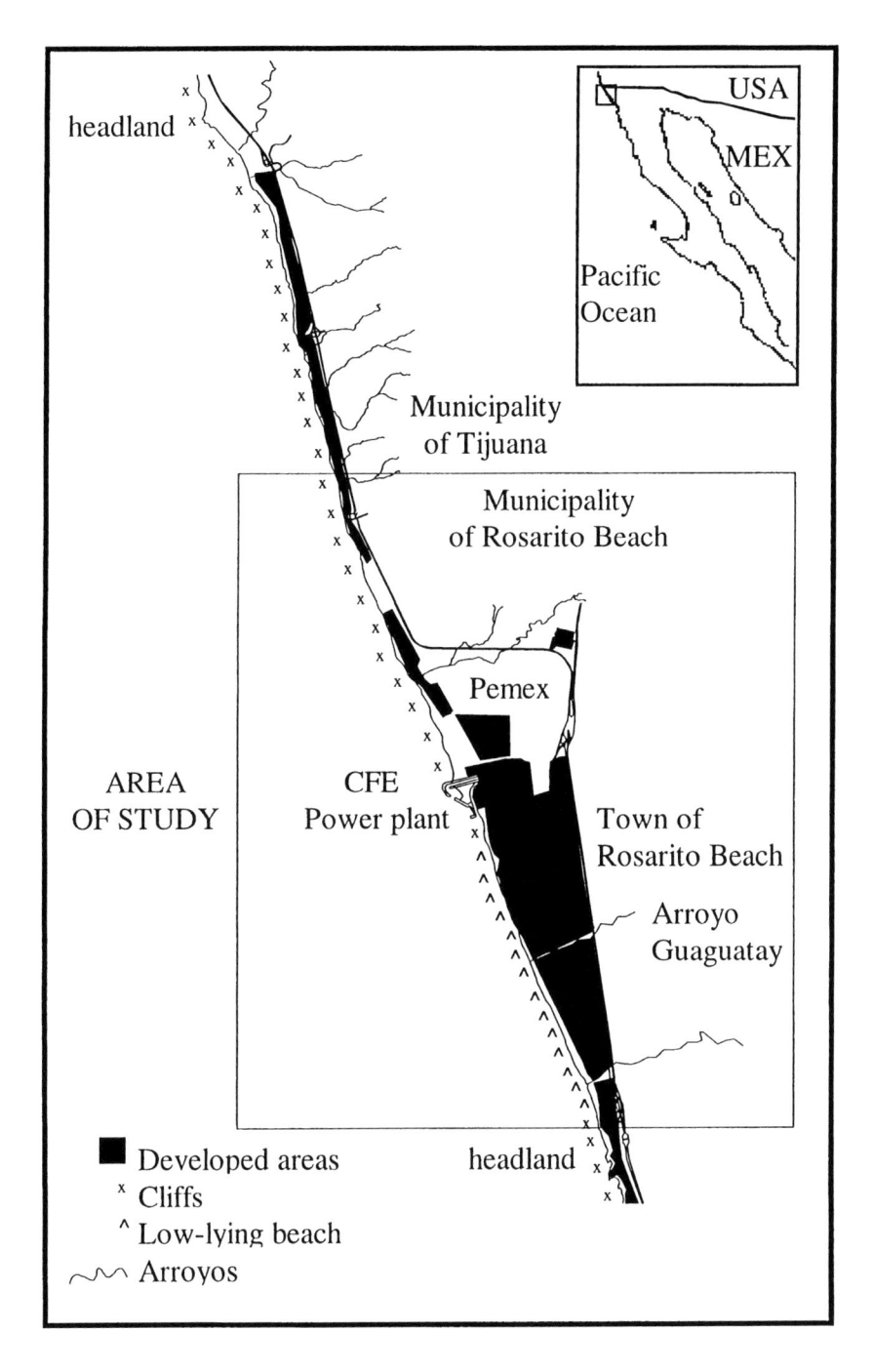

Figure 1. Map of the study area showing most relevant features.

3 Coastal processes

The waves that arrive to Rosarito Beach[7] are: 1) the winter waves (October-March) arriving at the coast from a westerly and northerly direction; 2) the summer waves (July-September) arriving at the coast from a south and southwestern direction ; and 3) a transitional period between winter and summer waves (April-June).

The sediment sources to the area are basically from the arroyos, cliff erosion and longshore sediment transport (LST) from adjacent areas. The arroyos combined have a drainage area of 235 km^2 and their contributions were estimated to be 15,000 m^3/yr following the methodology presented by Pou-Alberú & Pozos-Salazar[5]. The contribution of sand by cliff erosion was estimated to be 38,000 m^3/yr, considering the height and extension of the sedimentary cliffs and an estimated erosion rate of 0.5 m/yr. This rate closely resembles the rates presented by Sunamura[6], where he established an erosion rate between 10^{-1} and 10^0 for Pliocene deposits, as the ones present in this area. An estimation of net LST has been done by Appendini, et al.[2] and is in the order of 100,000 m^3/yr.

Other inputs and outputs of sand, i.e. the cross-shore transport, could not be established, although Lee and Osborne[4] have indicated that sand from the continental shelf up to a depth between 17-20 meters might be brought to the beach.

Considering the sources of sand, we have a sand supply of approximately 53,000 m^3/yr, so there is a deficit of 47,000 m^3/yr to satisfy the LST of 100,000 m^3/yr. This calculation shows that one of the problems in Rosarito Beach is the lack of sand in the system, leading to beach erosion. To give a broad idea of the erosion at Rosarito Beach, this sand deficit can be translated into an erosion rate of 0.21 m/yr using mass conservation; however, there are parts where the erosion may be higher or lower. Due to the lack of beach information these rates cannot be established for a long term.

Extreme erosion in Rosarito Beach exists in the short term resulting in a shoreline retreat up to 60 m in less than a month. This erosion is considered as a part of the beach cycle where the onshore summer transport later returns the sand to the beach. Although the beach show erosional-depositional cycles, extreme events as the ones in January and February of 1998 have transported the sand to depths more than 20 m, so it may be difficult for the beach to recuperate, exacerbating the sand deficit problem in the area.

4 Vulnerability to erosion

To establish erosion management strategies and policies it is necessary to assess the beach's vulnerability to erosion. The factors that were considered to define vulnerability were potential beach loss and probability of damage to infrastructure. The potential beach loss was established from beach width and morphology, while probability of damages was obtained considering land use, population density, and coastal protection. The combination of both factors gave a sense of the vulnerability toward beach erosion (Fig. 2). The percent length of shore subject to a high, medium and low beach loss potential, damage probability and vulnerability to erosion are shown in Table 1.

Table 1. Beach loss, damage probability and erosion vulnerability coast length percentages along Rosarito Beach.

	Beach loss potential	Damage probability	Vulnerability to erosion
High	62%	14%	25%
Medium	15%	25%	54%
Low	23%	61%	21%
Total	100%	100%	100%

From this analysis we see that most stretches of beach have a high potential to be eroded, which has a direct repercussion for the recreational capacity of the beaches, and thus on the tourism industry. The probability of damage is mostly low, since large stretches of coast show some type of protective structures, in particular in the north where tourism housing on cliffs is present. The northern part presents a medium vulnerability toward erosion in most parts, and although damage probability may be low, the loss of beach for recreation is likely to occur. Immediately south of the power plant, there is high vulnerability in an area extensively used for recreational activities (both local and tourism), and there are also local houses in a very vulnerable position to erosion. As a result, many of them where damaged during the El Niño waves in February 1998. The southern part presents medium and low vulnerability to erosion, in part, because this area still has a wide beach available to provide protection.

104 Environmental Coastal Regions

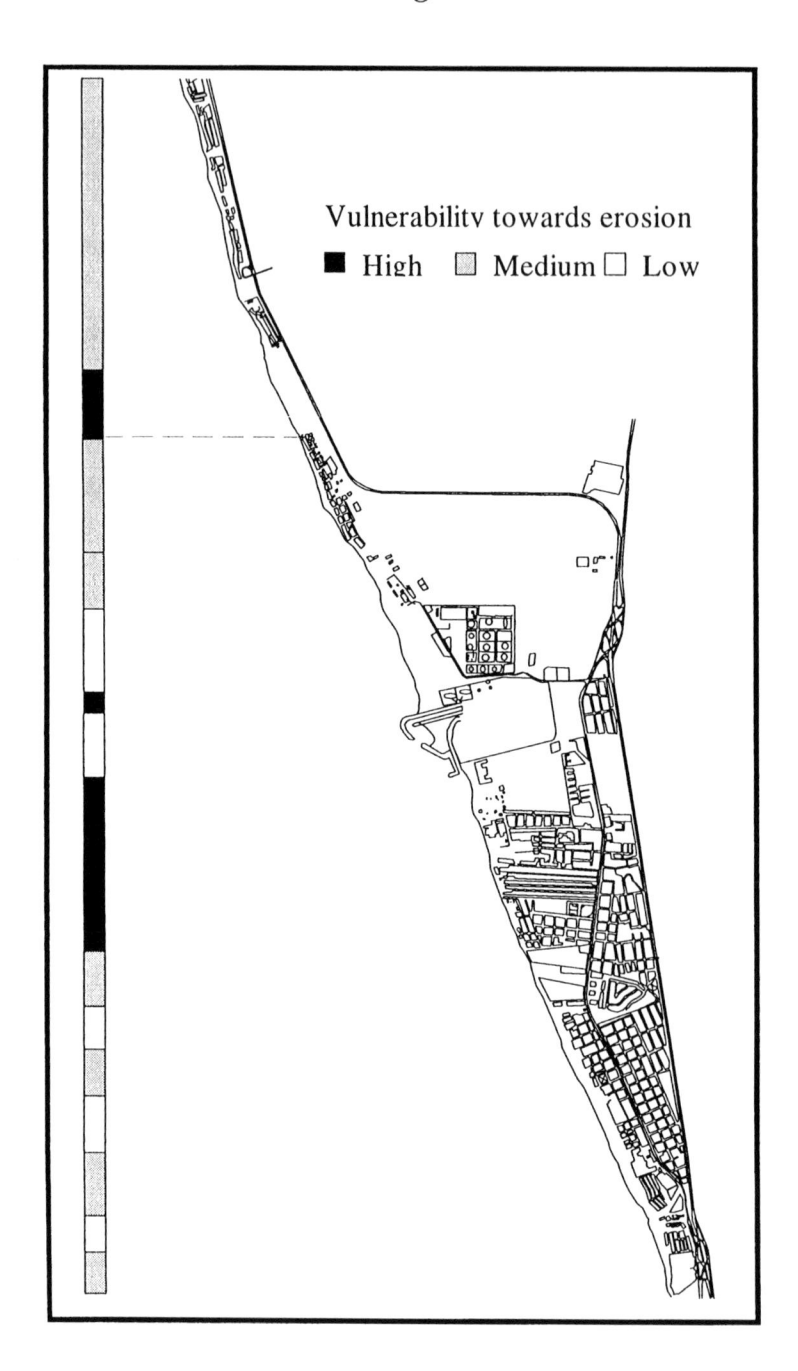

Figure 2. Erosion vulnerability on Rosarito Beach.

5 Management strategies

For the management strategies the area was divided in three reaches based on characteristics and erosion vulnerability: 1) the north reach, which goes from the northern boundary to the power plant, 2) the middle reach, which encompasses the beach south of the power plant to the Arroyo Guaguatay, and 3) the south reach, from Arroyo Guaguatay to the south (Figure 1). This division was established because each has different characteristics and vulnerability towards erosion. The beach at the north reach is not as important as the south reach beach for recreation and tourism, and it is not as important as the middle reach beach for protection and recreation.

Three basic problems exist in Rosarito Beach: 1) there is a long term shoreline erosion associated with a sand deficit in the littoral system; 2) there is localised shoreline erosion at the beach south of the power plant due to the direct effect of water intake structures[2]; and 3) there are severe impacts for infrastructure during storm induced erosion.

The possibility of a "do nothing" alternative could be unacceptable for Rosarito Beach because of the costs and public safety hazard associated with property and infrastructure damage during storm conditions and tourism decline. Thus, an erosion management plan is needed. As a first approach, three management strategies are indicated: 1) sand management, 2) protective devices and 3) regulations.

5.1 Sand management

Due to the sand deficit in the area, it is indispensable to increase the volume of sand in the littoral system. This may be attained by artificial beach nourishment or increasing natural deliveries. Cleaning of the arroyos' watershed and the water course itself as well as maintaining the cliffs' contribution north of Rosarito Beach (in the municipality of Tijuana) would be an important factor. Beach nourishment is the only management alternative that actually increases the sand volume in the littoral system. This alternative is more needed in the middle and south reaches, due to the beach's importance for tourism and protection. Also, this area is the one that historically has suffered more damage and is directly affected by the power plant water intake structures. Infrastructure in the southern part has not suffered damage, but the loss of the recreational beach is evident, requiring nourishment.

106 Environmental Coastal Regions

The use of opportunistic sand sources is needed because funding for dedicated sand is difficult. The possible sources of sand are: 1) the Rodríguez Dam in Tijuana which retains a vast amount of sand as it is the largest watershed area in the region; 2) the water intake structures of the power plant which generate shoaling rates in the order of 50,000 m^3/yr[2]; and 3) upland developments, where the spoil of construction may contain high amounts of sand due to the geology of the area.

5.2 Protective devices

The use of protective devices to retain the sand on the beaches was not considered appropriate for most of the beach reaches, although a groin at the southern headland may help in retaining sand with minimum downcoast effects since only pocket beaches exist that are not important as recreation sites. An alternate protection for the middle and south reaches would be the construction of artificial dunes and their stabilisation with vegetation, since they function as a protective device and increase the sand in the area. This option could be considered as a sand management strategy. Most of the northern reach is already protected with seawalls, and it is proposed to protect the remaining unprotected stretches of cliffs. Since their sand contribution to the system is small and the beach is not important for recreation in this area, the walling of this reach is of little importance.

5.3 Regulations

Because most of the coastline is already developed, the establishment of setback construction lines should be used as a measure for planned retreat. A minimum setback line in low-lying beach areas should be at least 70 m from the LMWL, which is the erosion observed during the extreme events in 1998 (60 m), plus a 50 year long term beach recession (10.5 m). Incentives for relocation of infrastructure should be set after severe damages occur to prohibit reconstruction. Construction (houses) in the beach areas provide too few benefits for the municipality (taxes), no public benefits, and do not contribute to beach preservation, so a planned retreat is needed. Buying out property by the local government is an alternative for a planned retreat, but most of the land in this area are concessions from the federal government (federal zone is 20 m from the mean HHWL) so this option may be politically difficult. However, it is an alternative that needs further exploration. Watershed management regulation is another necessity, since there are irregular settlements in the arroyos as well as illegal water detainment structures that may be contributing to sand loss on the beaches and pollution, increasing the public's safety hazard.

6 Conclusions

In Mexico there are no established policies for coastal management. Due to the economic importance of the beaches of Rosarito Beach, a management plan for handling erosion is highly needed. The sand deficit in its littoral zone clearly shows the need for beach nourishment. Protective structures are acceptable only in the north part where the beach is not an important recreational asset. Sand management, in particular nourishment, apparently is the most adequate option for the middle and southern beaches. It is clear that a need exists to establish policies for development in the beach areas and watershed management in order to sustain the beaches of the municipality as its most important natural and economic asset.

References

[1] Appendini, C.M. & Lizárraga-Arciniega, R., Development of a shoreline preservation strategy, *Proc. of the California and the World Oceans '97*, eds. O.T. Magoon, H. Converse, B. Baird & M. Miller-Henson, ASCE, Reston, pp. 1494-1498, 1998.

[2] Appendini, C.M., Lizárraga-Arciniega, R. & García-Krasovsky, R., Shoaling processes in Rosarito, B.C.: An exercise on longshore sediment transport modeling, *Proc. of the California and the World Oceans '97*, eds. O.T. Magoon, H. Converse, B. Baird & M. Miller-Henson, ASCE, Reston, pp. 1682-1693, 1998.

[3] Gobierno del Estado de Baja California (GEBC), Acuerdo y versión abreviada del programa regional de desarrollo urbano, turístico y ecológico del corredor costero Tijuana-Ensenada, *Periódico Oficial del Estado de Baja California, Mexicali, B.C., 2 de junio de 1995*, p. 75, 1995.

[4] Lee, A.C. & Osborne, R.H. Relative fluxes of sand for Southern California beaches: fourier grain-shape analysis, *Shore & Beach,* 63(3), pp. 9-19, 1995.

[5] Pou-Alberú, S. & Pozos-Salazar, G. Cantidad de sedimento drenado hacia la costa del Pacífico Norte en el Noroeste de Baja California, México. *Ciencias Marinas*, 18(3), p. 125-141, 1992.

[6] Sunamura, T. Processes of sea cliff and platform erosion, Chapter 12, *CRC Handbook of Coastal Processes and Erosion,* ed. P.D. Komar, CRC Press, Boca Raton, pp. 233-267, 1983.

[7] US Army Corps of Engineers (USACE). State of the Coast Report, San Diego Region, *Coast of California Storm and Tidal Wave Study*, USACE, Los Angeles District, 1991.

A digital elevation model of sections of the UK intertidal zone from the integration of satellite imagery and ocean height modelling

I.J. Davenport[1], D.C. Mason[1], G.J. Robinson[1], R.A. Flather[2], M. Amin[2], J.A. Smith[2], C. Gurney[3]

[1] NERC Environmental Systems Science Centre
[2] Proudman Oceanographic Laboratory
[3] National Remote Sensing Centre Ltd
EMail: ijd@mail.nerc-essc.ac.uk

Abstract

The conclusions of a four year study into the use of remotely sensing imagery and hydrodynamic sea height modelling to generate Digital Elevation Models (DEMs) of the intertidal zone, using the waterline method are presented. This technique involves determining the land-sea boundary in a number of images, and heighting it using a hydrodynamic tide-surge model. The quasi-contours derived are then interpolated to generate a DEM of the intertidal zone. Under the UK Natural Environment Research Council Land-Ocean Interaction Study this technique has been applied to a 100km stretch of the UK coastline between the Humber and the Wash on the east coast of England, and under the British National Space Centre Earth Observation LINK programme to Morecambe Bay on the west coast. A number of factors influence the accuracy and applicability of this technique to varying coastal zone types, such as beach slope, beach width, surface structure of the beach, and variability over time. The combination of these factors can determine whether this method is appropriate compared to alternative techniques such as airborne scanning LiDAR or geodetic levelling. The tools for coastal zone morphology study have advanced significantly in recent years, and we put this technique into the context of these changes, and consider the requirements for coastal zone protection and monitoring.

1 Introduction

The coastal zone is an important region to a number of different groups, both in the academic and commercial communities. The morphology of the intertidal zone is subject to the constant assault of water and suspended matter flow, and thus is uniquely susceptible to change, over the short term due to storm events, seasonal change, and long term erosion and accretion. Height maps, or DEMs of the intertidal zone are an essential component of hydrodynamic models of the coastal seas, and the associated flood warning systems. Endangered species reside on the coast, and many sites of special scientific interest lie on the foreshore, six in Morecambe Bay alone, the largest region of intertidal mud flats in the UK. Variability of form can also be studied for the causes of erosion or accretion and the effectiveness of sediment transport modelling. Some industries are also reliant on the inter-tidal zone, including tourism and shellfish fisheries.

Operationally, monitoring the morphology of the inter-tidal zone is important in two respects. Firstly in deciding where sea defences need to be installed to maintain the status quo, or to reclaim land, and what form of defence is most appropriate. Latterly, assessing the effectiveness of the defences also requires monitoring over a period of years. Whilst managed retreat is an alternative to coastal defence, there are regions of the world such as the Netherlands, where this is clearly not a viable option, and in any event a managed retreat also needs to be monitored. Until recently this has been only been possible using laborious techniques such as surveying, with its inherent hazards in flat intertidal regions where the tide can come in faster than walking pace. The advent of remote sensing techniques has presented some new possibilities appropriate to the intertidal zone, including aircraft based Light Direction And Ranging (LiDAR) and the system based on satellite data we describe in this paper, referred to as the 'waterline method'. More complete discussions of the various techniques can be found in [1]. Here we show the results that can be obtained using the waterline method in regions of the UK, and indicate what part it is best suited to play in the study of coastal morphology.

2 Technique description

The waterline technique is a means of deriving height information from coastal remotely sensed images, using the simple principle that the height of the land along the line where the land meets the sea, is the same as the height of the sea there. The two problems are then to delineate this line, and to determine the height of the sea at the time the image was acquired. Since the height of the sea varies over the distances involved, what we obtain by combining the line and the range of heights is a quasi-contour. Heighting the lines from a number of images at different tidal states we obtain a set of quasi-contours, at different times and heights. These provide some information about the form of the coastal zone and its evolution, indeed it is possible to observe dramatic movement of the channels in Morecambe Bay from this data alone as shown in Figure 9. However, interpolation of the height of the regions between these lines both in time and space can provide a DEM of the coastal zone covering the area between high and low tide, and showing the changes in time. Universal block kriging has been used to interpolate the region between the lowest and highest shorelines, due to its ability to interpolate in three dimensions, in this case the third being time, and its provision of an error map based on the local geostatistics.

3 Application to regions of the UK

This technique has been applied to three regions in the UK, highlighted in Figure 1, using ERS and RADARSAT SAR imagery. The Wash is a region of intertidal sand and mud flats on the east coast of the UK, with a typical slope of 1:500, and an intertidal area of 300km^2. The Lincolnshire coast north of the Wash, as far as the Humber, varies in slope between 1:100 and 1:30. Morecambe Bay is a region of sand flats on the west coast, of slope around 1:500 and intertidal area 350km^2, which can vary substantially even over a single season.

112 Environmental Coastal Regions

Figure 1 – The UK, highlighting the study areas

3.1 The Wash

For the Wash (Figure 2), ten ERS-1 SAR images covering the period 1991-4 were processed to determine the shoreline location in each one, as shown in Figure 3. A tide-surge model of the region based on a 240m grid, incorporating meteorological information and tide gauge data, was used to compute water levels and height each shoreline, creating a set of quasi-contours.

Figure 4 shows the results of interpolating the quasi-contours into a DEM. The low slope of the zone meant that a block size of 60m was used, since in most areas little change occurs over that distance, and the mean height standard deviation, verified by comparison with manually surveyed beach profiles, was 22cm. The error estimation, shown in Figure 5, increases with distance from shorelines, so that areas devoid of shorelines, such as exceptionally flat regions, or where interpretation of the sea edge from the SAR image is ambiguous, such as salt marshes, give rise to increased error.

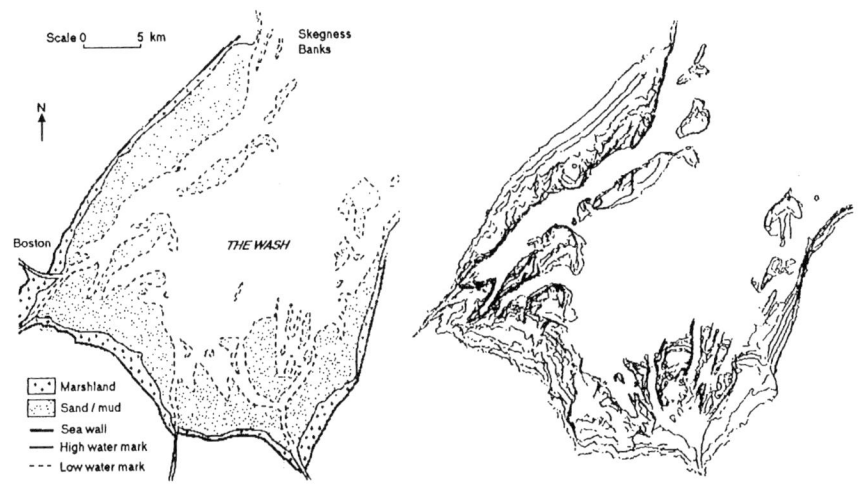

Figure 2 - OS map of the Wash

Figure 3 - Wash shorelines

Comparing this to the Ordnance Survey map of the low tide line in the 1970s, Figure 2, shows that some changes have occurred, particularly around the Skegness Banks, although in general the region seems to have remained largely static. Comparing DEMs generated in 1991 and 1994 using this technique showed that any change over the three year period was below the technique height measurement threshold.

114 Environmental Coastal Regions

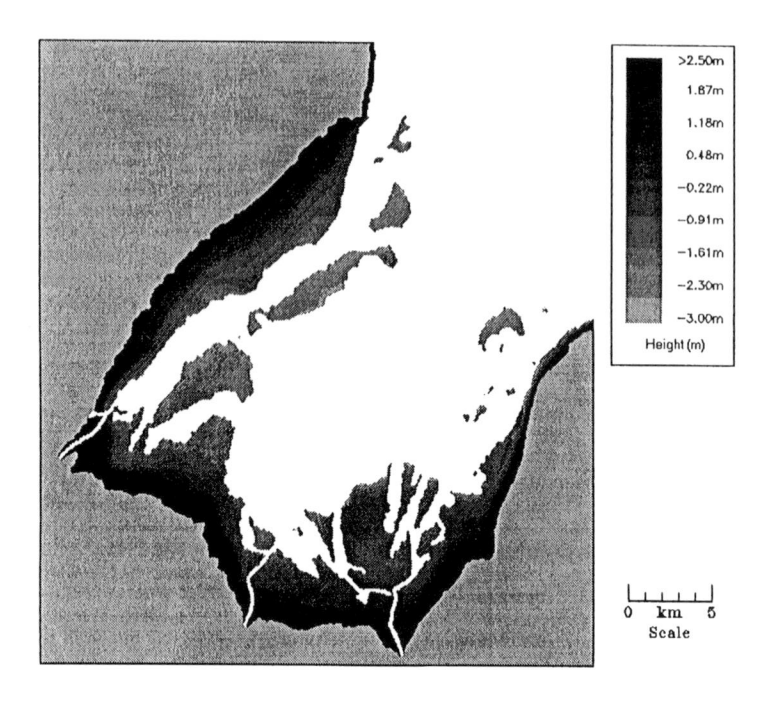

Figure 4 – Intertidal DEM of The Wash, UK

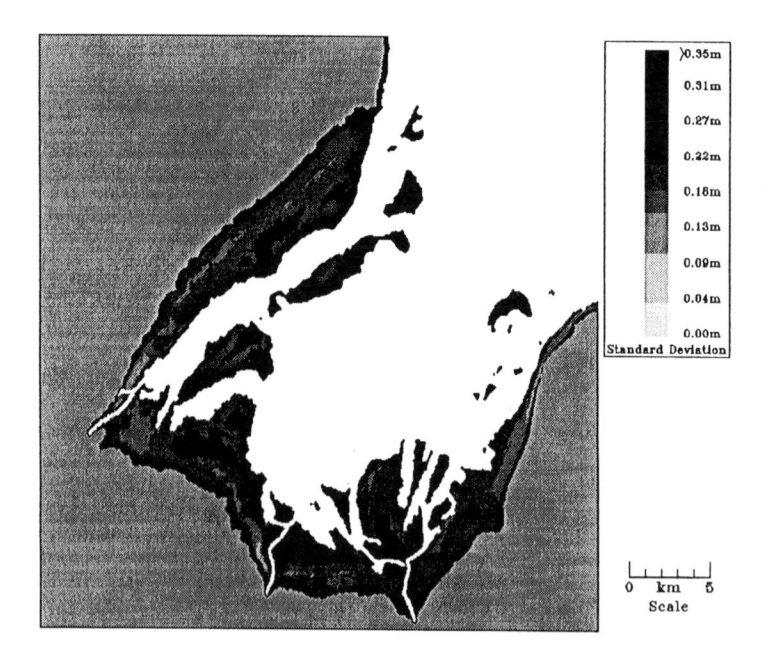

Figure 5 - Wash DEM error map

Environmental Coastal Regions   115

3.2 The Lincolnshire coast

Due to the steeper, narrower nature of the beaches found to the north of the Wash, and the considerable variability found within them, the beaches were divided up into a number of areas as shown in Figure 6. As might be expected, the accuracy of interpolation on the range of beaches varied depending on beach slope. Spurn Bight, with a slope similar to the Wash, but with slightly more fine structure, was interpolated onto a 30m grid, and yielded an average height error of 18cm. No independent survey data seems to exist for this region, however, so no verification has yet been established.

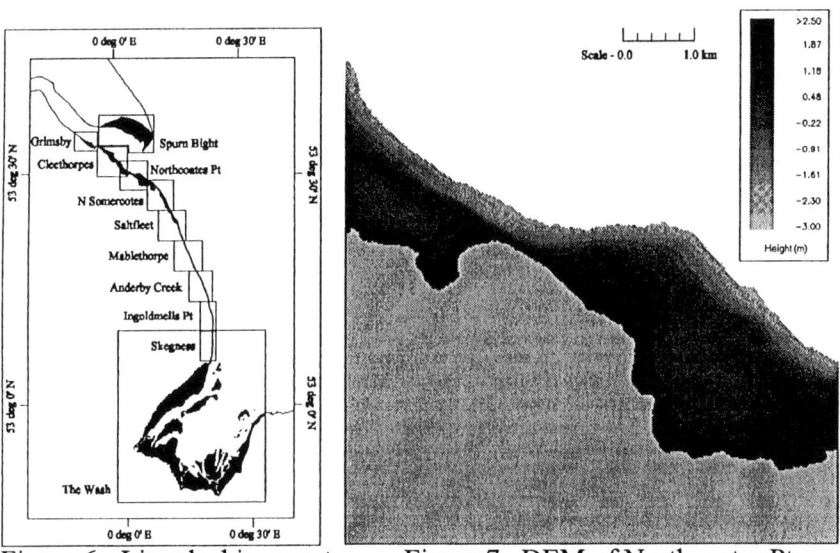

Figure 6 - Lincolnshire coast   Figure 7 –DEM of Northcoates Pt

The areas marked in the diagram as Grimsby through Mablethorpe, including the Northcoates Point region shown in Figure 7, have a beach slope of around 1:100, and the verified mean standard deviation for these regions was 27cm. The regions around high tide were frequently substantially flatter than the rest of the beach, and consequently shorelines were sparser, introducing greater error. Disregarding this static high tide region reduced the mean standard deviation to 23cm. The areas marked as Anderby Creek though to Skegness have a beach slope closer

116 Environmental Coastal Regions

to 1:30, and the mean standard deviation for these regions rose to 32cm, due chiefly to the increased height uncertainty imparted by a constant spatial uncertainty coupled with the increased slope. The complete set of DEMs for this region and a comprehensive analysis of the error sources are presented in [2].

3.3 Morecambe Bay

24 SAR images of the bay, an example of which is shown in Figure 8, between 1991 and 1997 were processed, and the resultant shorelines are shown in Figure 9. The bay differs from the other regions in this study in that it is highly dynamic, showing changes over even just a few weeks. Figure 10 illustrates the movement of the main Kent channel in the north-east of the bay between August 1991 and January 1993, indicating a shift of about 700m.

Figure 8 – Morecambe Bay SAR image

Figure 9 - Morecambe Bay shorelines

Interpolating these shorelines into DEMs is somewhat more complicated, due to the high temporal variability of the region, and the mean standard deviation of such DEMs, as shown in Figures 11 and 12 is about 40cm, although only crudely verified. At this error level it is possible to identify changes in the bay, and the generation of more detailed DEMs and a study of the movement of sediment in the bay will be the subject of further work.

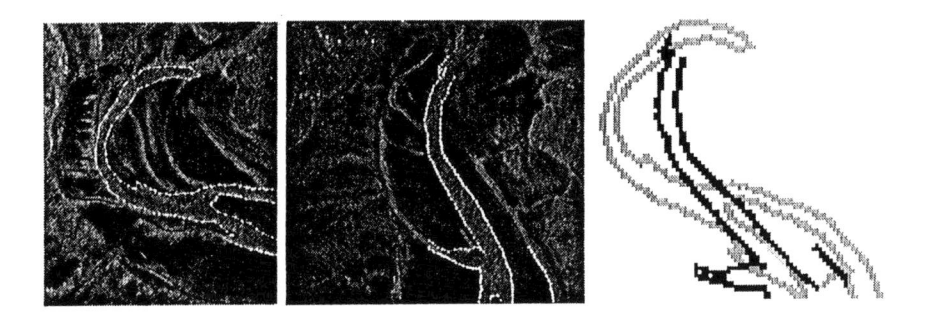

Figure 10 - SAR images of the Kent channel in August 1991 and January 1993, and the two derived shorelines showing the channel movement.

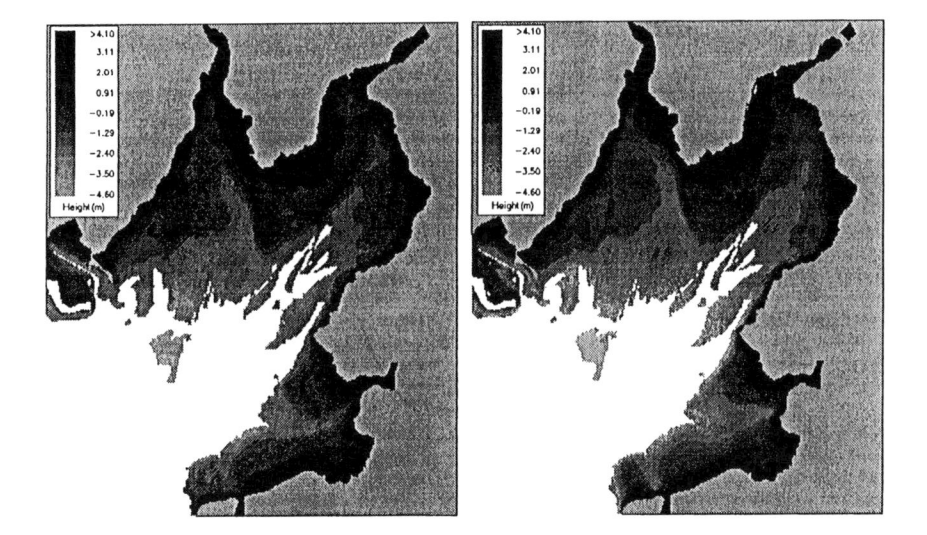

Figure 11 – Morecambe Bay DEM for 1991-1994

Figure 12 – Morecambe Bay DEM for 1995-1997

4 Alternative techniques

Another remote sensing based technique which achieves comparable heighting accuracy to the waterline technique is airborne LiDAR. This involves flying an aircraft over the area of interest. A laser beam scans a few hundred metre wide swath below the aircraft, recording the signal return time, and the geographical location from a differential GPS system based on one unit on the aircraft, and another nearby on the ground. These data can be processed to yield a DEM of the covered swath. At present the financial costs and height accuracy of the technique are unclear, due to its novelty, however current estimates are not substantially different from those for the waterline method. There are, however, clear problems and strengths with each technique which may indicate where they might be most appropriately applied.

LiDAR has a higher spatial resolution, of the order of 2 metres. The spatial resolution of the waterline technique is determined by the SAR pixel size, which for multi-look SAR is 12.5m. Smaller pixel size SAR is possible with systems such as single-look RADARSAT. However this increases data costs which form a significant fraction of the costs of the waterline method. An alternative means of using the waterline technique for higher resolution is to apply it to aircraft-derived images, and an attempt to do so is described in [4]. For narrow beaches, therefore, or where fine detail is required, LiDAR would appear to be a more appropriate option. Regions such as the Wash and Morecambe Bay described earlier have such a low slope that the height changes by only 2.5cm over a 12.5m distance, and clearly the waterline technique provides adequate resolution for these areas.

LiDAR swath width is limited to a few hundred metres. This necessitates a number of passes over beaches up to 5km wide. Such passes need to be carefully planned to avoid gaps between swaths, completed whilst the region below is above water, during the daytime, and in weather amenable to flight and the sensor. Whilst SAR images are sometimes unsuitable due to low wind speeds producing smooth sea surfaces, acquisition is independent of weather conditions, light level or absence thereof, and the waterline method may be more appropriate for building up a long-term model in wider flatter regions of the intertidal zone.

5 Conclusions

The waterline technique can be used to derive height information in the intertidal zone from remotely sensed images, and is most appropriate to large areas of low slope, where spatial resolution better than 12.5m is not required. In such regions it is possible to detect change over periods of months and years, and to generate height maps of height accuracy between 15cm and 40cm, dependent upon beach slope and temporal variability, and the number of images used.

Acknowledgments

This project has received funding under the LOIS Programme, the BNSC Earth Observation LINK Programme, the European Space Agency ERS Pilot Project Programme, and NERC contract F60/G6/12. Thanks are also due to the Environment Agency and the Meteorological Office. This is LOIS publication number 413. All SAR imagery is copyright ESA.

References

[1] Davenport, I.J., Mason, D.C., Flather, R.A. & Gurney, C., Foreshore study through shoreline delineation, *Proc. of the European Symposium on Remote Sensing III*, 1996.

[2] Mason, D.C., Davenport, I.J., Flather, R.A., Gurney, C., Robinson, G.J. & Smith, J.A., A sensitivity analysis of the waterline method of constructing a digital elevation model for intertidal areas in an ERS SAR scene of Eastern England, *Estuarine Coastal and Shelf Science* (submitted).

[3] Lohani, B. & Mason, D.C., 1998, Construction of a digital elevation model of the Holderness Coast using the waterline method and Airborne Thematic Mapper data. *Int. J. Remote Sensing* (submitted).

Section 3: Remote Sensing

Monitoring of coastal waters of the Baltic Sea by airborne imaging spectrometer AISA

T. Kutser[1], K. Eloheimo[2], T. Hannonen[2], P. Härmä[2],
T. Kirkkala[3], S. Koponen[4], J. Pullianen[4], T. Pyhälahti[2],

1. *Estonian Marine Institute, 1 Paldiski Road, Tallinn, EE-0001,Estonia, tiit@phys.sea.ee*
2. *Finnish Environment Institute, Kesäkatu 6, 00251, Helsinki, Finland*
3. *Southwest Finland Regional Environment Centre, PO Box 47, 20801 Turku*
4. *Laboratory of Space Technology, Helsinki University of Technology, Otakaari 1, 02150, Helsinki, Finland*

Abstract

Aim of the present work is to test what parameters of coastal waters could be quantitatively and operatively monitored with passive optical sensors from airborne and/or satellite measurements. Airborne imaging spectrometer AISA was used for remote measurements. Flight campaign was carried out in the Archipelago Sea, south-western coast of Finland. Simultaneous in situ measurements were carried out onboard research vessel Muikku and two outboard work boats. Suitability of more than seventy retrieval variables, proposed by different authors, was tested in estimation of concentrations of chlorophyll a, suspended matter and pheophytin, as well as water turbidity and Secchi disk depth.

1 Introduction

The Baltic Sea is a large semi-enclosed brackishwater basin surrounded by highly industrialized countries, inhabited by ca. 80 million people. The Baltic Sea (with Kattegat), has a surface area of 415,000 km^2, a volume of 21,600 km³ and a mean depth of 54 m. The hydrography of the Baltic Sea is largely regulated high riverine inputs and sporadic inflows of saline North Sea water through the Danish Sound. The Baltic Sea has several sub-basins with typical stratification patterns (temperature and salinity). Besides the vertical salinity gradient, the Baltic has large horizontal salinity variation from almost marine 34 PSU in the Kattegat to limnic 2 PSU in the northern part of the Gulf of Bothnia.

The high riverine inflow is reflected in the high concentrations, compared to open ocean waters, of yellow substances (Gelbstoff) in the Baltic Sea water. The average recidence time of water in the Baltic Sea is about 25 years.The Baltic Sea was previously oligotrophic. Due to antropogenic influence over several decades, primarily high nutrient inputs, The Baltic Sea is now considered to be eutrophic in the coastal areas.

The eutrophication is reflected in the phytoplankton primary production. Extensive growth of phytoplankton, algae blooming increases in frequency and intensity.

Reliable monitoring of the pelagic ecosystem has proved to be problematic because of its high temporal and spatial heterogeneity. Consequently, it often remains unobserved using the traditional sampling methods based on temporally sparse sampling at a few fixed stations. Furthermore, the traditional programs are usually unable to report rapidly the observations in case of exceptional events. Monitoring of water state could be more effective if to use satellite and/or airborne remote sensing. Using of spectral ratios or the same kind of algorithms in interpretation of remote sensing data is widely used, but the algorithms seems to have local and seasonal variability and special algorithms are needed for coastal and inland waters. Therefore we tested suitability of more than seventy retrieval variables, proposed by different authors, in case of coastal waters of the Baltic Sea. Possibility of using hyperspectral modelling in interpretation of remote sensing data was also used to elaborate interpretation methods depending less on specific characteristics of water body under investigation.

2. Equipment and measurements

2.1 The study area and measurement campaign

The sudy area was at the Archipelago Sea. This sea is part of the Baltic Sea, located between the Southwest part of the Finnish coast and the island of Åland (Fig 1). The Archipelago Sea, surface area of about 8,300 km² is heavily eutrophicated, semi enclosed sea area with 22000 islands.

The largest loads of nutrients are discharged to Archipelago Sea by diffuse loading from agriculture and pointsource loading from fish-farming.

In this remote sensing measurement campaing August 1997 we had three different type of measurement lines. The first line "Paimio" (line 1) is located in a estuary of the River Paimionjoki. The Second measurement line "Airisto" (line 2) is situated close to City of Turku and near this line was dreding work going on. The last line "Houtskär" (line 3) goes through the Archipelago Sea and the water exchange in many parts of this measurement line is very poor.

On August 14th, three field measurement lines (Fig 1) were flown with the aircraft (Short SC7 Skyvan) owned by the Helsinki University of Technology, Laboratory of Space Technology.

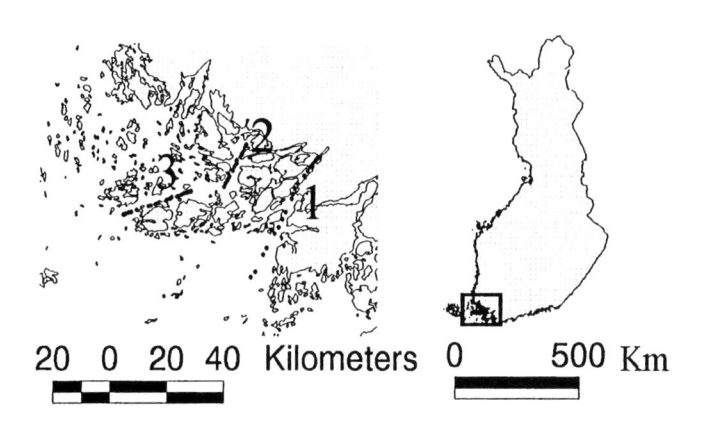

Fig. 1 Measurement lines in August 1997.

2.2 Water sampling and analysing methods

Water samples at each station were taken from the surface layer of the sea (0-0,5m) by using Ruttner-type water samplers. The Paimio measurement line (line 1, fig. 1) sampling was operated by the 28 metre research vessel Muikku with its two smaller motor boats. By using all three boats it was possible to take water samples from three different sampling stations of the Paimio at the same time as the aeroplane was flying over. The Airisto line (line 2, Fig.1) in situ measurements were done by a small outboard boat and the Houtskär line by R/V Aurelia.

The water quality variables measured and the standard methods were: water temperature, secchi disk transparency, chlorophyll a (ISO 10260 and pheophytins, suspended matter (filtered by Nuclepore polycarbonate 0,4 um), turbidity (FTU-units, EN 27027), water colour (comparison with standard platinium cobalt chloride disks, ISO 7887), total organic carbon (TOC, Carlo-Erba IR-analyzer), species of phytoplankton (preserved with Lugols solution and determined at the Finnish Environment Institute by the Zeiss inverted microscope using the Utermöhl [1] technique. All the other laboratory analyses were made at the Laboratory of Southwest Finland Regional Environment Centre. The chlorophyll-a and pheophytin samples taken at R/V Muikku (measurement line Paimio) were filtered immediately after sampling and then carried to a mainland laboratory for analysis. All samples from measurement lines Airisto and Houtskär were analysed in the mainland laboratory within 4-10 hours from sampling.

2.3 Water quality

The figures of the measured water quality date are shown in Figs. 2, 3 and 4. The phytoplankton species dominating at all three measurement lines were Cyanobacteria *Aphanizomenon spp*, and Cryptomonads *Teleaulax spp.*, and *Plagioselmis prolonga*.

Environmental Coastal Regions 127

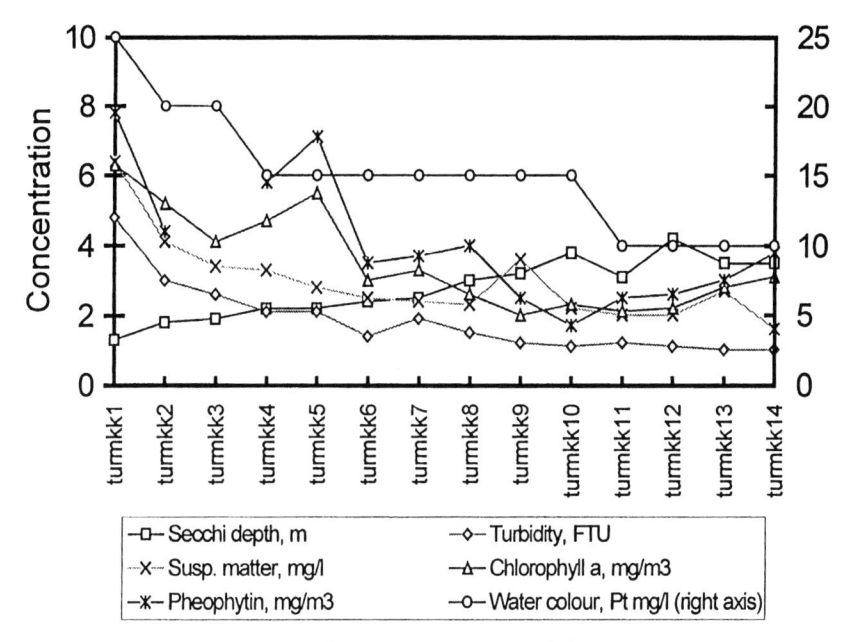

Fig. 2. Measured water quality parameters at Airisto 14.08.97.

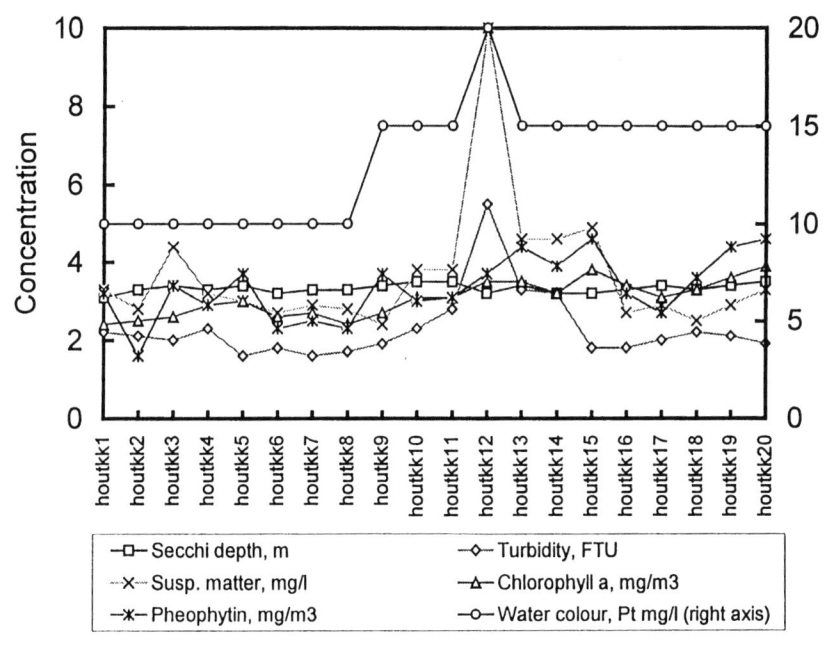

Fig. 3. Measured water quality parameters at Houtskär 14.08.97.

128 Environmental Coastal Regions

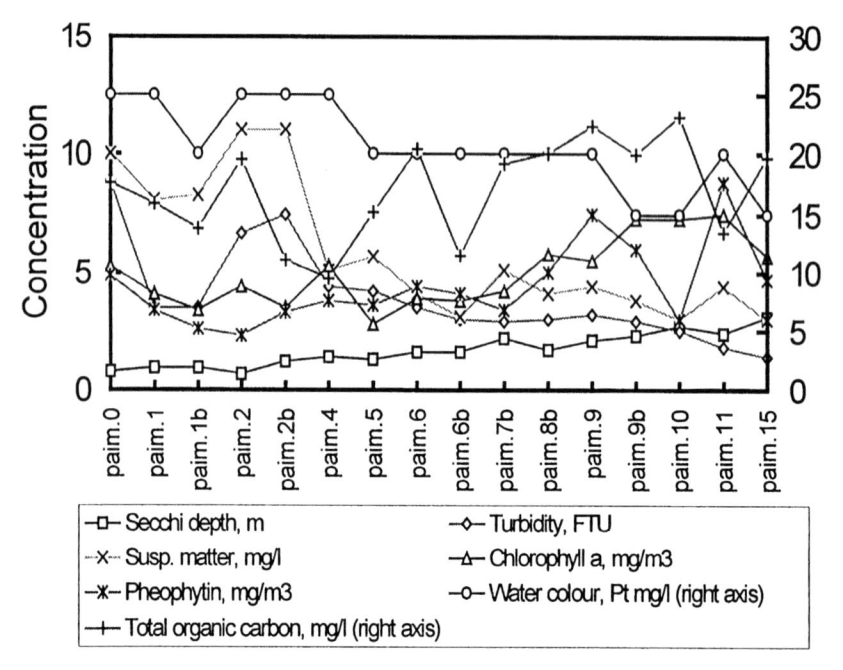

Fig. 4. Measured water quality parameters at Paimiolahti 14.08.97.

2.4 Imaging spectrometer AISA

Finnish imagine spectrometer AISA has been used during the measurement campaign in August 1997. AISA is a pushroom type instrument using a charge-coupled device (CCD) sensor matrix. AISA operates in the range of 450-900nm that is divided into 288 channels. The channel width is programmable from 1.6nm to 9.4nm. The number of spatial pixels is 384.

During the campaign the flight altitude has been 1000 meters. The instantaneous field of view of the CCD (AISA) is 1 milliradian and so the spatial resolution across the track has been 1 meter. The resolution along the track has been 4.5 meters using the integration time of 80 ms and the velocity of 110 knots.

The AISA can be used in one of four modes:
1. all of the information on the CCD is stored - a long integration time is needed (min. 350 ms),
2. all of the data is stored in the spatial direction, but only for a small number of selected wavelength bands (spectral channels),
3. all of the spectral information is stored at selected spatial locations,

Environmental Coastal Regions 129

4. the user can select a set of wavelength bands where full spatial information is stored and a set of spatial location where full spectral information is stored (Mäkisara et al., [2]; Mäkisara, [3]).

During the campaign the second mode has been used on the coastal sites. Some flight lines have been also flown using the third mode for atmospheric correction development purposes. The used channels are shown in Fig. 5.

The geometric correction software makes a geocoded image using the raw AISA data, navigation data and information given by the operator (Mäkisara, [3]). Software for geometrical and radiometrical pre-processing is introduced in (Mäkisara et al., [4]) and (Mäkisara et al., [5]). The geocoded images of the campaign have been re-sampled to the resolution of 2m x 2m.

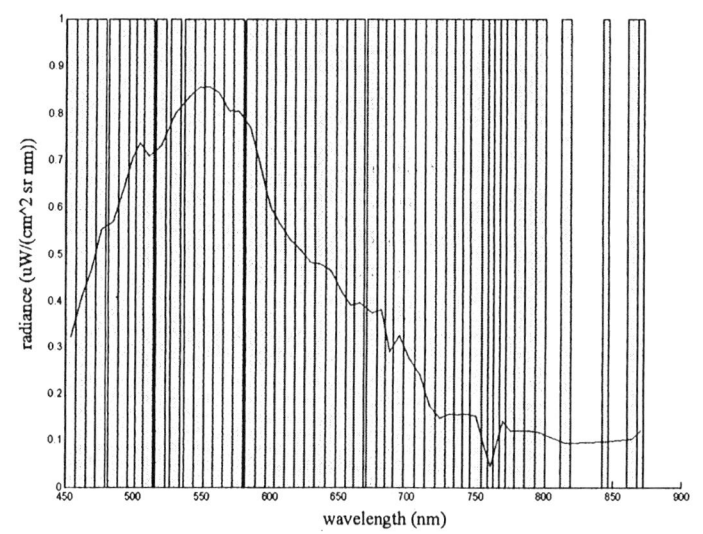

Fig. 5 Spectral channels used in August 1997 campaign.

2.5 Atmospheric correction

An atmospheric correction system was developed for the AISA airborne imaging spectrometer. The correction method is based on the method developed by de Haan and Kokke [6]. The method is based on fitting a single spectral in-situ measurement to a measured AISA spectrum. A number of sets of correction parameters are calculated with MODTRAN using different visibility and humidity values and the best fitting coefficients are searched in an automated iterative process. An averaged reference reflectance spectrum was measured from water at the exact time of aeroplane measurements.

In the correction method by de Haan and Kokke [6] the forementioned phenomena can be simplified to a following equation:

$$R_{APP} = \frac{c_1 + c_2 L_{rs,t} + c_3 L_{rs,b}}{c_4 + c_5 L_{rs,b}}.$$

The R_{APP} is the atmospherically corrected reflectance over the surface, $L_{RS,t}$ is the measured radiance from the targets direction and $L_{RS,b}$ is the mean radiation from the target's neighbourhood. c1 is atmospheric path radiance, c2 and c3 account for the adjacency effect. It should be noticed that if c2 = 1 (in which case c3 = 0) the reflectances are uniform in the neighbourhood of the target and the adjacency effect vanishes. c4 is proportional to the product of the transmittances from sun to target and from target to the sensor. c5 is the spherical albedo for illumination from below.

The coefficients are calculated by running the MODTRAN code three times in the considered situation: The albedo of the surface is set to 0.0, 0.5 and 1.0 respectively. In this way the equations for the software may be calculated as follows:

$$c_1 = -L_{path}(A_{app} = 0.0)$$

$$c_2 = 1 + \frac{L_{path}(A_{app} = 0.5) - L(A_{app} = 0.0)}{L(A_{app} = 0.5)}$$

$$c_3 = 1 - c_2$$

$$c_4 = (1-c_5)\{L_{ground}(A_{app} = 1.0) + L_{path}(A_{app} = 1.0) - L_{path}(A_{app} = 0.0) - c_5 L_{path}(A_{app} = 0.0)\}$$

$$c_5 = \frac{2L_{ground}(A_{app} = 0.5) - L_{ground}(A_{app} = 1.0)}{L_{ground}(A_{app} = 0.5) - L_{ground}(A_{app} = 1.0)}.$$

In our automatic parameter estimation procedure the measured reference reflectance spectrum was compared to corrected AISA spectrum. In each iteration those atmospheric parameter values were selected which gave minimum sum of squared errors. System enables also a "recalibration" of AISA wavelenghts based on atmospheric absorption peaks.

3 Results and discussion

Spectral ratios, their combinations or similar kind of retrieval variables are widely used in interpretation of remote sensing data. We tested suitability of algorithms proposed by different authors ([7]-[15]) in the estimation of various water quality parameters.

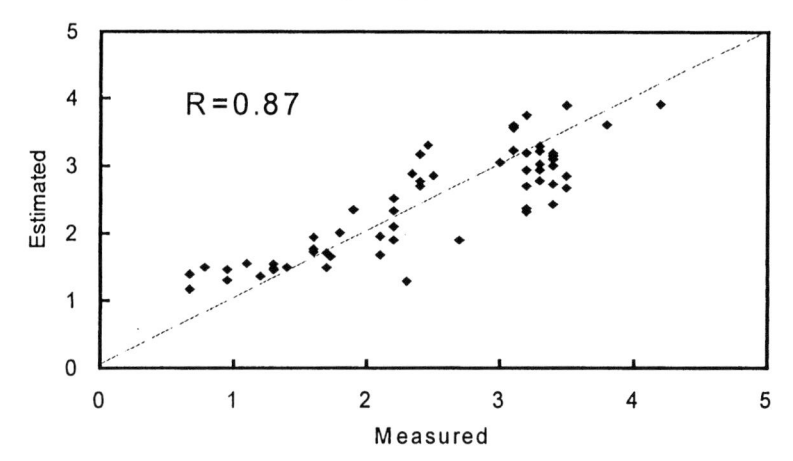

Fig. 6. Correlation between measured in situ and estimated from AISA data Secchi disk depths (in meter). Estimated Secchi depths are calculated by formula: SD=-41.46F$_2$+45.96, where F$_2$=Lu(549)/Lu(527).

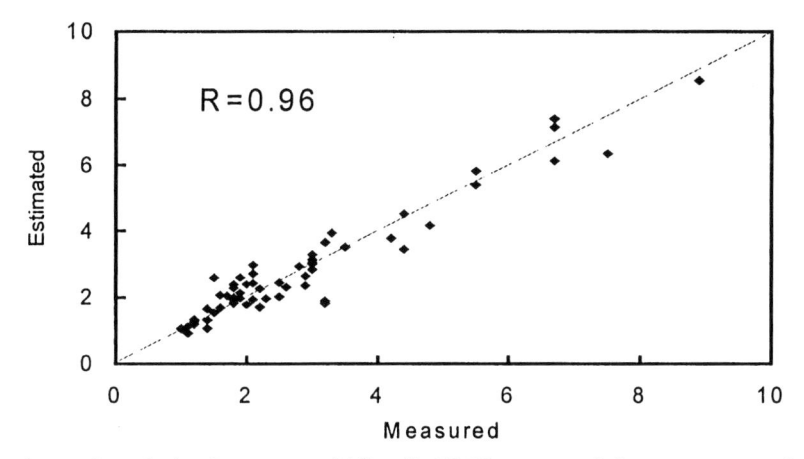

Fig. 7. Correlation between turbidity (in FTU) measured from water samples and estimated from remotely measured data. Estimated turbidity is calculated from AISA spectra using formula: T$_{FTU}$=0.634De$_4$-4.70, where De$_4$=6.34+6.99[Lu(454)-Lu9753)]+4.23[Lu(556)-Lu(753)]+10.7[Lu(622)-Lu(753)]+10.9[Lu(673)-Lu(753)]

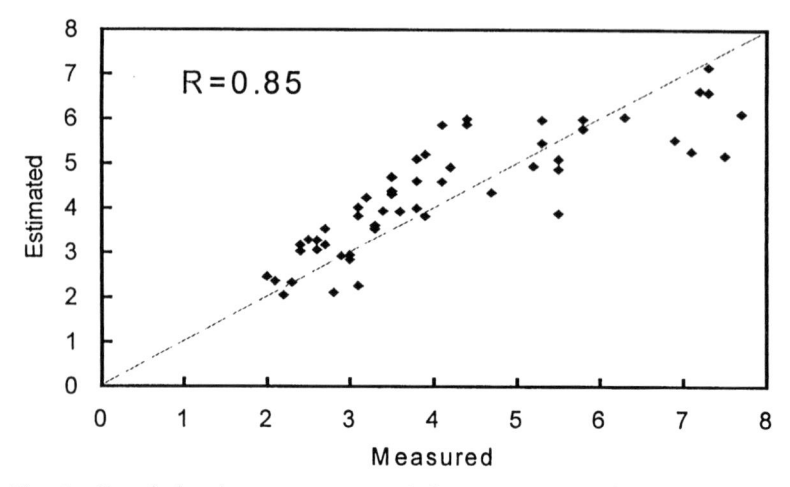

Fig. 8. Correlation between measured from water samples and estimated from AISA data concentrations of chlorophyll-a (in µg/l). Estimated chlorophyll concentrations are calculated using power function: CHL=1.2De$_{24}^{-7.78}$, where De$_{24}$=[Lu(520)-Lu(753)]/[Lu(629)-Lu(753)].

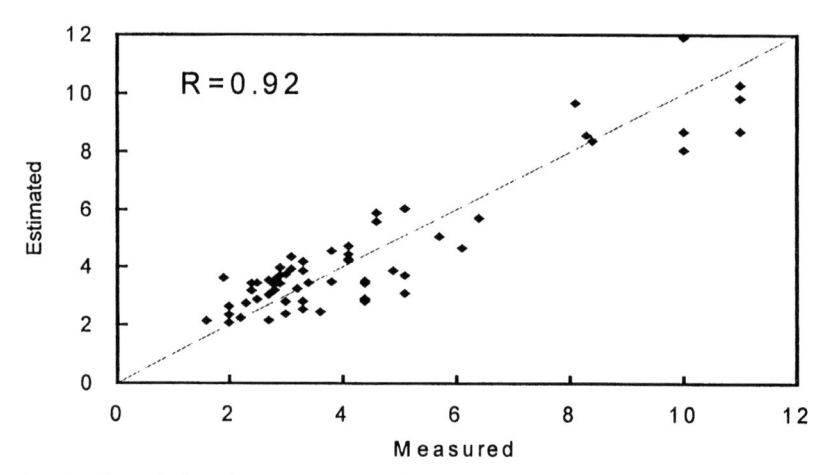

Fig. 9. Correlation between suspended matter concentrations measured from water samples and estimated from AISA data by algorithm: SUS=2.72De$_2$-10.65, where De$_2$=-3.97-3.19[Lu(563)-Lu(753)]+9.38[Lu(622)-Lu(753)].

Our results show that the chlorophyll a, and suspended matter concentrations as well as water turbidity and transparency can be estimated from airborne spectral data. Some algorithms elaborated for new satellite (SeaWiFS, MERIS) were also suitable for estimation of the same water characteristics. It suggest that these parameters can also be estimated from the satellite sensors.

Results obtained in interpretation of AISA data collected on Finnish lakes show that it is possible to estimate concentrations of dissolved and total organic carbon, pheophytin and total phosphorus from AISA data. We can assume that these parameters can be estimated also in case of coastal water whereas the difference between lake and coastal waters of the Gulf of Finland (differences in salinity, dominating algae species, bottom type) is small.

The Baltic Sea is 70% of the year totally or partly covered with clouds therefore the most effective way for monitoring of water state in the Baltic Sea could be organised by combining satellite images, airborne data and automatic in situ measurements from commercial ships and monitoring buoys.

Acknowledgements

The study was organised by the Finnish Environment Institute and carried out in co-operation with the Helsinki University of Technology, Laboratory of Space Technology, Finnish Forest Research Institute, Estonian Marine Institute and Southwest Finland Regional Environment Centre. The authors wish to tank to research and laboratory staff of the Southwest Finland Regional Environment Centre and the crew of R/V Muikku.

The study was financed by the Technology Development Centre (TEKES), Finnish Environment Institute and Helsinki University of Technology, Laboratory of Space Technology.

References

[1] Utermöhl, H., Zur Vervollkommnung der quantitativen phytoplancton-metodik. Mitteilungen der Int.Ver.Limnol. 9, p 1-38, 1958

[2] Mäkisara, K., Meinander, M., Rantasuo, M., Okkonen, J. Aikio, M., Sipola, K. Pylkkö, P., Braam, B., Airborne Imaging Spectrometer for Applications (AISA), *IGARSS Digest*, Tokyo, pp.479-481, 1993.

[3] Mäkisara K. *The AISA Data User's Guide*, To be published in "VTT Research Notes" in 1998.

[4] Mäkisara, K., Kärnä J.-P., Lohi, A., Geometric correction of airborne imaging spectrometer data, *IGARSS Digest*, Pasadena, USA, pp. 1503-1505, 1994.

[5] Mäkisara, K. Kärnä, J.-P., Lohi, A., *IGARSS Digest*, Pasadena, USA, pp. 851-853, 1994.

[6] de Haan, J. F., Kokke, J. M. M. 1996. Remote Sensing Algorithm Development, Toolkit I: Operationalization of Atmospheric

Correction Methods for Tidal and Inland Waters, RWS-Survey Department, Netherlands Remote Sensing Board (BCRS). 91 p.
[7] Afonin E., Kravtsov, G., *Modern apparatus and methods of passive remote sensing of the ocean and atmosphere in the visible part of spectrum*, Sevastopol 1985, (in Russian).
[8] Althuis I., Vogelzang, J., Wernend, M.R., Shimwell, S.J., Gieskes, W.W.C., Warnock, R.E., Kromkamp, J., Wouts, R., Zevenboom, W., *On the colour of case 2 waters. Particulate matter North Sea. Part I: Results and conclusions*, Delft, ISBN 90 5411 170 4, p. 163, 1996
[9] Arst H., and Kutser, T., Data processing and interpretation of sea radiance factor measurements, *Polar Research*, **13**, 1994, pp. 3-12.
[10] Gitelson A., Garbuzov, G., Szilagyi, F., Mittenzwey, K.-H., Karnieli, K., Kaiser, A., Quantitative remote sensing methods for real-time monitoring of inland waters quality, *Int. J. Remote Sens.*, **14**, 1993, pp.1269-1295.
[11] Dekker A.G., Hoogenboom, H.J., *Operational tools for remote sensing of water quality: A prototype toolkit*, Vrie Universiteit Amsterdam, Netherlands Remote Sensing Board report 96-18, ISBN 90 5411 215 8, p. 52, 1997.
[12] Kutser T., Arst, H., Mäekivi, S., Estimation of water quality by passive optical remote measurements, *Sensors and Environmental Applications in Remote Sensing*, ed. J. Askne, Balkema, Rotterdam, pp.281-288, 1995.
[13] Kutser T., Arst, H., Mäekivi, S., Kallaste, K., Estimation of the water quality of the Baltic Sea and some lakes in Estonia and Finland by passive optical remote measurements on board a vessel, *Lakes and Reservoirs; Research and Measurements*, **3**, pp. 53-66, 1998.
[14] Kutser T., "Estimation of water quality in turbid inland and coastal waters by passive optical remote sensing," Dissertaciones Geophysicales Universitas Tartuensis, vol 8, 1997, p. 161.
[15] Tassan S., Local algorithms using SeaWiFS data for the retrieval of phytoplankton, pigments,suspended sediment, and yellow substance in coastal waters," *Appl. Optics*, **33**, pp.2369-2378,1994.

Study of the land-sea interface in the Barcelona Area with lidar data and meteorological models

Cecilia Soriano and José M. Baldasano
Universitat Politècnica de Catalunya (UPC)
Avd. Diagonal 647, planta 10, 08028 Barcelona-SPAIN
soriano@pe.upc.es; baldasano@pe.upc.es

Abstract

This contribution analyzes the circulatory patterns of air pollutants in Barcelona (Catalonia, Spain), an area with strong coastal and orographic influence, during a typical summer-time situation. Special emphasis is put on the development of the sea-breeze circulation and the penetration of its front in the terrain. The analysis was carried out using data from an elastic-backscatter lidar and results from the application of a mesoscale meteorological model. Vertical scans from the lidar revealed a multilayer arrangement of the aerosols above the city, which is related to the sea-breeze circulations and the mountain and valley winds that originate in the region. The formation of a thermal internal boundary layer above the city, as cold and stable air from the sea flows over the heated surface, was also captured with the lidar. The non-hydrostatic meteorological model MEMO was applied to the Barcelona Area. A dispersion simulation, using CO as a tracer, was also carried out, whose aim is the identification of the atmospheric circulatory patterns in the region. Results from the model helped to understand the information acquired with the lidar to make a full description of the circulatory patterns of air pollutants in the Barcelona air basin.

1 Introduction

Being located near the Mediterranean coast, the city of Barcelona and its area of influence provide a privileged scenario for the study of the land-sea-air interactions. Regions with complex orography and a strong coastal influence develop a very wide range of meteorological phenomenology that makes this kind of domains very attractive to the study. The main processes going on coastal environments are daytime convective vertical mixing, sea-breeze circulations, and circulations produced by mountain and river valley thermal and mechanical effects [1], [2], [3].

These mesoscale phenomena develop better when the synoptic conditions prevailing at large scale is weak. When the mesoscale conditions prevail, circulatory patterns will be mainly determined by the characteristics of the terrain in the area. A region's unique orography often requires the development of unique air-quality schemes at a regional scale [4]. That is evidenced in the frequent fact that air-quality remediation schemes developed for one region often prove useless when applied to other areas.

The mesoscale circulatory patterns of air pollutants in the region of Barcelona were described in a study that combined experimental measurements with numerical simulations from meteorological and dispersion models [5]. The application of these two approximations proved very useful for this kind of study, since they compliment each other. Measurements consist mainly of elastic-backscatter lidar data. The lidar data were acquired in July 1992 during a collaborative campaign carried out between Los Alamos National Laboratory (LANL), Los Alamos, NM (USA), and the Technical University of Catalonia (UPC), Barcelona (Spain). The lidar provided information about the distribution of aerosols and the prevailing winds.

In this contribution, we will focus in the dispersion study rather than in the meteorological component of the simulation with the mesoscale model, and we will put the stress on the development of the sea-breeze circulation and the penetration of its associated front in the terrain. The role that mountains in the area have in the establishment of the circulatory patterns will also be studied.

2 Lidar observation of the mesoscale circulations

Vertical scans of range-corrected elastic-backscatter from the lidar can map the distribution of aerosols within the atmosphere, and can be used to infer the patterns of air pollutant circulation in the monitored region. This capability follows from the fact that aerosols trace flow and mass motions within the atmosphere, and have been used in the description of the circulatory patterns of air pollutants in Barcelona mentioned above. Figure 1 shows the evolution of the Mixing Layer (ML) as captured by the lidar, as a function of range from the lidar (x-axis) and m.s.l. altitude (y-axis). Light colors indicate high content of aerosols, whereas dark colors indicate low presence of aerosols. As we move away from the lidar we are approaching the sea, which is at about 6 km from the lidar's position.

Environmental Coastal Regions 137

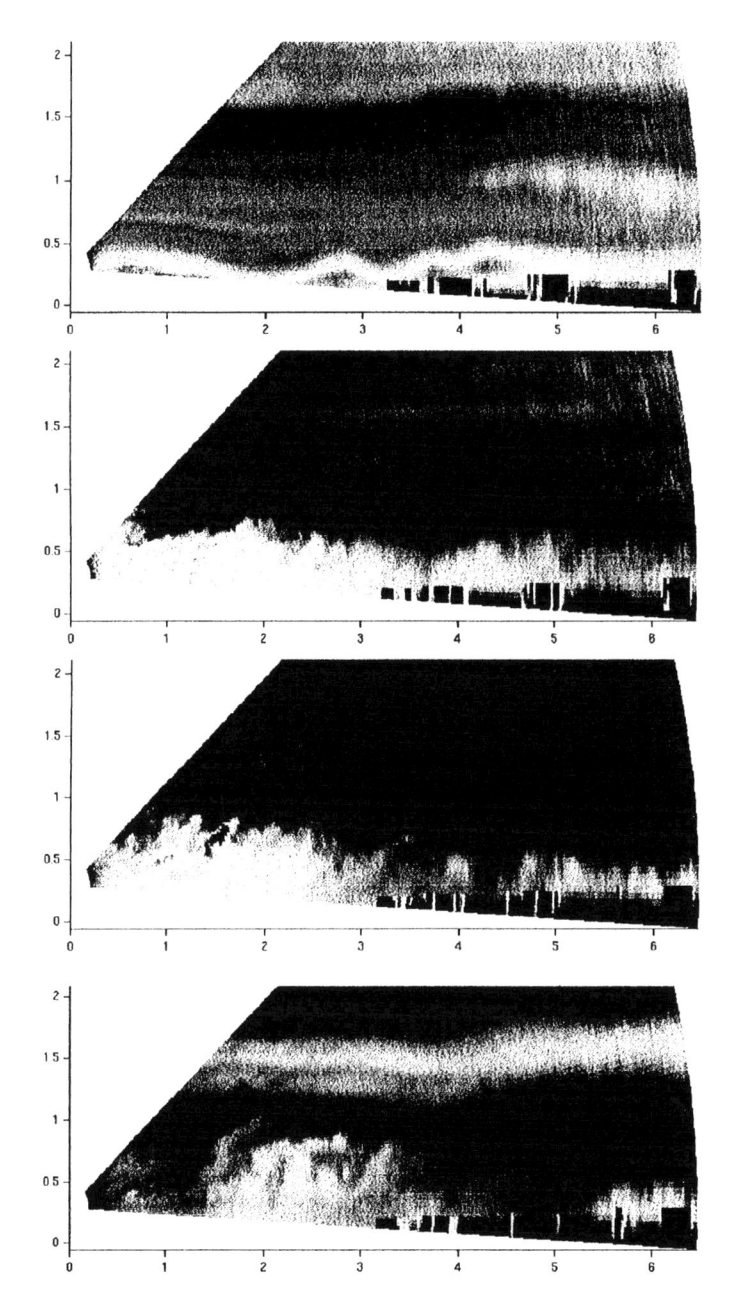

Figure 1: Lidar vertical scans of elastic backscatter acquired in Barcelona, on July 28, 1992, at (from top to bottom) 5:21, 11:36, 13:35 and 14:34 LST. (Y-axis is altitude and X-axis is range from lidar, both in km. Light/dark colors indicate high/low content of aerosols).

The scans show the structure of the first 2.5-km of the atmosphere in Barcelona. Scan at the early morning shows a shallow Nocturnal Boundary Layer (NBL) of about 300-400 m. During nighttime we talk of NBL rather than ML because mixing is reduced. Offshore flow (moving away from the lidar position) is evident in this layer, and produce waves in the frontier between clean and dirty air. Daytime scans show a development of the ML, evident in the lidar scans as a highly convoluted layer with origin in the surface, where aerosols (and pollutants) are emitted. The maximum depth of the ML is observed at about 13 LST. Under conditions of light onshore winds and strong insolation, the ML becomes a Thermal Internal Boundary Layer (TIBL) in coastal zones ([6], [7]), which shows a continuos rise from its over water depth to a continental depth. This rise of the TIBL is clear in the scan acquired in Barcelona at 13:35, where we can see that it has an altitude of 400 m at 6 km from the lidar position (next to coast-line), and rises up to 800 m inland. Finally, the scan at 14:34 shows an elevated layer of aerosols at about 1.5 km of altitude. Those aerosols are positioned aloft by the return flow of the sea-breeze circulatory cell and the injection produced by the Coastal Mountains, the mountain range limiting Barcelona city in its interior side and where the lidar was installed during the campaign. Convection is still evident on the ML, and we can also identify some Kelvin-Helmholtz billows. These kind of circulatory cells are formed as air masses with different density (sea-air and terrain-heated air) find each other.

3 Dispersion Simulation

A dispersion simulation has been carried out with the mesoscale model MEMO [8]. The simulation was performed CO, a very low-reactive pollutant, to be used a tracer of the atmospheric movements. The simulation domain is a region of 80x80 km^2 centered in the city of Barcelona, with a cell resolution of 2x2 km^2. The CO inventory for Barcelona was extracted from the emission inventory for year 1990 [9]. Results from that inventory were that 90% of the total emissions of CO in the region are from traffic.

The simulation day chosen was 28 July 1992. The weak synoptic situation for that day allowed the development of the mesoscale phenomena that takes place in the region. The evaluation of the meteorological model with data from surface station and remotely measured winds with elastic lidar was shown in [5]. We will show here the results of the dispersion simulation. Figure 2 shows the horizontal fields (at 10 m above the surface) of CO calculated by the model for 6 LST and 18 LST. Two different scales have been used, as the increase of the ML during daytime provokes a dilution of

Environmental Coastal Regions 139

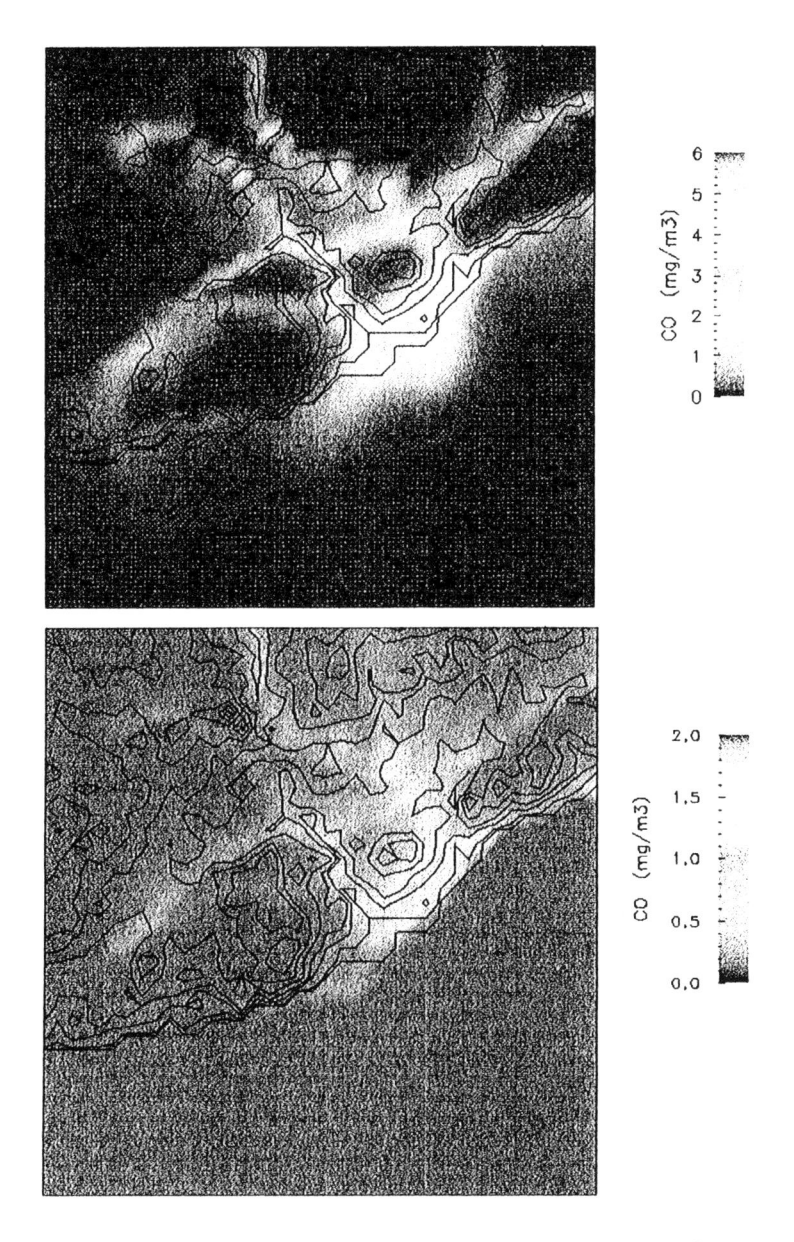

Figure 2: Horizontal cross-sections of the CO concentration (in mg/m^3) at 7 LST (top) and 18 LST (bottom) on July 28.

the emissions in the afternoon. The CO distribution reveals the main aerosols sources in the region (Barcelona City and main roads). At 7 LST the offshore flow produce a migration of the CO emitted in the land towards the sea. The

140 Environmental Coastal Regions

situation reverses as sea breeze starts to blow, and provokes a penetration of the pollutants inland. We can see this penetration in the plot for 18 LST in the afternoon. The rise of the ML decreases surface concentration levels of CO (and pollutants in general), compensating higher emissions at that time of the day.

The vertical cross-section of CO concentration at 11 LST is shown in figure 3. We can identify the TIBL that is formed above the land. It is shallow above the sea and next to the coastline (right side of the plot) and gets deeper as we move inland (depths of about 600 m at that time of the day). The plot is in agreement with the aerosol distribution captured by the lidar and represented in the form of vertical scans.

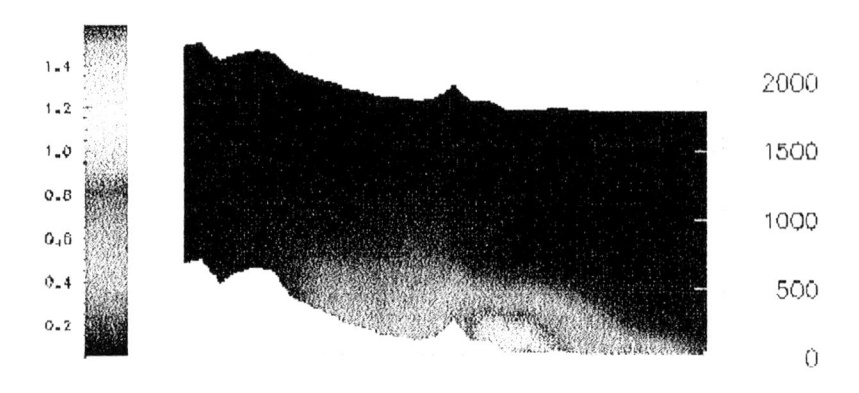

Figure 3: Vertical cross-sections of the CO concentration (mg/m^3) simulated by the model at 11 LST, on July 28, 1992. Altitude is in meters.

Cross-sections of CO concentration for next times (not included here) show the development of the ML (to a maximum depth of about 900 m) and some transportation of CO to high altitudes, specially the injection produced by the mountain ranges that we had identified in the lidar scans. However, this injection is reproduced later in the afternoon by the model, and only in the Pre-Coastal Mountain range.

4 Evaluation of the dispersion simulation

The evaluation of the dispersion simulation has been carried out by comparison with data from surface stations in the domain. Results of this comparison are shown in figure 4.

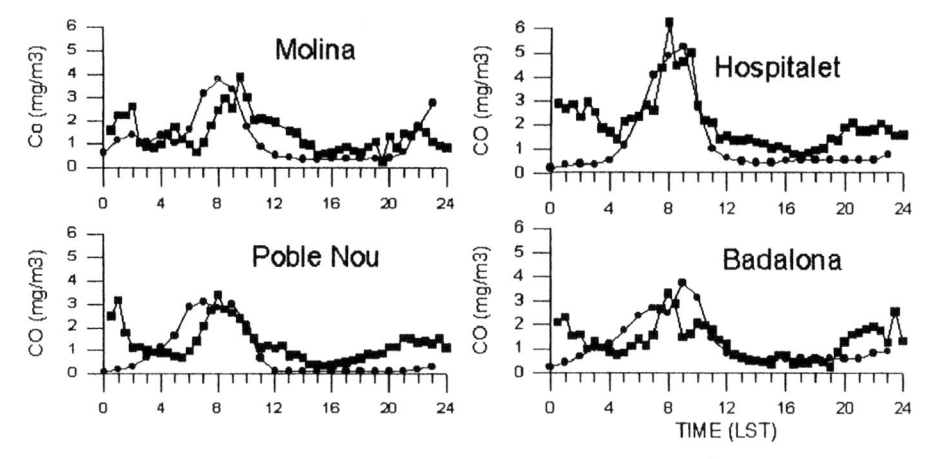

Figure 4: Comparison of the concentration of CO (in mg/m³) simulated by the model (circles) with measurements from surface stations (squares) for July 28, 1992.

As we can see, the agreement between measured and simulated data is quite good. The model was able to simulate very well, both in time and magnitude, the position of the maximum of the CO concentration. This maximum takes place between 8 and 10 LST, when higher emissions (rush hour time) coincide with a still shallow ML, since solar radiation is still weak at this time of the day. Even the higher pick registered at the station of Hospitalet was reproduced by the model.

5 Conclusions

We have studied the penetration of the sea breeze front in the region of Barcelona. Lidar scans have captured the development of a TIBL (Thermal Internal Boundary layer) above the terrain as cold and stable air from the sea flows over the terrain. A dispersion simulation has been carried out for the region, using CO as a tracer of the atmospheric movements. Results have shown how the dispersion of pollutants is highly determined by the daily cycle of the sea breeze circulation and the mesoscale circulations originated by the complex surrounding orography in the region. Evaluation with surface stations has given excellent results, and the model has reproduced the formation of the TIBL.

Bibliography

[1] Wakimoto, R.M. and McElroy, J.L. Lidar Observation of Elevated Pollution Layers over Los Angeles. *Journal of Climate and Applied Meteorology*, **25**, 1583-1599, 1986.

[2] Melas, D. and Kambezidis, H.D. The Depth of the Internal Boundary Layer over an Area under Sea-Breeze Conditions. *Boundary-Layer Meteorology*, **61**, 247-264, 1992.

[3] Mc Kendry, I.G., Steyn, D.G., Lundgren, J., Hoff, R.M., Strapp, W., Anlauf, K., Froude, F., Martin, J.B., Banta, R.M. and Olivier, L.D. Elevated Ozone Layers and Vertical Down-mixing over the Lower Fraser Valley, BC. *Atmospheric Environment*, **31**, 14, 2135-2146, 1997.

[4] Hoff, R.M., Harwood, M., Sheppard, A., Froude, F., Martin, J.B. and Strapp, W. Use of an Airborne Lidar to Determine Aerosol Sources and Movement in the Lower Fraser Valley (LFV), BC. *Atmospheric Environment*, **31**, 14, 2123-2134, 1997.

[5] Soriano, C., Baldasano, J.M. and Buttler, W.T., On the application of meteorological models and lidar techniques for air quality studies at a regional scale. *Proceedings of the Twenty-second NATO/CCMS International Technical Meeting on Air Pollution Modeling and its Application.* Clermont-Ferrand, France, 2-6 June 1997, 413-420.

[6] Nakane, H. and Sasano, Y. Structure of the Sea-breeze Front Revealed by Scanning Lidar. *Journal of the Meteorological Society of Japan*, **64**, 5, 787-792, 1996.

[7] Hayden, K.L., Anlauf, K.G., Hoff, R.M., Strapp, J.W., Bottenheim, J.W., Wiebe, H.A., Froude, F.A., Martin, J.B., Steyn, D.G. and McKendry, I.G. The Vertical Chemical and Meteorological Structure of the Boundary Layer in the Lower Fraser Valley during Pacific'93. *Atmospheric Environment*, **31**, 14, 2089-2105, 1997.

[8] Moussiopoulos, N. MEMO-A Non-Hydrostatic Mesoscale Model. In: *The EUMAC Zooming Model: Model Structure and Applications.* Editor: Moussipoulos, N., EUROTRAC-155, Garmish-Patenkirchen, pp 7-22, 1994.

[9] Costa, M. and Baldasano, J.M. Development of a Source Emission Model for Atmospheric Pollutants of the Barcelona area. *Atmospheric Environment*, **30A**, 2:309-318, 1996.

Fluorescence Bands From Georeferenced LIDAR Spectra of Marine Water Mapped as RGB Images

P.C.C. Barbosa; R.A. Nunes; R.H. Tabares & S. Paciornik
Pontifícia Universidade Católica do Rio de Janeiro.
pccb@fis.puc-rio.br

Abstract

Brazil extracts 70% percent of its oil production from Campos Basin. Petrobrás, a Brazilian oil company, is promoting an environment pre-monitoring study of this area. In addition to traditional analysis techniques, we are using LIDAR spectroscopy in order to compare it with these well established techniques. In this paper we present our first results using an intuitive and flexible visualization tool for spatial distribution of different fluorescent materials in water, as the phytoplankton and other pigments, humic compounds or petrogenic hydrocarbons. A Nd:YAG LIDAR coupled with a GPS receiver is used to acquire georeferenced fluorescence spectra of marine water. Emission spectra excited with a 532nm (second harmonic) pulse were associated to each georeferenced point. Spectra were normalized by the water Raman line. Chlorophyll-a and dissolved organic matter (DOM) fluorescence bands of each point were extracted and weighted. The bands were attributed to the "primary" components of colored pixels which are superimposed on the map of the measured region creating a RGB image. Pixel colors between measurement points are obtained through interpolation.

1 Introduction

The "Grupo de Instrumentação e Materiais" (Instrumentation and Materials Group, GIM - PUC-Rio, Brazil) has mounted a compact and transportable system of Laser Induced Fluorescence (LIF-LIDAR)[1].

During the last few years several experiments were carried out in order to employ this equipment on water[2], vegetation[3] and soils[4] studies.

More recently, with the sponsorship of Petrobrás (Brazilian oil company), an expedition was programmed to the Campos Basin region for testing of the system and for validation of a methodology that allows the generation of maps containing information about water conditions in this region. This work is part of a multitask effort inserted in a institutional program called "Campos and Cabiúnas Basins Environmental Preliminary Stage Monitoring". The LIF-LIDAR system was installed in a container in the deck of a tug, and was used to map the water fluorescence spectra along the tug's path.

The wide penetration depth variation of a laser beam in different samples, even for near positions, was the major problem for utilization of this kind of equipment in direct measurement and comparison of fluorescent materials concentration between samples. Those variations are caused by the optical attenuation properties change in, due to the concentration of fluorescent materials and other compounds like suspension sediments. In 1981[5], laser induced water Raman emission was successfully applied for normalizing the fluorescence signals from the chlorophyll-a.

The fluorescent materials present in natural waters are characterized by an emission band around λ= 350nm, referring to aminoacids and proteins, and two other bands with peak values between 390nm and 450nm, that are related to the humic compounds[6]. Other important fluorescence occurrences are those from the phycoerythrin in 580nm, from the chlorophyll-a in 680nm and from the chlorophyll-b in 720nm. After 1981[6] some studies associated the normalized emission of fluorescence from the chlorophyll-a with the phytoplankton concentration. In 1985[7] the relation between the chlorophyll fluorescence emission and other accessories pigments was investigated with the goal of obtaining a spectral signature for each phytoplankton composition.

In this work we present the necessary modifications to the original system so that georeferenced data acquisition would be possible. In order to visualize these data in an integrated fashion, a new methodology for the visualization of fluorescence signals through the association of these signals to the intensity variation of the Red, Green and Blue (RGB) color primaries is developed.

2 Experimental Procedure

The most important technical features of the LIF-LIDAR system used in this expedition are shown on table 1.

The laser with a Newtonian telescope, the optical detection sub-system and the control electronics were mounted in a rack. The power supply units and the cooling system were mounted in a separated rack so that there would be more flexibility when transporting of the system.

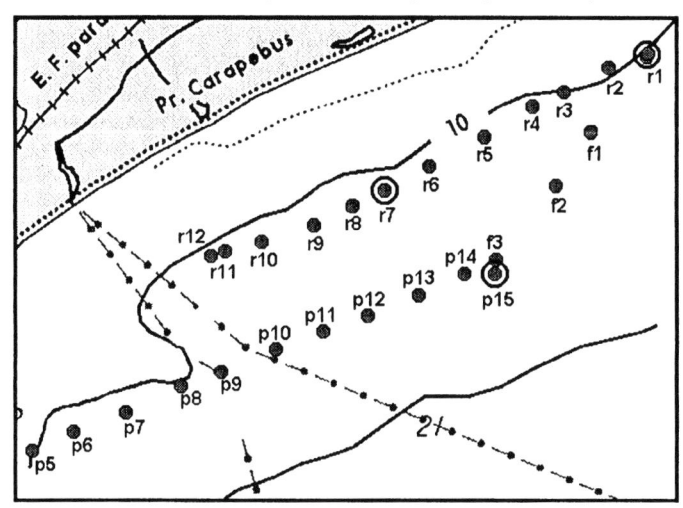

Figure 1 - Measuring sites map. Measurements were made between 10m and 21m Isobaths (black lines)

The location of the region where this mission took place is shown in Figure 1. The upper left point of the rectangle is at 22°21'39''W latitude, 41°32'31''S longitude and the lower right point at 22°16'00''W, 41°42'13''S.

In order to improve signal-to-noise ratio and minimize effects generated by ship movement and surface water ripple, measurements were assigned to the mean value obtained by shooting the laser 32 times. The background illumination was subtracted from the fluorescence signal. This was achieved by making a measurement without laser excitation for every laser shot measurement made. These mean values were then associated to their georeferenced positions.

Two representative bands in these spectra were chosen to generate the RGB map. These bands are associated to Dissolved Organic Matter (DOM) and Chlorophyll-a. DOM and chlorophyll intensities were extracted from bands centered at 560 nm with 20 nm width and 685 nm with 10 nm, respectively. These spectra were normalized by the water Raman signal (λ = 650 nm). Deconvolution pre-processing techniques were applied to mitigate instrumental broadening.

Table 1 - LIDAR Characteristics

Polychromator: grating 300 lines/nm, range 400 - 900 nm, slit 150 mm
Image Intensifier: gain - 10^4
CCD camera: 1024 linear array
Control System: 14 bits A/D converter, 10 photons/bit resolution
Telescope mirrow: 15cm diameter

Nd:YAG laser characteristics:
     probe pulse duration   8 ns,
     pulse repetition rate   10Hz
     pulse energy:
     1st harmonic (λ = 1064nm)   600mJ
     2nd harmonic (λ = 532 nm)   280mJ
     beam divergence   0,5 mrad
     operation mode   active Q-switch

Power consumption: 1kW
Cooling system: closed
Windows compatible control Software

3 Results

In Figure 2 we present 3 of the 25 collected spectra. These spectra were selected because they clearly exhibit variations of band intensities related with DOM and chlorophyll-a. As it can be seen in this Figure, comparison of bands of these three spectra are somewhat difficult. With a higher number of spectra the comparison becomes even more confusing. Also, the lack of association between spectrum and its physical position in this kind of graph makes it more difficult to have the perception of space changes.

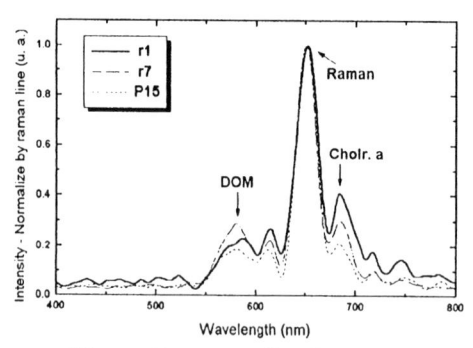

Figure 2 - normalized spectra

Environmental Coastal Regions   147

Table 2 shows fluorescence bands intensities DOM and chlorophyll-a in the considered sites (shown in figure 1), both normalized by the water Raman emission. Water Raman intensities are proportional to water light-attenuation coefficients and to effective signal-integration depth (which is defined as the greatest depth from which light emission reach the remote receiver[8]) and are also listed in this table. To display data related to all these measurement points coherently in a single graphic, 2-D or 3-D plots should be used.

By associating the intensities of fluorescence to color we obtain a quick visualization of distribution over an area as we can see in figure 3 and 4. In this paper, we use shades of gray to simulate color change once we are limited to black and white print. Color version of these graphics are available in the Internet at http://www.ctc.puc-rio.br/gim.html .

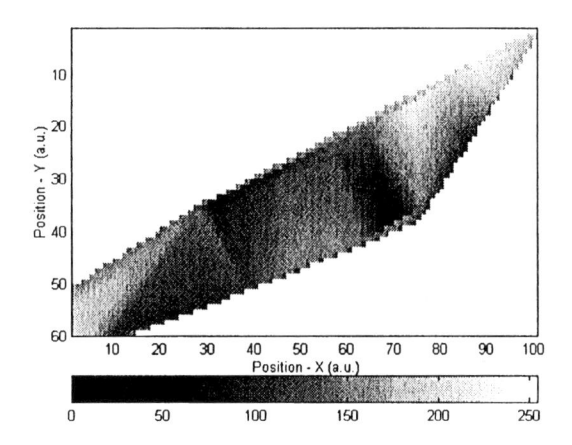

Figure 3 - Chlorophyll-a mapped as color intensity

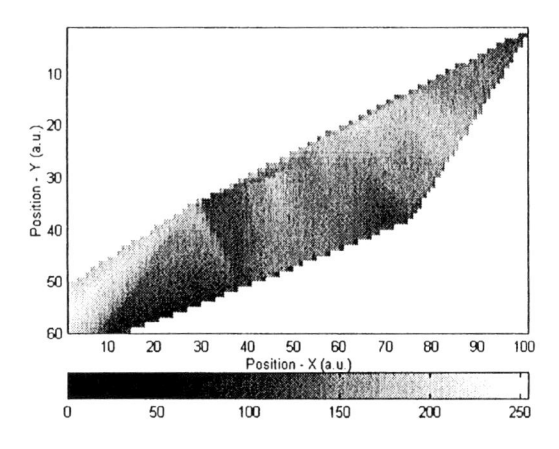

Figure 4 - DOM mapped as color intensity

Although spatial reference could be achieved this way, correlating both pigments concentrations is still difficult. With a 3-D graphic it is also possible to get spatial reference of a variable as we can see in figure 5. But two sets of isolines are unpractical in the same graphic. In order to put two pigment intensities in the same map we use RGB color decomposition.

Table 2 - Normalized fluorescence band intensities and water Raman emission.

Point	DOM	Chlorophyll-a	Raman
p6	0.242	0.265	4815
p7	0.268	0.257	4508
p8	0.247	0.264	4308
p9	0.294	0.292	4357
p10	0.357	0.339	3704
p11	0.350	0.360	3575
p12	0.318	0.332	4159
p13	0.272	0.336	4206
p14	0.323	0.328	3695
p15	0.289	0.281	3758
f3	0.265	0.172	5510
f2	0.423	0.252	1992
f1	0.356	0.386	1275
r1	0.315	0.409	1319
r2	0.303	0.468	1366
r3	0.372	0.521	1627
r4	0.345	0.451	1341
r5	0.410	0.512	2529
r6	0.447	0.321	2629
r7	0.396	0.303	2835
r8	0.354	0.303	3532
r9	0.401	0.299	3424
r10	0.297	0.244	3993
r11	0.296	0.221	378
r12	0.424	0.317	1409

Color in CRT equipment can be described as composed of three "primaries": Red, Green and Blue. The linear combination of intensities of each "primary" color gives its gamut, i.e., all colors that the equipment can reproduce[9]. By choosing a "primary" intensity to represent a pigment concentration it's possible to visualize two or three variables in the same map. The question here is how sensitive is the

Environmental Coastal Regions   149

human eye to differentiate between these colors, in order to create a simple standard to compare data.

If we use three variables, the color of each point varies in a RGB cube. Even if these colors are perfectly distinguishable, a pattern to understand its meaning would be difficult to generate. But if we set one "primary" with a fixed value, two variables can be well represented in a color square. We fixed the blue color at mid-range, in order to generate a good color contrast when changing chlorophyll as Green and DOM as Red. The result can be seen in figure 6.

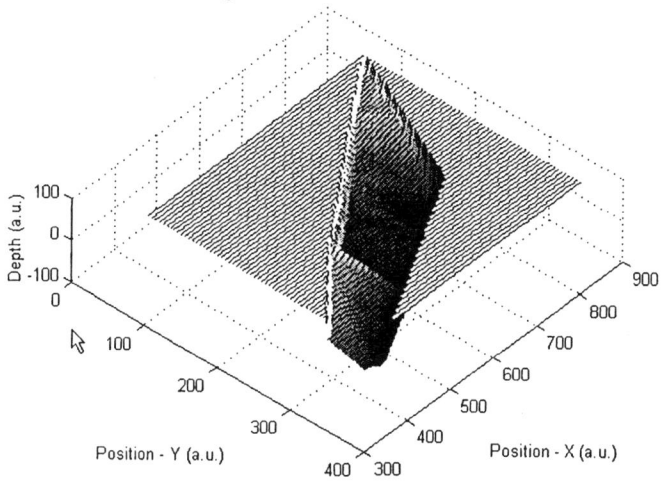

Figure 5 - Water Raman intensities

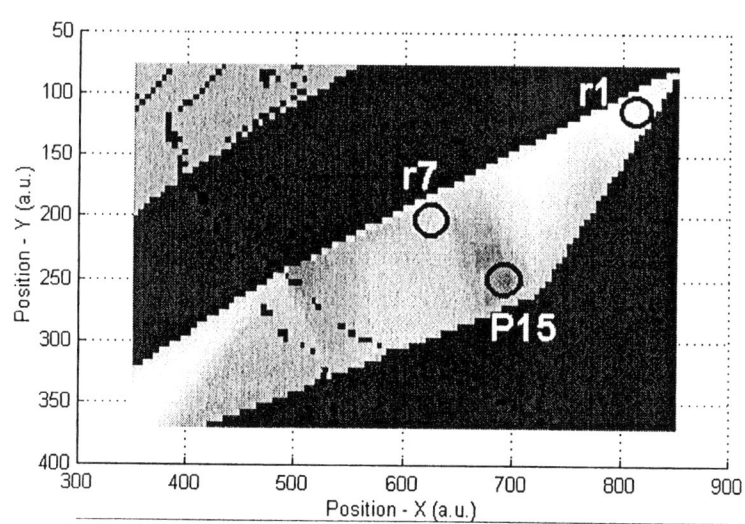

Figure 6 - DOM and Chlorophyll-a mapped as R and G pixel intensities

150   Environmental Coastal Regions

Three sites were chosen to illustrate the usefulness of this approach. To clarify differences, colors of these points are plotted on a CIE chromaticity diagram[8] (figure 7).

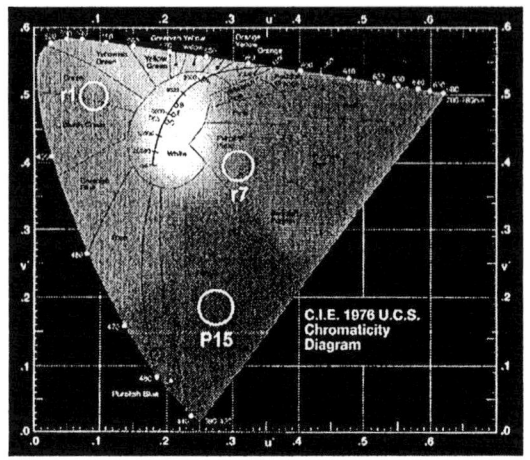

Figure 7 - a CIE 1976 U.C.S. chromaticity diagram

In Figure 8, combination of color decomposition and 3-D map was made. Z coordinate has the same meaning of Figure 5, i.e., local water transparency measured by water Raman band intensities.

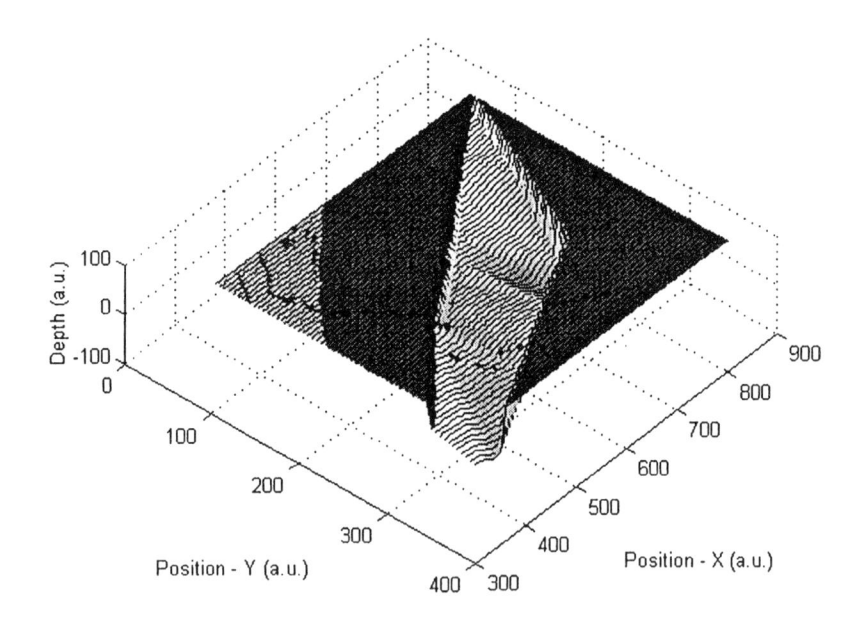

Figure 8 - DOM and Chlorophyll-a mapped as R and G pixel intensities with Water Raman intensities as depth

4 Conclusions

An easier identification of the bidimensional variation of georeferenced parameters as DOM and Chlorophyll-a concentrations can be achieved with the use of trichromatic space. The association with R, G and B color intensities allows visualization of more than one parameter in only one map, which is very useful for a quick historical evaluation of a region. The same representation is not possible for isolines graphics. Besides, the addition of a third z coordinate associated to the effective signal-integration depth can show in the same graph information about local water transparency. However it is necessary to adjust weights and intervals for each RGB component to choose the fluorescence intensity range that better represent the phytoplankton and DOM distribution in the region. This methodology should yet be tested for different seasons and water quality conditions.

References

[1]. Nunes, R.A.; Barbosa, P. C. C.; Tabares, R. H.; Scavarda do Carmo, L. C.; Bunkin, A. & Pershin, S.; "Utilização de Sistema LIDAR Transportável para Sensoriamento Ambiental" - XX ENFMC, 1997.

[2]. Nunes, R.A.; Tabares, R. H.; Bunkin, A. & Pershin, S.; "Compact LIDAR for Remote Sensing of Water Pollution"; Water PollutionIV: Modeling Measuring and Prediction - Computational Mechanics Publications, 1996.

[3]. Nunes, R. A.; Carvalho, I. M.; Cremona, M.; Bunkin, A.; daSilva, F. C.; "Fluorescência Induzida por Laser aplicada à identificação remota de espécies e estresse na vegetação" - SIAGRO, 1997.

[4]. Nunes, R. A.; Tabares, R. H.; Carvalho, I. M.; Bunkin, A.; Pershin, S.; da Silva, F.C.; Palmieiri, F.; Millioli, V.S.; Ramalho, A. L.; "Caracterização *in situ* de solos por espectroscopia induzida por laser". Congresso Brasileiro de Ciência do Solo; CD-ROM, 1997.

[5]. Bristow, M.; Nielsen, D.; Bundy, D. & Furek, R; "Use of water Raman emission to correct airborne laser fluorosensor data for effects of water optical attenuation"; Appl. Opt. 20, 2889(1981).

[6]. de Souza Sierra, M.M.; Giovanela, M.; Donard, O.F.X. & Blein, C.; "A utilização da espectroscopia de fluorescência no estudo da matéria orgânica dissolvida nas águas naturais: evolução e perspectivas"; Química Nova, 19(3) (1996).

[7]. Yentsch, C.S. & Phinney, D.A.; "Fluorescence Spectral Signatures for Studies of Marine Phytoplankton". *in* Mapping Strategies in Chemical Oceanography, p. 259 (1985).

[8]. Poole, L. R. & Esaias, W. E.; "Water Raman normalization of airborne laser fluoresensor measurements: a computer model study". Applied Optics, vol. 21, n. 20 (1982).

[9]. Computer Graphics Principles and Practice, Foley, J. D.; van Dam, A.; Feiner, S. K. & Hughes, J. F.. 2nd ed. (1996), Addison-Wesley P.C..

Section 4: Atmospheric Pollution

Fraction of the overall ozone levels explained by meteorological variables in the Bilbao area

G. Ibarra[1], I. Madariaga[2], A. Elías[3], J. Caamaño[4,] M. Albizu[5] E. Agirre[6], J. Uria[7],

The following authors are at UPV-ETSII y de IT. Alda. Urkijo s/n. 48013 Bilbao. Spain:

[1] Dpto. I. N. y Mecánica de Fluidos. e-mail: inpibbeg@bi.ehu.es
[2] Dpto. I. Electrónica y Telecomunicaciones.
[3] Dpto. I. Q. y del Medio Ambiente.
[4] Dpto. E. G. y Proyectos de Ingeniería.
[5] Dpto. Medio Ambiente. Gobierno Vasco. Gran Via, 89. 48013. Bilbao. Spain.
[6] Dpto. de Matematica Aplicada. UPV. La Casilla, s/n. Bilbao. Spain.
[7] HOSPITAL DE BASURTO. 48013. Bilbao. Spain.

Abstract

This paper shows the work developed in order correctly to assess the fraction of the total ozone levels that meteorology variables account for in the coastal area of Bilbao. Following the methodology developed by Kolmogorov-Zurbenko, the low-pass KZ filters have been applied to daily data of ozone as well as meteorology variables. According to this methodology, time series of daily values of the studied variables have been decomposed into trend, high frequency or random noise component, and long cycles term. The conclusions show that unlike previous works, in a coastal zone such as the Bilbao area temperature itself does not account for great fractions of the overall variability of ozone. The results show that nearly 85% of the overall variability of ozone time series can be attributed to meteorological variables, the most important being wind speed, radiation and temperature.

1 Introduction

Bilbao city is located in North Central Spain (Figure 1) and the whole area, as many industrial towns in the world, has air quality problems. In the recent past years, the nature of these problems has shifted from SO_2 due to industrial activities, to photochemical smog (O_3, NO_2, NO). A recent report to the Basque Government [1] suggested that traffic played an important role in the formation of ozone.

Figure 1. Bilbao is located in North-Central Spain.

For monitoring and control purposes, an air pollution and meteorological network is operated in the whole area by the Basque Government since 1977. The network, in its 1993-94 configuration, measured several meteorological parameters at 13 locations and air pollution variables at 26.

Urban ozone formation is a complex phenomenon since this pollutant is not emitted into the atmosphere directly but it is produced owing to the interaction of meteorology and NO_x and VOC`s [2]. Therefore, several works have focused on trying to remove the impact of meteorological effects from ozone data in order to detect changes in ozone precursors emissions [3, 4]. With the same aim of moderating or filtering out the effects of meteorology in ozone time series, low-pass filters have been applied to daily maximum ozone data. To that end, the Kolmogorov-Zurbenko (KZ) low-pass filter has been used [5, 6, 7] to remove the effects of cycles below one month and analyze the apportion of periodicities over one month plus long- term trend to the overall variability of the daily maximum log ozone time series at a given location. Joint application of KZ filter to ozone, surface temperature and dew point temperature time series [7] has made possible to assess the relative importance of these variables and their contribution to the ozone formation process.

In this paper, the long-term periodicities of ozone time series are analyzed and explained with the long-term components of meteorological variables.

2 Database

The database used for this study have been hourly data measured in the Bilbao air pollution and meteorological network during 1993 and 1994 (Figure 2). Daily mean values of ozone and meteorological variables for years 1993 and 1994 (730 cases) constituted the database used in this research. In the case of ozone, the daily means of its hourly logarithmic values were used.

Meteorological variables were chosen from Feria, at sea level, and Banderas, located 200 metres above sea level (Figure 2). The height difference between these two locations allows detection of thermal inversion. The wind speed vector, originally measured as a direction plus a module in Feria, was projected along the Nervion river axis and its perpendicular obtaining two components, V_x and V_y, easier to handle for a statistical analysis (Figure 2). The reason for choosing the river axis was that it is well known in the Bilbao area [8] that the most important air circulations take place forth and back along Nervion river. The variables chosen were, daily maximum thermal contrast between Feria and Banderas and daily means of temperature at Banderas, and temperature at Feria, V_x, V_y, radiation, absolute humidity and dew point temperature at Feria.

158   Environmental Coastal Regions

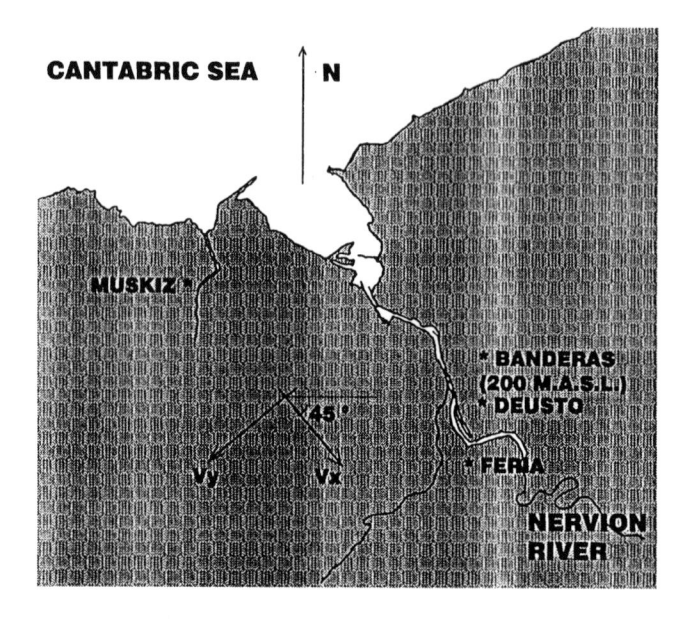

Figure 2. Bilbao area, location of sensors in the zone and projected axis.

3 Methodology

A low-pass filter applied to a certain variable separates, on the one hand the contributions of periodicities lower than the length of the filter, and on the other hand the contribution of periodicities higher than the length of the filter plus long-term trend. The Kolmogorov-Zurbenko ($KZ_{m=29,p=3}$) is a low-pass filter produced by a moving average of length 29 and obtained after three iterations [5, 6, 7]. This means that applied to the variables mentioned above, the filtered time series represent the contributions of periodicities equal or higher than one month plus long-term trend to the overall variability of each variable. If X_t is the original time series of variable X, the filtered time series will be denoted as X_{KZ} while short-term variations (X_t-X_{KZ}) will be denoted as X_{WT}.

$$X_{KZ} = 3 \text{ ITERATIONS (m=3) of} \quad Y_t = \frac{1}{29} \sum_{z=-14}^{z=14} X_{t+z}$$

In this work, the KZ $_{m=29,p=3}$ filter was applied to the following variables of the Bilbao area: mean daily values of log ozone (O_3), temperature at Feria (FER), temperature at Banderas (BAN), maximum daily thermal gradient between both (GRD), daily mean wind speed (as V_X and V_Y), absolute humidity (HUM), dew-point temperature (DPT), and radiation (RAD).

The correlations of filtered values (X_{KZ}) and residuals (X_{WT}) of ozone and meteorological variables can be seen in Table 1. It shows that long-term componentes of ozone are not correlated with short-term components of meteorological variables, while long-term components of ozone show significant correlations with the long-term components of meteorological variables.

	VXKZ	VYKZ	GRDKZ	HUMKZ	RADKZ	FERKZ	BANKZ	DPTKZ	PREKZ
O3KZ	0.0886	-0.0341	0.4536	0.2512	0.3903	0.3531	0.3777	0.294	-0.232
O3WT	0.0052	-0.0081	-0.0422	-0.0156	-0.0142	-0.0098	-0.0079	-0.012	-0.0355
	VXWT	VYWT	GRDWT	HUMWT	RADWT	FERWT	BANWT	DPTWT	PREWT
O3KZ	0.0066	-0.0124	-0.0227	-0.0109	-0.0089	0.0129	0.0059	0.001	-0.0523
O3WT	0.0049	-0.0102	-0.0637	-0.1847	0.2607	0.2113	0.175	-0.2147	-0.3001

Table 1. Correlation matrix of meteorological long-term and short-term components with O_{3KZ} and O_{3WT}.

Furthermore, Table 2 shows that short-term (X_{WT}) and long-term components (X_{KZ}) are uncorrelated for all these variables and long-term components account for most of the variability of original variables (Table 3). In the case of ozone, the long-term component, includes [5, 6, 7] the apportion of periodicities above one month and the long-term trend in ozone emissions.

VARIABLE	V_X	V_Y	GRD	HUM	RAD	FER	BAN	DPT	PRE
CORR. X_{WT} X_{KZ}	0.0555	0.0358	0.0510	0.0573	0.0315	0.0578	0.0665	0.0612	0.1159

Table 2. Correlations between short-term (X_{WT}) and long-term or filtered component (X_{KZ}) of variable X.

VARIABLE	V_X	V_Y	GRD	HUM	RAD	FER	BAN	DPT	PRE
KZ	0.7942	0.7764	0.1431	0.9792	0.8132	0.9660	0.9667	0.9640	0.999
WT	0.2058	0.2236	0.8569	0.0208	0.1868	0.034	0.0333	0.036	0.001

Table 3. Fraction of the overall variability in variable X attributable to long-term component (X_{KZ}) and short-term component (X_{WT}).

In previous works, the effect of periodicities of ozone above one month has been explained with long cycles of meteorology [5, 6, 7]. Following the same approach for Bilbao, in order to remove the effects of meteorology on ozone, the filtered values of ozone and meteorology, were used to build a regression equation, being O_{3KZ} the dependent variable and the meteorological variables the independent ones. The Multiple Regression Analysis was performed using a stepwise regression procedure so that only meaningful variables are included in the equation.

4 Results

The results of the Multiple Regression Analysis can be seen in Table 4. In this table, it can be seen the coefficients of the regression equation and also the absolute and relative increase in R^2 due to the inclusion of the variables at the different steps of the regression process.

The fraction of the overall variabilty of O_{3KZ} explained by meteorology (METKZO3) is 87.3% . The most relevant variables whose inclusion in equation originates a higher increase in R square are V_{YKZ}, RAD_{KZ}, BAN_{KZ} and FER_{KZ}. Therefore, the apportion of periodicities above one month to O_{3KZ} can be attributable to the effects of long cycles of meteorology. The residuals are far from being white noise and represent the long-term trend in ozone.

Step	Rsq	ΔRsq	%Δ Rsq	Variable
1	0.1523	0.1523	17.45	RAD_{KZ}
2	0.6984	0.5416	62.56	VY_{KZ}
3	0.7397	0.0413	4.73	FER_{KZ}
4	0.7642	0.0245	2.81	DPT_{KZ}
5	0.7856	0.0214	2.45	HUM_{KZ}
6	0.8019	0.0163	1.87	GRD_{KZ}
7	0.8518	0.0499	5.72	BAN_{KZ}
8	0.8685	0.0167	1.91	PRE_{KZ}
9	0.8729	0.0044	0.5	VX_{KZ}
R Square			0.87291	
Standard Error			0.07662	

Dependent Variable = O_{3KZ}

Independent Variable	Coefficient B
VX_{KZ}	-.040693
VY_{KZ}	-.371748
GRD_{KZ}	0.277609
HUM_{KZ}	-.226045
PRE_{KZ}	-.015480
RAD_{KZ}	3.629567
FER_{KZ}	0.058063
BAN_{KZ}	-.314052
DPT_{KZ}	0.336874
Constant	20.655837

Table 4. Results of applying Multiple Linear Regression to O_{3KZ} (dependent) and VX_{KZ}, VY_{KZ}, GRD_{KZ}, HUM_{KZ}, RAD_{KZ}, FER_{KZ}, BAN_{KZ}, DPT_{KZ}, PRE_{KZ} (independent) using a stepwise procedure.

Taken into account that O_{3KZ} and O_{3WT} are uncorrelated and O_{3KZ} accounts for 97.9% of the overall variability of the daily ozone data, it can be concluded that long cycles (above 1 month) of meteorology are responsible for approximately 85% (0.873 x 0.979) of the overall variabilty of ozone.

Unlike previous works [10, 11, 12] when it was used in this equation surface temperature (FER_{KZ}) as the only independent variable, the maximum R square obtained was as low as 0.125. No increases in R square were obtained when using the temperature shifted by any number of days. The same Multiple Linear Regression analysis when applied to raw daily data of ozone and all the meteorological variables, but without filtering, yielded a maximum R square of 0.21382. In this case, the stepwise process only chose as meaningful V_y, GRD, PRE and RAD with a constant in the equation of value 24.

5 Conclusions

Meteorology plays a crutial role in ozone formation. Application of low-pass filters makes it possible to separate long-term components (long cycles+trend) and short- term components. In the case of Bilbao, 85% of the overall variability in the daily ozone time series can be attributed to meteorology.

The component due to long-cycles can be explained by meteorological variables but comparison with previous works suggests that explanatory variables and its relative importance is different for each place. In a coastal area like Bilbao, located near the sea, the most relevant variables have turned out to be wind speed, radiation and temperature.

Acknowledgements

This work has been performed under financial support from the Education, Universities and Research Department of the Basque Government, *Eusko Jaurlaritza*. The authors wish to thank the Environmental Department of the Basque Government for providing data for this study.

References

1. TEKNIMAP AMBIENTAL. *Bases técnicas para el plan de saneamiento atmosférico en la comarca del Alto Ibaizabal.* Report to the Basque Government. 1995.

2. Finlayson-Pitts B.J.; Pitts J.N. *Atmospheric Chemistry: Fundamentals and experimental techniques.* Wiley, New York. 1986.

3. Bloomfield, P.;Royle, J. A.; Steinberg, L.J.; Qing Yang. Accounting for meteorological effects in measuring urban ozone levels and trends. *Atmospheric Environment* .1996, *30*, 3067-3077.

4. Cox, W. M.; Shao-Hang Chu; Assesment of interannual ozone variation in urban areas from a climatological perspective. *Atmospheric Environment* .1996, *30,* 2615-2625.

5. Rao, T. S.; Zurbenko, I.G. Detecting and tracking changes in ozone air quality. *Journal of the Air & Waste Management Association.* 1994, *44,* 1089-1092.

6. Rao, T. S.; Zalewsky, E; Zurbenko, I.G. Determining temporal and spatial variations in ozone air quality. *Journal of the Air & Waste Management Association.* 1995, *45,* 57-61.

7. Flaum J. B.; Rao, T. S.; Zurbenko, I.G. Moderating the influence of meteorological conditions on ambient ozone concentrations. *Journal of the Air & Waste Management Association.* 1996, *46,* 35-46.

8. Ibarra, G. Ph. D. Thesis, University of the Basque Country, 1993.

Fog Formation Prediction In Coastal Regions Using Data Mining Techniques

Nelson F. F. Ebecken
COPPE/UFRJ - Federal University of Rio de Janeiro
Caixa Postal 68506
21945-970 Rio de Janeiro - Brazil
e-mail: nelson@ntt.ufrj.br

Abstract

The recent growth of many scientific and business databases created a need of a new class of tools and techniques for automated search the knowledge hidden in these databases. For this reason it became necessary to integrate data mining and database technology on scalable parallel platforms.

This paper discusses a parallel implementation of a data mining strategy developed for MIMD platforms: objected related data mining technology, that enables a specific form of data mining optimized for better performance and more complete results.

Our application is related to observed meteorological conditions at the International Airport of Rio de Janeiro. The main objectives are to understand the parameters that have influence on visibility/weather conditions (fog formations), and to compare with current methodologies to predict future situations with some antecedence. In our analysis we have considered the last 20 years to build a model and extract the knowledge to make predictions.

1 Introduction

With recent explosion of information, and the availability of cheap storage, it has been possible to gather massive data during the last decades. The eventual goal of this massive data gathering is the use of this information to gain competitive advantages, by discovering previously unknown patterns in data, which can guide the decision making.

Recent years have shown that it is becoming difficult to analyze data using only OLAP tools, showing the need of an automated process to discover interesting and hidden patterns in data. Data mining techniques have increasingly been studied e.g. Mark[1], especially in their application in real-world databases. One typical problem is that databases tend to be very large, and these techniques often repeatedly scan the entire set. A solution that has been used for a long time is the extraction of a training sample, so that the data mining algorithm could be run on a smaller portion of data. In this case, subtle differences among sets of objects in the database will become less evident.

The need to handle large amounts of data implies a lot of computational power, memory and disk I/O, which can only be provided by parallel computers.

A variety of data mining algorithms have been proposed addressing parallel processing. Most of them use flat files and exploit parallelism through parallel programming.

2 The Data Mining Strategy

In the present work the neural approach was used. A large number of papers have been presented: e.g., Lu & Liu[2], Bigus[3], Schurmann[4]. The neural network based data mining approach consists of tree major phases:

- *Network construction and training.* This phase constructs and trains a three layer neural network based on the number of attributes and number of classes and chosen input coding method.

- *Network pruning.* The pruning phase aims at removing redundant links and units without increasing the classification error rate of the network. A small number of

Environmental Coastal Regions   167

units and links left in the network after pruning enable us to extract concise and comprehensible rules.

- *Rule extraction.* This phase extracts the classification rules from the pruned network.

The rule extraction algorithm used was suggested by Lu & Liu[5] and can be summarized as:

- Apply a clustering algorithm to find clusters of hidden node activation values.

- Enumerate the discretized activation values and compute the network outputs. Generate rules that describe the network outputs in terms of the discretized hidden unit activation values.

- For each hidden unit, enumerate the input values that lead to them and generate a set of rules to describe the hidden units' discretized values in terms of the inputs.

- Merge the two sets of rules obtained in the previous two steps to obtain rules that relate the inputs and outputs.

3 Scalable Parallel System

Our parallel algorithms have been implemented on an IBM POWER parallel System SP from NACAD, the High Performance Computing Nucleus of COPPE/UFRJ, the Engineering Graduate Center of the Federal University of Rio de Janeiro, with the following characteristics:

- 2 wide nodes (spn1 and spn2): RAM memory 1GB(spn1)/512MB(spn2), Disk8GB, processor type POWER2 model 590.

- 2 thin nodes (spn3 and spn4): RAM memory 128MB, Disk 2GB, processor type POWER2 model 390.

- TOTAL: 1064Mflops, RAM 1.75Gbytes, Disk 20Gbytes.

The parallel implementation of the backpropagation algorithm has two approaches:

- partitioning the training data sets, that distributes the data sets over nodes, and

- the partitioning of the network itself, that distributes the neurons over processing nodes.

The approach for partitioning the data has the advantage of reducing the required communication to produce the network output. Each node executes the forward and backward passes independently on a different subset of total data. This leads to a performance gain and low complexity of the implementation. The network may be copied in each processing node and perform the training over its training data set.

In a MIMD architecture, each node can have different power processing, local memory and disks. The mapping algorithm of processes to nodes must take into account these differences and choose the suitable procedure of distribution. Another issue of this implementation is the workload balance on processing nodes. On systems where dynamic workload balance is present the mapping algorithm should use this feature to specify the proceedings. But it's hard to determinate the best parameter to be used in remapping the network, leading us to a complexity growth and some undesired situations.

Our first implementation works with static mapping of nodes and partitioning of the training data sets, keeping copies of the entire network on each node.

4 Oracle Parallel Server

The present work was conducted under ORACLE Parallel Server environment.

Oracle software was installed in CWS (Control Workstation). A nfs mounted directory (/oracle) is then imported by each of the nodes of RISC/6000 SP2.

Environmental Coastal Regions   169

The advantages are: centralized installation (easier to do software upgrades) and it saves disk storage needed, but it does not offer high availability, because the CWS represents a single point of failure.

There is a single database, which is accessible from any of the SP2 nodes (transparent to users). The files which store the database are distributed among all the 4 nodes of the system. All datafiles are accessible from all nodes through the VSD (Virtual Shared Disk), a device driver layer above the AIX Local Volume Manager. To guarantee data consistency, each node of the system runs an Oracle software called DLM (Distributed Lock Manager). The DLM is responsible for locking services of shared resources of the system (in this case, the datafiles).

To improve performance, Oracle Parallel Server uses a Parallel Cache (each node has its own local data cache, which is available from all other nodes through DLM and HPS).

The Parallel Query Option provides scalability to most common query operations. It dramatically improves response times for very large complex queries. Scans, sorts, joins, distinct and groups can all be performed in parallel.

The application developed to access data in parallel is exactly the same that would access data sequentially (transparent parallelism to application). Oracle Parallel Server requires all the disks in the system to be accessible from all nodes (equivalent to a Shared Disk approach). When a query is processed in parallel, Oracle knows which data are on the local disks of a given node. The nodes which are local to the requested data are primarily used for query processing to increase performance.

The parallel processing can be decomposed in three phases:

Phase I: Shared-Everything - the query is divided into subtasks, which are sent to the nodes where data reside.

Phase II: Shared-Disk - each of the processing nodes does purely local processing to retrieve the requested data.

Phase III: Shared-Disk - nodes which have finished early their processing can be borrowed to take over tasks from nodes which have not finished processing yet.

170   Environmental Coastal Regions

The workload can be redistributed dynamically since all nodes have access to all data. The processing power of all nodes can be used to process the query as effectively as possible.

Oracle Parallel Server for the RS/6000 SP provides a number of advanced system administration tools, making system management of an MPP system as easy as any other system. System administration tasks can be performed from a single point of control.

5 The Studied Application

Our application is related to observed meteorological conditions at the International Airport of Rio de Janeiro. The main objectives are to understand the parameters that have influence on visibility/weather conditions (fog formations), and to compare with current methodologies to predict future situations with some antecedence.

In our analysis we have considered the last 20 years to build a model and extract the knowledge to make predictions. The weather condition is stored in the 12th field, and it is varying in the range 0-9. The fog condition corresponds to the value 4.

In this way our problem can be stated as:

- What are the relevant parameters and rules that imply in value 4?

- What are the relevant parameters and rules, occurred "n" hours before, that imply in value 4?

The database stores for each hour (24 hours per day) the following values:

1. register code
2. register block
3. register location
4. year
5. month
6. day

19. clouds at 3rd level: number
20. clouds at 3rd level: type
21. clouds at 3rd level: height
22. clouds at 4th level: number
23. clouds at 4th level: type
24. clouds at 4th level: height

7. hour
8. clouds total
9. wind direction
10. wind velocity
11. horizontal visibility
12. weather condition
13. clouds at 1st level: number
14. clouds at 1st level: type
15. clouds at 1st level: height
16. clouds at 2nd level: number
17. clouds at 2nd level: type
18. clouds at 2nd level: height
25. altimeter correction
26. temperature (Dew-Celsius)
27. blast velocity
28. atmospheric pressure - sea level
29. atmospheric pressure-runway level
30. cloud direction
31. pressure trend
32. pressure - last 3 hours variation
33. air temperature - dry
34. air temperature humid
35. precipitation
36. relative humidity

6 Results

The primary metric for evaluating classifier performance is classification accuracy: the percentage of test samples that are correctly classified. The other important metrics is classification time (speed-up) and the structure of the extracted knowledge (accurate and compact). The obtained error is 5.96%, and the sensitivity analysis is shown in Table 1.

A great number of informations were extracted by rules. The fog prediction can be characterized by:

$$IF\{(-7796 \leq 21.x1 + 32.x2 + x3 - 14.x4 - 84.x5 \leq -5880)$$

and

$$-10460 \leq 27x1 + x3 - 25.x4 - 107.x5 \leq -8060)\}$$

These results will be confronted to current practice methodology and evaluated by expert opinion.

The obtained speed-up corresponds to a performance factor of 2.2 between the 4 nodes solution and the best 1 node solution.

Table 1. Sensitivity

Field Name	Sensitivity
Precipitation	42.9
Hour	9.0
Clouds 2nd level type	5.1
Wind direction (x2)	4.8
Temperature (dew) (x4)	4.4
Clouds 2nd level: number	4.3
Pressure-sea level	3.4
Pressure trend	3.4
Pressure 3hours variation	3.3
Month (x1)	3.1
Clouds 1st level height	3.0
Wind velocity	2.9
Clouds total	2.9
Clouds 1st level number	2.6
Clouds 1st level type	2.0
Clouds 2nd level height (x3)	0.6
Relative humidity (x5)	0.5

7 Conclusion

This work is an attempt to study the connectionist approach to data mining to generate classification similar to that of decision trees. The main issue is to reduce the training time of neural networks, developing fast algorithms and database implementations on parallel scalable arquitectures.

Although the parallel environment permits a fast initial training of a network, it is desirable to have incremental training and rule extraction during the life time of an application database: the accuracy of rules extracted can be improved along the change of database contents. Another possibility to improve the performance is to reduce the number of input units of the networks by feature selection.

The present data mining technology adds the following improvements:

- Better performance and reduced cycle times because compute-intensive tasks are executed in parallel, and closer to the data source.

- The analysis is more interactive and involves fewer process steps when performed online, directly against the relational data source.

- Ability to mine much larger data sets, not just the data in flat files or extracts..

- More complete results - without the need to extract a target data set, all potentially valuable information is available for mining.

References

[1] Mark, W., Special Issue on Data Mining, IEEE Expert, Vol. 11, nº. 5, October, 1996.

[2] Lu, H., Setiono, R. and Liu, H., Neurorule: A Connectionist Approach to Data Mining, Proc. VLDB'95, pp 478-489, 1995.

[3] Bigus, J. P., Data Mining with Neural Networks, McGraw-Hill, 1996.

[4] Schurmann, J., Pattern Classification: A Unified View of Statistical and Neural Approaches, John Wiley, 1996.

[5] Lu, H., Setiono, R. and Liu, H., Effective Data Mining Using Neural Networks, IEEE Transactions on Knowledge and Data Engineering, Vol. 8, no. 6, December, 1996.

[6] Fayyad, U. M., Piatetsky-Shapiro, G., Smyth, P. and Uthurusamy, R., editors, Advances in Knowledge Discovery and Data Mining, AAAI Press/MIT Press, 1996.

[7] Agrawal, R. and Shafer, J. C., Parallel Mining of Association Rules, IEEE Transactions on Knowledge and Data Engineering, Vol. 8, no. 6 December, 1996.

[8] ORACLE Parallel Server, ORACLE Coorporation, 1996.

Estimation of acidifying deposition in Europe due to international shipping emissions in the North Sea and the North East Atlantic Ocean

S.G. Tsyro & E. Berge
Research and Development Department, Norwegian Meteorological Institute, P.O.Box 43, Blindern, N-0313 Oslo, Norway
Email: svetlana.tsyro@dnmi.no

Abstract

The contribution of sulphur (SO_2) and nitrogen oxides (NO_x) emissions from international shipping in the North Sea and the north-eastern (NE) Atlantic Ocean to acidifying deposition across Europe is evaluated for 1990, 1995 and 2010 with the EMEP (European Monitoring and Evaluation Programme) two-dimensional trajectory model. A new ships emission data set for the North Sea and the NE Atlantic Ocean is implemented based on information from the Lloyd's Register of Shipping. The new emissions of SO_2 and NO_x are a factor of 2.2-2.8 larger then the earlier estimates used. Therefore, the presented calculations give considerably higher ships' contribution to acidification in Europe. Ship emissions from the North Sea and the NE Atlantic Ocean in 1990 exceeded 100-150 mg m^{-2} a^{-1} and accounted for more than 10-15% of the total acidifying deposition in many coastal areas. That is comparable with the contributions of large industrial countries.

The considerable decrease of the total European emissions of SO_2 and NO_x by 26% and 15% respectively between 1990 and 1995 was mainly due to the reduction of land-based emissions, whilst ship emissions are thought to remain comparatively stable. Consequently, a relative increase of importance of international shipping to acidification is registered. Our estimates also suggest that the importance of ship emissions in the North Sea and the NE Atlantic Ocean will increase in 2010 providing that European countries have reduced their national emissions according to the current reduction plans and no abatement measures with respect to international shipping are taken.

1 Introduction

A number of recent estimates (e.g. Streets, Carmichael & Arndt, 1997, Corbett & Fischbeck, 1997) shows that ship traffic presents a significant source of sulphur dioxide and nitrogen oxides on a global scale. Ship emissions contribute to acidification of soils and fresh waters hundreds kilometres inland. An attempt to combat this problem by approving the global emission limits was made by the International Maritime organisation in 1997. An initiative was also taken to establish a SO_x emission control area for the Baltic Sea and the North Sea (The North Sea Ministerial Conference, 1995).

Through the work of the EMEP under the Convention on Long Range Transboundary Air Pollution (LRTAP), transport and deposition of transboundary acidifying species has been calculated since 1985. Emissions from international shipping were first introduced in model calculations in 1990 and derived from Bremnes (1990). A new ships emission data set for the NE Atlantic Ocean and the North Sea have been made available to the Meteorological Synthesizing Centre-West (MSC-W) for EMEP from the Lloyd's Register of Shipping in 1996 (Lloyd's Register of Shipping, 1995). The year 1990 was chosen as a base year for emission estimate, as this year was considered to be a 'typical year' with respect to shipping movement. The obtained estimates of SO_2 and NO_x ships emissions for this area appeared considerably higher than the previous estimates used by EMEP.

The new ship emission data aggregated to the EMEP 150x150 km grid gives a total sulphur emission of 1080 ktonnes as SO_2 and a total nitrogen emission of 1550 ktonnes as NO_x in 1990. In earlier EMEP calculations emission figures of 490 and 541 ktonnes respectively were used (Barrett and Berge, 1996). The evidence of a factor of 2.2-2.8 larger emissions than previously assumed asked for a renewed assessment of acidifying effects of the ship emissions in Europe. The purpose of this work is, thus, to present the updated calculations on long range transport of air pollution from international shipping in the North Sea and in the NE Atlantic Ocean.

In the present paper, firstly, we analyse the modelled depositions of sulphur and nitrogen originating from the ship emissions in 1990. Furthermore, the total European emissions of SO_2 and NO_x declined markedly from 1990 to 1995 by approximately 26% and 15% respectively largely due to the reduction of landbased emissions. The relative importance of international shipping was, therefore, expected to increase assuming comparatively stable ship emissions. In order to study this, contribution of the ship emissions to the total deposition in Europe in 1995 was estimated. Finally, a scenario calculation for the year 2010 is presented.

2 Data description

2.1 The dispersion model

The calculations presented in this paper were carried out with use of the EMEP/MSC-W routine acid deposition model (Eliassen & Saltbones, 1983; Barrett & Berge, 1996). The model is one-layer, receptor-oriented Lagrangian with 150 km spatial resolution. It employs official estimates of SO_2, NO_x and NH_3 national emissions and real-time meteorological data derived from the Numerical Weather Prediction model of the Norwegian meteorological Institute or analysed from observations. 4-day backward trajectories are calculated every 6 hours. Mass balance equations for chemical concentrations within air parcels following the motions in the Atmospheric Boundary Layer include emissions, chemical processes, dry and wet deposition. Exchange with the free troposphere is parameterised.

2.2 Ship emissions in the North Sea and the NE Atlantic Ocean

Spatial distribution of the annual international shipping emissions of SO_2 and NO_x in the North Sea and the NE Atlantic Ocean (not included here) shows highest emissions of the both compounds in the regions of heavy shipping activity in the southern North Sea, the English Channel, the western part of the Biscay, and in the Strait of Gibraltar.

The SO_2 and NO_x emissions adopted in the calculations are given in Table 1.

Table 1: Total emissions from international shipping in the North Sea and the NE Atlantic Ocean in the EMEP 150 km grid in 1990 (ktonnes).

	SO_2	NO_x
the NE Atlantic	641	911
the North Sea	439	639
Total	1080	1550

Emissions from ships in the NE Atlantic Ocean and the North Sea within the EMEP calculation domain constitute around 2.6% of the total SO_2 and 6.5% of the total NO_x emissions officially reported in Europe in 1990. Ship emissions of SO_2 in the North Sea, which contains some of the

178 Environmental Coastal Regions

busiest ports in the world, are 2-7 times the national emissions in 1990 of some adjacent countries, e.g. the Netherlands, Denmark, Sweden and Norway. Emissions of NO_x are lower than only the national emissions of 8 main country-emitters in Europe.

If we assume that international ship emissions have not declined noticeably since 1990, then shipping in the NE Atlantic and the North Sea accounts for 3.5 and 7.5 % of the total SO_2 and NO_x emitted in Europe in 1995. The projected further reduction of the land-based emissions will result in even larger portions of ship emissions (5.2 and 10.5% respectively) in 2010.

3 Contribution from ship emissions to the acidifying deposition in Europe

This section presents the results of estimation of the contribution from the international shipping in the NE Atlantic and the North Sea to the total deposition of oxidised sulphur (SO_x) and oxidised nitrogen (NO_x) on land.

3.1 Sulphur and nitrogen depositions from shipping in 1990

Figure 1 presents the modelled deposition field of SO_x due to international shipping emissions in the NE Atlantic and the North Sea in 1990. The modelled deposition field of NO_x deposition (not shown here) has a similar pattern, but the maximum depositions are noticeably lower. The fate of the emitted SO_2 and NO_x is determined by the meteorological and chemical conditions in the atmosphere. Shortly after emission a certain fraction of the emitted species is deposited in the vicinity of the sources as a local deposition. The remaining part of the emissions undergoes long range transport with the winds, chemical transformation, and dry and wet deposition to the surface. The largest SO_x and NO_x depositions coincide well with the areas of highest emissions. SO_2 is, in particular, effectively deposited close to the emission source and has an enhanced dry deposition velocity over sea. Depositions of SO_x in excess of 200 mg(S) m^{-2} a^{-1} are found on the both side of the English Channel, in the narrow coastal areas of England, Denmark, Germany, the Netherlands, Belgium, and northern France. Maximum depositions of NO_x, however, do not exceed 100-150 mg(N) m^{-2} a^{-1} and occur in the Netherlands and most of Belgium. The southern parts of the England, Norway and Sweden, Denmark, the north-western parts of Germany and France, entire Belgium and Luxembourg, Denmark, Germany, the Netherlands, Belgium, and northern France.

Environmental Coastal Regions   179

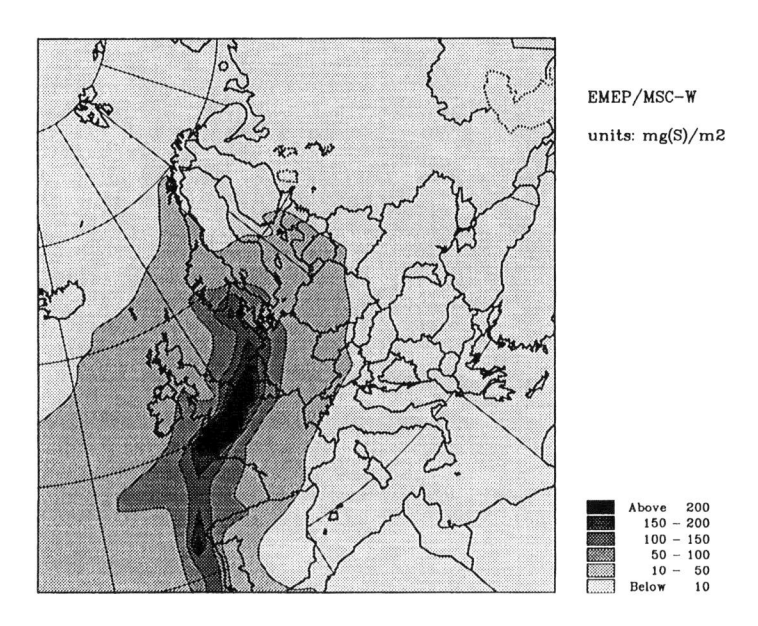

Figure 1: Deposition of oxidised sulphur from international shipping in the North Sea and the NE Atlantic Ocean in 1990.

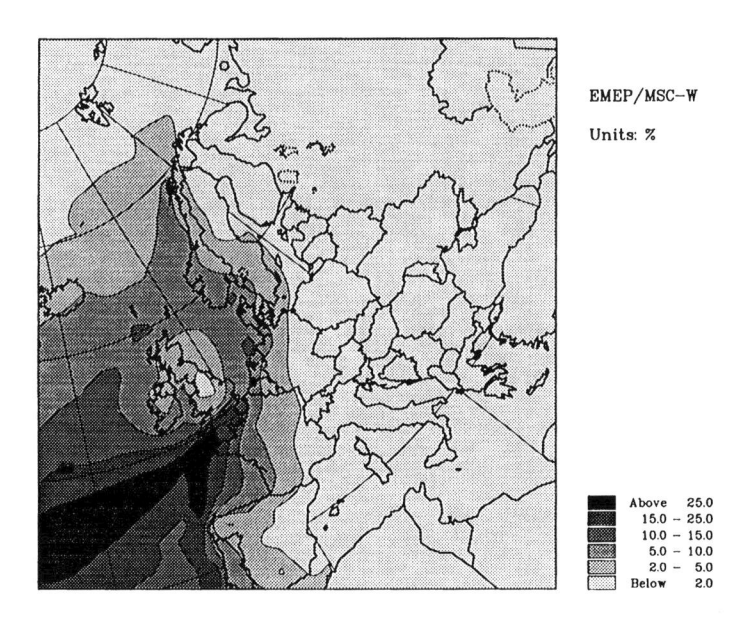

Figure 2: Relative contribution of international shipping in the North Sea and the NE Atlantic to the total SO_x deposition in 1990.

180 Environmental Coastal Regions

Maximum depositions of NO_x, however, do not exceed 100-150 mg(N)m^{-2} a^{-1} and occur in the Netherlands and most of Belgium. Southern parts of Norway, Sweden and England, Denmark, the north-western parts of Germany and France, entire Belgium and Luxembourg, the Netherlands, and the costs of Portugal and Spain received depositions of the both components of 50 -100 mg(S/N) m^{-2} a^{-1} in 1990.

Figures 2 and 3 show the relative contribution of SO_2 and NO_x emissions originating from the North Sea and the NE Atlantic Ocean to the total on-land depositions in 1990. The maximum SO_x contribution from ships (Figure 2) is above 15% and attributed to the large emissions in the English Channel. Contributions of 5 to 10% out of the total deposition is seen in quite an extensive area covering southern England and Ireland, Portugal, and a wide coastal zone from Spain to Norway.

The fraction of total anthropogenic NO_x deposition due to ships is, in general, greater than that for SO_x (Figure 3), with maximum contribution of 15-20% in the south of England and Ireland, and the north-west of France. Ship contributions in excess of 10% are found in this case over a considerably larger area, as NO_x is rather slow deposited in its primary phase and, hence, can be transported to longer distances.

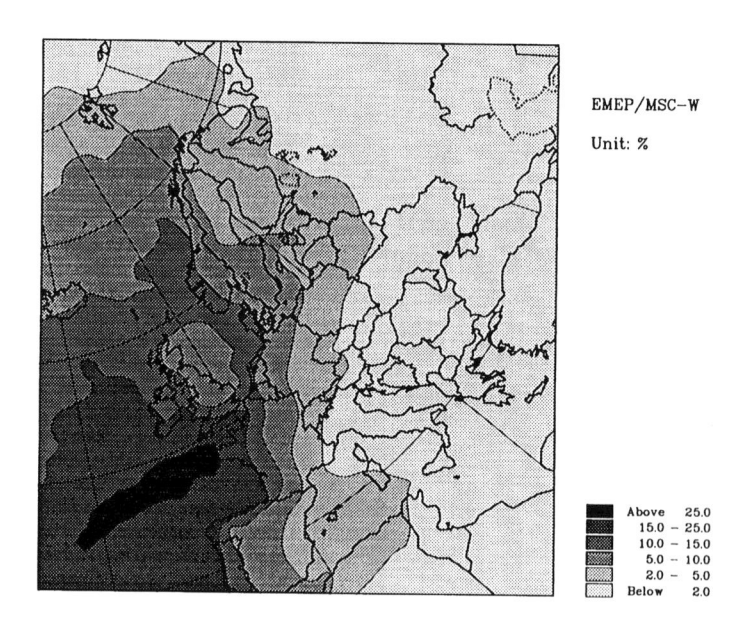

Figure 3: Relative contribution of international shipping in the North Sea and the NE Atlantic to the total NO_x deposition in 1990.

Environmental Coastal Regions   181

Using the ability of the Lagrangian model to allocate the modelled deposition in any receiving area to certain emission areas, it is feasible to distinguish the regions of influence of the North Sea and the NE Atlantic.

It was found that emissions of SO_2 and NO_x from the North Sea plays the most important role in acidifying depositions in the Netherlands (contributions of 7 and 9% relatively), Belgium (5 and 9%), Denmark (5 and 9%), Norway (4 and 7%) and England (2 and 6%). Whilst international shipping in the NE Atlantic Ocean affects considerably Portugal, contributing with 12% to SO_x and 24% to NO_x depositions, Ireland (6 and 15%), Spain (3 and 10%), and England (2 and 7%).

On the whole, a greater importance of shipping fleet to acidification in Europe than estimated by EMEP earlier is found (see e.g. EMEP/MSC-W Report 1/96, Jerre *et al.* (1994)). That is, even more European countries appear to be affected by ship emissions, and the relative ship contributions to acidifying depositions in the countries are found to be larger than all earlier estimates.

It should be pointed out that about 90% of the total SO_2 and NO_x emissions in the North Sea is found to originate from a zone of approximately 50 nautic miles from the coast line associated with the heavy traffic routes in its western and southern parts and, particularly, in the English Channel. Consequently, this zone is the dominant contributor to SO_x and NO_x depositions in the most environmentally sensitive areas located in Scandinavia, England, Germany and the Netherlands. Such estimates may be of a particular interest, when considering emission reduction strategies in limited, most heavily trafficked shipping areas. Besides, reducing emissions is clearly the more effective the closer intensive emission source is to sensitive areas.

3.2 Contribution of ship traffic to SO_x and NO_x depositions in 1995

As it was discussed above, the relative role of international shipping to acidification in Europe was expected to be increasing since 1990 as a consequence of the land-based emissions reduction during this period.

A measurable increase of relative ships contribution to the total SO_x and NO_x deposition in many locations in 1995 is modelled. The area with relative SO_x deposition of 10-15% is noticeably larger then in 1990 (not shown) and covers now broad coastal regions of France, Belgium, the Netherlands, almost entire Denmark, and southern Norway. Even more pronounced is the grown importance of the NO_x ship emissions. The area of relative contributions of 10-15% has extended to include practically entire Norway and the Netherlands, considerable parts of Germany and Sweden. Portugal, Denmark and western Norway receive up to 25% of

182 Environmental Coastal Regions

their NO_x depositions in 1995 from ships.

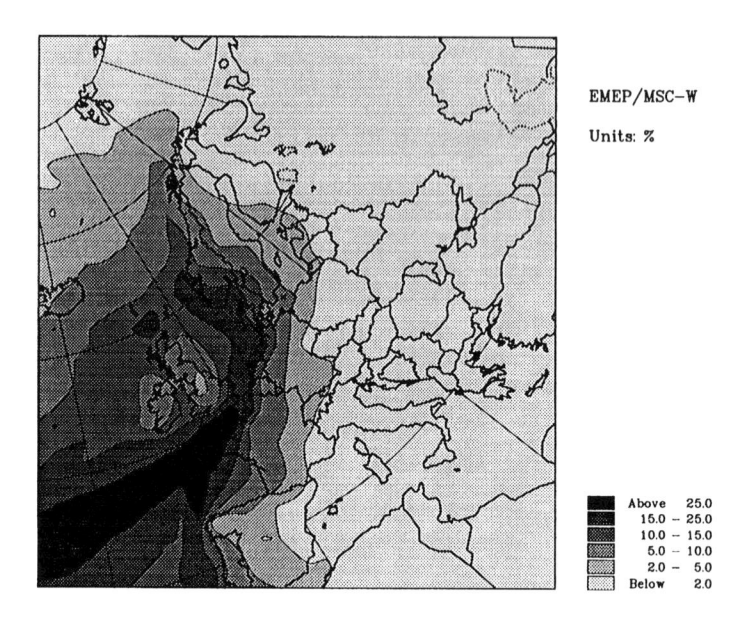

Figure 4: Relative contribution of international shipping in the North Sea and NE Atlantic to the total SO_x deposition in 2010.

3.3 Scenario calculation of acidifying deposition in Europe due to ship emissions in 2010

An assessment of ship traffic contribution to the total SO_x and NO_x deposition in 2010 is made, assuming ship emissions remain unchanged at 1990 levels and projected national emissions for land-based sources after implementation of the current reduction plans. Figure 4 shows clearly that international shipping in the North Sea and the NE Atlantic could become a major cause of acidification and eutrophication in a number of European countries, accounting for 15-25% of the total SO_x deposition in southern Norway, Denmark and coastal areas of Germany, the Netherlands, Belgium and France. For NO_x depositions the respective area appears even larger (not shown here) and includes also Portugal and Spain. Ship contributions may reach 40% on French and English coasts in the vicinity of the English Channel, and in northern Spain (for NO_x). Table 2 illustrates the increasing relative ships impact to acid depositions in Europe unless actions to abate emissions from international shipping are taken.

Table 2. Relative ships contribution compared with some countries to the on-land depositions in Europe (in %)

country	SO_x 1990	1995	2010	NO_x 1990	1995	2010
Germany	6	12	43	23	36	38
United Kingdom	16	27	55	47	55	105
France	31	47	61	55	62	62
Belgium	111	155	180	211	247	247
Norway	1020	1489	1531	558	515	710

4 Summary and conclusions

New data on international shipping emissions for the North Sea and the NE Atlantic Ocean from the Lloyd's Register of Shipping was employed by the EMEP/MSC-W to assess the impact of ship traffic to acidification in Europe. These data are based on a comprehensive refined methodology for quantification of ship emissions (Lloyd's Register, 1995), and, therefore, considered to be of a higher quality than earlier estimates used in the EMEP calculations. The new emissions are a factor of 2 larger for SO_2 and about a factor 3 larger for NO_X. The EMEP Lagrangian acid deposition model was used in order to analyse the long range transport of the SO_2 and NO_x emissions from ships.

The calculation results and their discussion presented in this study clearly points out the importance of international shipping emissions as a source to the deposition of acidifying compounds in Europe. Oxidised sulphur and nitrogen depositions resulting from these ship emissions are estimated to exceed 10-15% and 15-20% respectively in 1990 in narrow coastal regions and are comparable with the contributions of large industrial countries. Countries exposed most to the acidifying depositions from ships in the North Sea are Belgium, Denmark, the Netherlands and Norway. The NE Atlantic is found to be a major contributor to acidifying depositions in Portugal, Ireland and Spain. The increased impact of ship emissions to the acidification in Europe in 1995 is assessed. Moreover, our estimates show the appreciable further increase of ships importance as a source of acidifying emissions in 2010 taking into consideration the current reduction plans for land-based emissions, and if no measures are taken to limit ships emissions. Their relative contribution to the acidifying deposition may exceed 20-25% (up to 40%) in the areas effected most, including those where ecosystems are especially sensitive to acidification.

References:

[1] Barrett, K. & Berge, E. (eds.), *Transboundary air pollution in Europe. MSC-W Status Report 1996.* Part 1: Estimated dispersion of acidifying agents and of near surface ozone. EMEP/MSC-W, Report 1/96, Norwegian Meteorological Institute, Oslo, Norway, 1996.

[2] Bremnes, P.K., Calculation of exhaust gas emission from sea transport. Methodology and results. In *EMEP Workshop on emissions from ships*, Oslo, 7-8 June 1990 (ed.by Gabriel Kielland). State Pollution Control Authority, Oslo, pp. 60-84, 1990.

[3] Corbett, J.J. & Fischbeck, P., Emissions from Ships, *Science*, **278**, pp. 823-824, 1997

[4] Eliassen, A. & Saltbones, J., Modelling of long range transport of sulphur over Europe: a two-year model run and some model experiments. *Atmospheric Environment*, **17**, pp. 1457-1473, 1983.

[5] Jerre, J., Barrett, K. and Styve, H. (1994). *The contribution from ship emissions to acidification in the North Sea countries. The North Sea as a Special Area - Phase II.* Report No. 94-3437. The Norwegian Veritas Industry, Høvik, Norway.

[6] Lloyd's Register of Shipping, *Marine Exhaust Emissions Research Programme.* Lloyd's Register of Shipping, London, 1995.

[7] Streets, D.G., Carmichael, G.R. & Arndt, R.L., Sulfur dioxide emissions and sulfur deposition from international shipping in Asian Waters, *Atmospheric Environment*, **31**, pp. 1573-1582, 1997.

[8] *The North Sea Ministerial Conference*, Esbjerg, 1995.

[9] The Swedish NGO Secretariat on Acid Rain, The European Federation for Transport and Environment & European Environmental Bureau, *Cleaner shipping - a cheap way to reduce acidification in Europe.* Visby, Sweden, 1997.

Alternative Models for the Disposal of Flue Gases in Deep Sea

P. N. Tandon & P. Ramalingam
Universiti Brunei Darussalam
Negara Brunei Darussalam
e-mail : tandonpn@ubd.edu.bn

Abstract

Substantial amounts of flue gases are released into atmosphere from industries as well as from coal and gas fired power plants. Considerable concern has been expressed about the environmental impact of flue gases such as sulfur dioxide and carbon dioxide which contribute to the problems of acid rain and global warming respectively. Out of all the alternatives proposed in literature, it is concluded that sea water is a convenient absorption medium for flue gases emitted from many industrial sites. In sea water, flue gases such as SO_2, HCl, CO_2 are converted into dissolved harmless sulphate, chloride and bicarbonate ions. These ions may be discharged into the ocean in considerable amounts and remain there for several hundred years. The capture and disposal of CO_2 from fossil-fueled power plants near the bed is technically feasible. The increase in density resulting from the dissolution of CO_2 in sea water may be sufficient to transport the dissolved gas to lower depths towards the bottom. We have used CORMIX models to investigate the process as this constitutes a negatively buoyant plume under the sea. Several cases of flue gases from coal and gas fired power plants have been studied and it has been concluded that it is a possible solution. Particularly, for CO_2 the dilution reaches to a level so as to combine with carbonates already present near the bed of same dilution. We have studied two examples: one from coal fired power plant and the other using oil.

1 Introduction

Several authors[3,6,7,16,17] have considered production, capturing and sequestering of CO_2 from flue gas previously. Furthermore, the mode of release of CO_2 in deep sea has not been considered previously in detail. It has created uncertainty about the ultimate fate of the released CO_2. Study of assessment of the feasibility, energetics and economics of CO_2 sequestering varied between authors[7]. The capture and disposal of CO_2 from flue gas of fossil-fueled power plants is technically possible. It requires a significant fraction of the energy produced and additional equipment which involves a large capital expenditure. Oceans have enormous capacity of holding CO_2 at the bottom for several hundreds of years. It has been rightly said if all the carbon is dumped into sea as CO_2 solution it will increase only by 3% of its value already present in the sea. Therefore, an attempt is made in this paper to assess the possibility of transporting the dissolved CO_2 into sea. We also present various alternatives for the disposal of flue gases and conclude that of all the alternatives available at present, the oceans are the best reservoirs for the time being till other alternatives are developed.

Herzog et al.[7] have investigated five processes for capturing CO_2 from flue gas of coal fired power plants:
1. Combustion of coal in an atmosphere of O_2 and recycled flue gas.
2. Scrubbing of flue gas with a recyclable solvent (monoethanolamine).
3. Cryogenic CO_2 fractionation of flue gas.
4. Separation of CO_2 by selective membrane diffusion.
5. Scrubbing of flue gas with sea water.

Marchetti[13,14] was the first to propose the idea of disposing CO_2 in deep ocean. He considered both the processes of scrubbing and burning of coal in pure oxygen for the separation of CO_2 from flue gas. Marchetti[13] also analyzed the disposal of flue gas CO_2 in the ocean. Mustacchi et al[16] considered the use of alkalis, salts or amines to strip CO_2 from flue gas and dispose in deep ocean. In the first four processes, the CO_2 from flue gas is captured, compressed, and released as a subcritical fluid in deep sea whereas the fifth process ie., involving sea water scrubbing and disposing through a diffuser, results in a saturated solution of CO_2 in sea water. While this solution may not be required to be disposed at great depth, it certainly requires huge flow rates of sea water. Other alternatives such as depleted oil and gas fields could be used as short term storage option, storage as buried mass or as solid CO_2 near the South pole also seem to be less attractive. Thus the last case of injecting a solution of compressed CO_2 at a depth in sea water. As the density of this solution is greater than sea water, a plume of negatively buoyant carries the solution to the bottom where it spreads due to local turbulence. This injection at a depth greater than 500m, is to avoid bubble formation. This solution introduces carbon from flue gas into the terrestrial carbon cycle, where the rate at which it is taken up will be determined by the rate of circulation between deep ocean and well-mixed surface layer and therefore it can remain at the sea bed for hundreds of years[7,9]. Haugen and Drange[6] have shown that the increase in density as a result of the solubiltiy of CO_2 may be sufficient to transport the dissolved gas to lower depths even in the case when it is released at shallow depths of 200-400m from the sea level. If the CO_2 is injected near the shore, the gravity currents will carry the rich CO_2 solution along the bottom slope into deep water. Shallow injection will be less expensive in terms of energy and capital than deep ocean injection. For example, along the coast of Norway[5], near the vicinity of oil and gas fields and power plants, negatively buoyant CO_2 enriched sea water would transport the flue gas from emission site into deep ocean. Actual site location and diffuser design can be predicted through a series of alternatives and trial results using CORMIX 1 and 2. Through optimization of these results, one can design the diffuser so as to safeguard any harmful effects of such measures on marine life and the environment. In this paper, we have tried various alternative designs with the help of CORMIX 1 and 2 and the conclusions are being presented.

2 Mathematical and computer models

A large number of mathematical and computer models have been developed in recent past based on Euler's approach and Lagrangian concepts[15]. Euler models are accompanied by spurious oscillations, which depict non-physical representation of transport phenomenon. In Eulerian approach pollutant concentration is determined at certain fixed points. The numerical solutions by finite difference and or finite element method introduces numerical diffusion and hence may result in negative concentration. Thus, it does not depict conservation of the conservative pollutants. Lagrangian approach follows the plume element along the course of its trajectory for example JETLAG, developed by Lee and Cheung.[12] They consider round buoyant jet directed into an uniform cross flow. The jet discharges the effluent at angle θ with respect to the horizontal plane and σ with respect to the current. The trajectories in turbulent flow field is determined by considering advection by flow and dispersion of the substance in water at desired time intervals by using local balance equations.[12] This approach describes exact mass conservation and absence of numerical diffusion. A series of models using length scales are

Cornel Mixing Zone Expert Systems(CORMIX 1, 2, and 3). CORMIX 1[4] is intended for submerged single port discharges. CORMIX 2[2] considers the dilution and plume formation for any of the three diffuser types: staged, unidirectional and alternating. CORMIX 3[11] is meant for surface discharges of effluents from canal, open channels or open pipes.

3 Statement of the problem

In this paper, we have studied the alternative methods of flue gas disposal from two types of power plants: coal fired and gas/oil fired. Examples have been taken from literature[5,7]. The flue gas at atmospheric or elevated pressure is brought into contact of sea water in a vertical column. Assuming the inlet water contains no CO_2 then the out let water becomes saturated if 413,000 kg/s water is used to absorb about 90% of CO_2 from a 500 MW power plant. This includes the cooling water from the power plant. The savings in capital cost in reduced flow rate will approximately be required in pressure elevation for dissolution. However, at an elevated pressure of 10 atmospheres the requirement of sea water flow rate will be about 41,000 kg/sec as compared to that mentioned above for 1 atmosphere.

In Norway, a 1200MW power plant using 270 tons/hour of oil releases an effluent of 50m³/s has been considered as a second example in this study. The flue gas from the stack is passed through a scrubber with a filter at the inlet to remove any particulate matters. Sulfur dioxide removal efficiency of the scrubber is assumed to be 100%. Only 27% of the impurities levels of oil (20 ppm. Nickel and 60 ppm. Vanadium) is assumed to be collected by the scrubber[5].

In the first case of study, a negatively buoyant plume of rich solution of CO_2 in sea water having a density of 1032 kg/cm³ is released through a pipe with a single port. The effluent is released at various angles to the direction of the current at a depth of 700 meters from the surface and 300 meters above the bottom. Various cases have been studied by varying the diameter of the port and or the angle of the diffuser with the current. The smaller diameter pipes reduce the cost of transport but it may require thicker pipe which may increase the cost. Variations of the concentrations and dilution for different angles and port diameter, geometrical presentation of plume central line with x - axis as well as with z - axis have been presented through graphs and tables. The second case deals with the distribution of the concentration of Vanadium, as an example, in the near field region for different properties of the effluent. Such a study is useful so as to protect certain areas from higher concentration of Vanadium and/or Nickel after the near field.

4 Results and conclusions

It has been asked several times whether by this process we transmit a pollutant from the gas to sea. The answer is no. The pollutants in the gas are SO_2 and CO_2 which are converted as sulphate, totally dissolved as ions in the scrubber and carbonates and CO_2 in the sea water. Sea water contains enormous amounts of sulphates naturally. This process increases the concentration of sulfur only by a small amount as compared to the quantity already present. It has been commented that if all the sulfur present in the atmosphere is dumped into the sea, it will form a thin surface layer of sulfur (thickness of a paper).

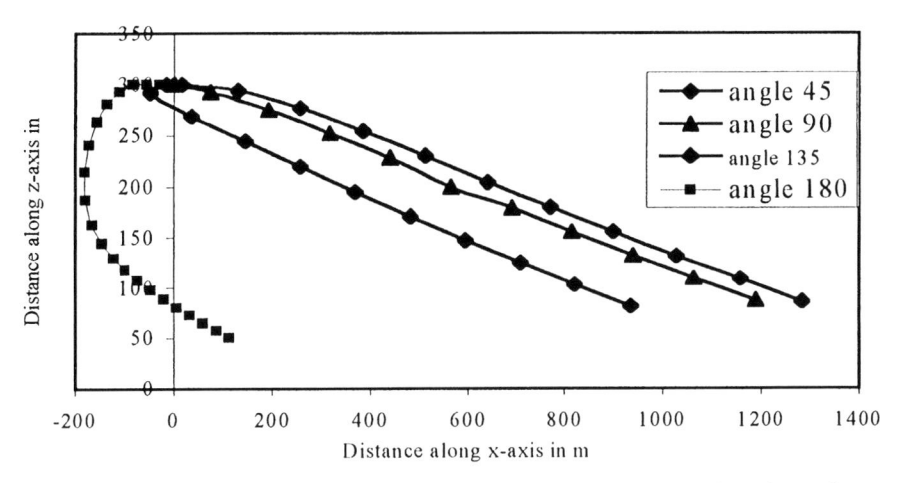

Figure 1. Variation of central line for different angles σ to direction of current.

We have collected the data and by introducing in model CORMIX 1 and obtained the results for various possible alternatives. The results of our analysis are presented figure 1 to 5 and tables 1 and 2. Figure 1 shows the concentration profiles for different angle σ. It may be observed that near field region decreases with increasing angles.

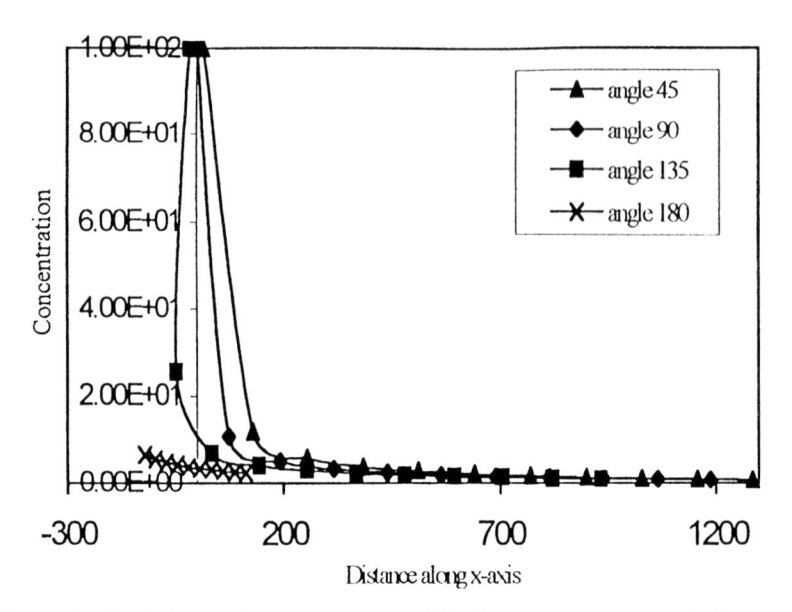

Figure 2. Variation of concentration with distance along x-axis for different angles σ

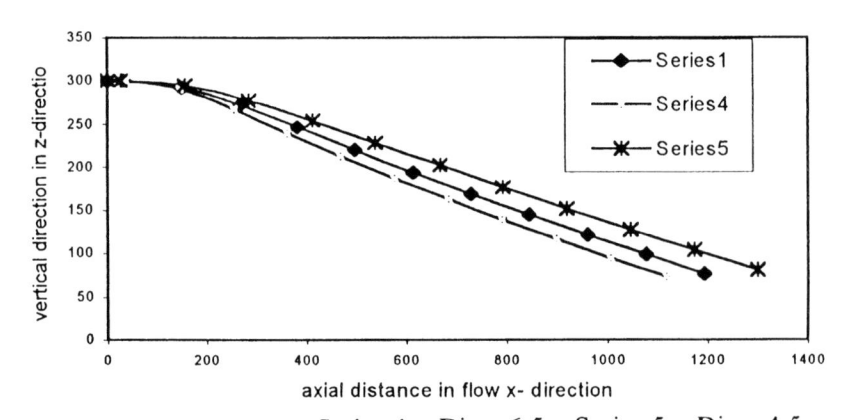

Series 1 = Diam 5.5m, Series 4 = Diam 6.5m Series 5 = Diam 4.5m

Figure 3. Central line profiles for different diameters of the diffuser port.

Figure 2 describes the variation of concentration along the flow direction whereas figure 3 shows the effect of diameter of the ports on the central line. Only single port has been considered in this study. Region of interest ie. near field region decreases with increasing the diameter of the exit.

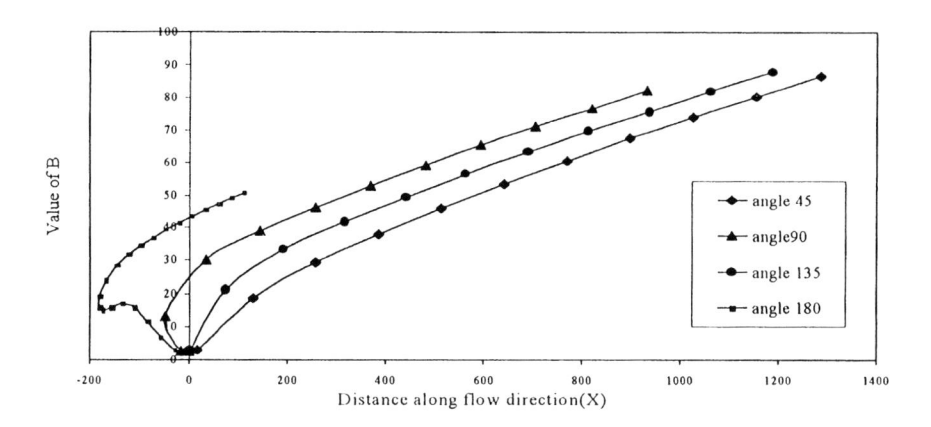

Figure 4. Variation of radius of the plume with distance along x-axis for different angle σ.

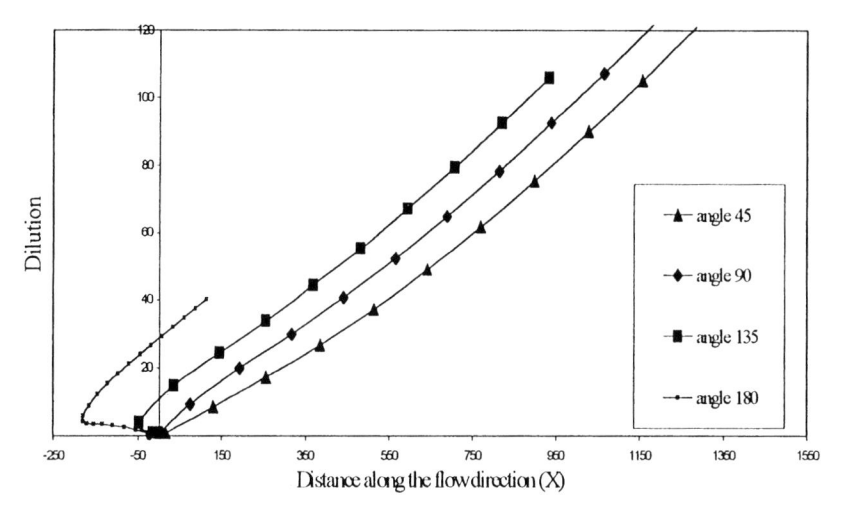

Figure 5. Dilution along the central line for different angles σ.

Figure 4 concludes that the plume area (dilution) increases with decreasing values of angle(σ). Figure 5 describe the central line dilution profiles for different angles. These results are consistent with the concentration profiles as shown in figure 2.

Tables 1 and 2 describe the variation of the plume parameters for different values of the diameter of the port. The table 2 describes the second case under the assumption that the density of the effluent is greater than sea water and it constitutes a negatively buoyant plume.

Table 1. Effect of Variation of Port diameter on various characteristics of Negatively-Buoyant Plume

No. of ports = 1, θ = 0, σ = 0, Region of Interest : 100,000 m., Wind vel. : 3m/s, Flow rate: 401.8 cm^3/s. Uniform current along x-axis : 1.35 m/s.

Port Diam. (m.) Refer fig 3.	Location of Edge of NFR (m.)	Poll. Conc. (%)	Dilution at the edge of NFR	Plume half width at NFR (m.)
4.5 Series 5	1625.3	0.4772	211.4	177.4
5.5 Series 1	1460.3	0.5506	181.6	164.4
6.5 Series 4	1345.6	0.6159	162.4	155.5
7.0	1301.1	0.6236	155.3	152.3
7.5	784.6	0.6254	144.6	143.7

Table1 presents results of this study. One may observe that the near field region decreases with increasing diameter of the pipe. Variation of the pollution concentration and dilution depicts the desired results. Variation of plume half width ie. plume area normal to the central line decreases with increasing diameter of the pipe. These results help in optimal design of the diffuser.

Consolidated results for the second example are shown in table 2 which describes the variation of important parameters with the port diameter port. In this case, we have considered the distribution of an important pollutant Vanadium found in oil. As mentioned above the parameters have been taken from a power plant in Norway. The effluent after the scrubber is released into the sea at Skagerak.

Table 2. Near field characteristics for negatively plume containing metals(Vanadium) as impurity

Initial concentration : 43.8 ppm. No. of ports = 1, θ = 0, σ = 0, Region of Interest : 20,000m., Wind vel.: 2m/s, Flow rate: 50 m^3/s. Uniform current along x-axis : 1.0 m/s. Density of sea water = 1020 kg/cm^3. Density of effluent = 1028 kg/cm^3.

Port Diam. (M.)	Location of Edge of NFR (m.)	Poll. Conc. ppm	Dilution at the edge of NFR	Plume half width at NFR (m.)
3.00	269.13	0.6482	67..50	41.10
4.00	232.58	0.8539	51.20	35.81
5.00	207.91	1.0270	42.60	32.65

All results present that the effluent is sufficiently diluted so that it is not at all harmful to the aqua cultural growth in the bottom of the sea. Particularly, the dilution of carbon dioxide reaches to a level where it can easily combine with carbonates present and forms dissolved solution of bicarbonates. The dilution at the near field region has also reached to less significant levels[18].

Further, this analysis is capable of locating the disposal site as well as designing suitable diffusers for optimal results. It can be designed in such a way so as to safeguard the sensitive region and take care of any prescribed dilution at the edge of near field region by changing different effluent flow rate, diameter of the pipe and or diffuser and other parameters after fixing the environmental parameters. Hence any practical risk can be taken care off [10,19].

References

[1] Abraham, B. M., J. G. Asbury, E. P. Lynch and A. P. S. Teotia, "Coal Oxygen Process Provides CO_2 for Enhanced Recovery", Oil & Gas Journal, 80(1982), 68 – 75.

[2] Akar, P. J. and Jirka G. H., An Expert System for Hydrodynamic Mixing Zone Analysis for conventional and Toxic Multiport diffuser discharges, Report No. EPA/ Report No. EPA/600/3-90/073, ERL - U.S. EPA, Athens, Georgia (1991).

[3] Arnold, D.S., D.A.Barrett and R.H. Isom, "CO_2 can be produced from flue Gas", Oil and Gas Journal, 80(1982), 130 – 136.

[4] Doneker, R.L. and Jirka G.H., Expert System for Hydrodynamic mixing Zone analysis of conventional and toxic submerged single port Discharges (CORMIX I), Report No. EPA/600/3-90/012, ERL - U.S. EPA, Athens, Georgia (1990).

[5] Hagen, R. I. and Kolderup, H., Flue gas desulphurisation pilot study Phase I : Survey of Major Installations -- Sea Water scrubbing & flue gas de-sulphurization Process, NATO-CCMS Study Phase I, Appendix 25-D, US department of Commerce, National Technical Information serviced, PB – 299005, (1979)

[6] Haugan, Peter M. and Helge Drange, Sequestering of CO_2 in deep ocean by shallow injection, Nature 357(1992), 318 – 320.

[7] Herzog, H, Golomb, D., and Zemba, S., "Feasibility, Modeling and Economics of Sequestering Power Plant CO_2 Emission in the deep ocean" Environmental Progress, 10(1) (1991), 64 – 74.

[8] Hopson, S., "Amine inihibitor copes with corrosion" Oil & Gas Journal, 83(1985), 87 – 92.

[9] Ishitani H. and Matuhashi R., Possibility and Evaluation of CO_2 Disposal in Deep Ocean" in "Global environmental Security" edited by Yuji Sujuki, Kazuhiro Ueta and Shunsuke Mori, Springer-Verlag Berlin Heidelberg, 1996.

[10] Jacquij G. Ganoulis, Engineering Risk Analysis of Water Pollution, VCH Weinheim. N.Y.(1994)

[11] Jones, G. R., CORMIX 3: An Expert System for the Analysis and Prediction of Buoyant Surface discharges, M.Sc. thesis Cornel University, Ethaca, NY (1990).

[12] Lee J. H and Cheung, V, Discussion on marine Outfall Design – Computer Models for initial Dilution in a Current and Reply by Authors, Proc. Instn. of Civ. Engineers, Part I, No.88(1990), 481 – 486.

[13] Marchetti, C., "On Geoengineering and CO_2 problem in Climatic Change", I. D. Reidel Publishing Company, Dordrecht, Holland (1977).

[14] Marchetti, C., "Constructive solutions to CO_2 problem", in Man's Impact On Climate, W. Bach, J. Pankrath and W. Kellogg (eds) Elsevier Scientific Publications(1979).

[15] Muellenhoff, W.P., Soldate, Jr., A. M., Baumgartner, D.J., Schuldt, M. D., Davis L.R. and Frick, W. E., Initial Mixing Characteristics of Municipal discharges, Vol. I and II, EPA-60/3-85-073a, U. S. Environmental Protection Agency, New port, Oregon.

[16] Mustacchi, C., P. Armenante and V. Cena, "Carbon dioxide Removal from Power Plant Exhausts" Environment International, 2(1979), 453 – 456,

[17] Pauley, C. R., "Recovery of CO_2 from flue gas", Chem. Engg. Progress, 80(1984), 59 – 62.

[18] Tsanis, I. K., and Valeo C., Mixing Zone Models for the Submerged discharge, Computational Mechanics Publications, Southampton Boston(1994)

[19] Wood, I. R., Assymptotic Solutions and behaviour of Outfall Plumes, J. Hydraulic Engg.,119(5) (1993), 555 - 580.

Section 5: Water Pollution

Prevention of hydrocarbons sea pollution: Sensitivity Index Maps for the Venice Lagoon as integral component of oil-spill contingency planning and response

F. Cinquepalmi, D. Schiuma, D. Tagliapietra, C. Benedetti and A. Zitelli[*]

University Institute of Architecture of Venice (I.U.A.V.) - S. Croce 1957, 30135 Venice, Italy. Italian Ministry of the University and Scientific and Technological Research (MURST); Venice Lagoon System Project (DAEST-IUAV). EMail: cinpalmi@iuav.unive.it, donna@iuav.unive.it, davidet@iuav.unive.it, cbenedet@ronet.it, [*]andreina@iuav.unive.it (Coordinator)

Abstract

In Italy the international regulations governing the transport of dangerous liquids by sea are strictly enforced and there are specific sets of safety rules for each port. Oil-spill risk management procedures could be further improved by the application of Environmental Sensitivity Index mapping principles to the assessment of coastal vulnerability.

The aim of this project is the development of an Environmental Sensitivity Index (ESI) for the Venice Lagoon and the creation of Sensitivity Maps using GIS technologies.

1 Introduction

Venice stands on a group of islands in a lagoon formed 6000 years ago in the north-east Adriatic. The present-day brackish lagoon covers about 570 sq.km. and is connected with the sea through three inlets at Chioggia, Malamocco and the Lido; these are demarcated by long jetties on either side and are kept practicable by outer breakwaters. Over 75% of the lagoon is less than 1m deep and navigation channels vary in depth from 5m to 20m at the Malamocco inlet. Since the first half of the 20th century intensive economic and industrial development has transformed the lagoon; about a third of the mudflats have gradually been reclaimed, channels serving the industrial port have been deepened and a vast industrial zone focused on energy production and the petro-chemical industry has grown up along the mainland side of the central part of the lagoon. The inestimable historical, landscape and environmental values embodied in the Venice Lagoon co-exist with increased anthropic pressure along its mainland shore due to the expansion of the chemical and petro-chemical industries in Porto Marghera.

Currently, approximately 20 million tonnes of crude oil and bulk shipments of dangerous liquid chemicals are transported across the lagoon every year; in addition there are about 10,000 motorized craft of all sizes operating permanently within the lagoon basin. The situation constitutes a high-risk potential for a catastrophic oil spill and the certainty of continuous sub-acute spillage.

2 An integral component of oil-spill contingency planning and response: Sensitivity Index Maps for the Venice Lagoon

Environmental Sensitivity Maps have been used since as far back as 1979 as an integral component of measures for the prevention of oil spills and for planning contingency response mechanisms in case they do occur.

Environmental Sensitivity Index (ESI) atlases have been prepared, for example, for most of the American coast, including Alaska and the Great Lakes [5], and for representative sections of the coast of several other countries, including Canada [1], France [2] and Germany [4].

In Italy the international regulations governing the transport of dangerous liquids by sea are strictly enforced and there are specific sets of safety rules for each port.

Oil-spill risk management procedures could be further improved by the application of Environmental Sensitivity Index mapping principles to the assessment of coastal vulnerability.

The Venice Lagoon and the Upper Adriatic are particularly vulnerable areas given the intensive traffic of oil and dangerous cargo vessels. Previous studies have provided an analysis of the historical series of oil spills in the Venice Lagoon and of the probability and possible consequences of a full-scale catastrophe[6]. They also developed a plan for controlling oil tanker traffic in the Adriatic in order to prevent accidents and consequent pollution of the sea[3]

2.1 The basic elements of a sensitivity mapping system

The aim of this project is first to identify the parts of the lagoon system that are most vulnerable to oil spill damage and to establish the basic elements - the criteria and scale of the survey and of the representation - for the development of an Environmental Sensitivity Index (ESI) for the Venice Lagoon and for the creation of Sensitivity Maps using GIS technologies.

The sensitive aspects of each of the constituent elements of the lagoon system were identified and assessed in terms of risk, impact and damage. Assessment of the sensitivity of the Venice Lagoon is based on criteria that are usually considered fundamental:

- the risk factor, determined in terms of the morphological, physical, chemical, sedimentological and topological characteristics of the lagoon ecomosaic (natural and anthropic components) and in relation to the difficulties attached to Clean-up;
- the biological and/or anthropic value factor, which represents the potential damage that the system could suffer.

A third specific element was taken into consideration as an amplification of the potential damage factor:

- the anthropogenic factor, i.e. the socio-economic and historical-cultural values that are specific to the city of Venice and its lagoon and make them unique.

It was decided to use a scale of representation that could include all the main constituent parts of the lagoon, from the sea to the sandy coastal shore, the inlets, the morphological features of the intertidal transitional zone, the historic island settlements and the fishfarms.

3 The lagoon ecosystem: natural and historical features

The natural morphological structure of the Venice Lagoon depends on its relationship with the sea, from which it is separated by a discontinuous sandbar of dunes and beaches.

Being a micro-tidal coastal basin, subject to a twice-daily tidal cycle, the lagoon has acquired characteristic morphological features such as tidal flats, mudflats, salt marshes, marshes and islands, which began to support human settlements from the 4th century BC.

3.1 Natural features

The transition wetlands and intertidal environment areas represent very important habitats and form the natural landscape of the Venice Lagoon.

In constructing the ESI matrices we considered the following natural morphological features:
- Coastlines (table 1, matrix 1):
  - seabed close to shore;
  - beach/shore.
- Natural intertidal elements (table 1, matrix 2)
  - low marshes;
  - high marshes;
  - islets;
  - mudflats;
  - brackish marshes.

3.2 Stretches of water

Navigable channels of varying depths constitute the main communications network; there are water-based economic activities such as aquaculture and fishfarms, while cultural and recreational activities include rowing and fishing. It was decided that stretches of water with such significance should be identified in the Sensitivity Index and should be distinguishable on the maps.

In constructing the ESI matrices we considered the following stretches of water (table 2, matrix 3):
- major navigational channels;
- tidal channels/marsh creeks;
- mudflats;
- open waters;
- brackish basin;
- secondary channels;
- fish farms.

3.3 Artificial features

The *insulae* of the city are entirely built up with *palazzi*, churches and lesser buildings of inestimable historical and architectural value, built and modified over many centuries; their façades rise directly from the water of the canals and so are subject to the risk of erosion and soiling.

The phenomenon of *acque alte* brings the flooding first of the city streets and squares and then of the ground floors of the buildings. Paving is mainly of stone slabs and brick laid at least two centuries ago, and fine marble flooring is often literally irreplaceable in that the quarries it came from are now worked out; all this is at severe risk in case of oil spillage.

Environmental Coastal Regions 201

MATRIX 1 — Coastlines

ESI FOR NATURAL MORPHOLOGICAL FEATURES

	CLASSES	Risk classes	Standardized weighting	beaches	seabed close to shore		
Tide height	>100 cm	1	6,7				
	50/100 cm	2	13,3				
	25/50 cm	3	20,0				
	-25/25 cm	4	26,7	26,7			
	-25/-50 cm	3	20,0		20,0		
	-50/-100 cm	2	13,3				
		15	100				
Angle of slope	15°><45°	1	16,7				
	5°><15°	2	33,3		33,3		
	<5°	3	50,0	50,0			
		6	100				
Wave movement	low	1	16,7				
	average	2	33,3				
	high	3	50,0	50,0	50,0		
		6	100				
Risk value				**127**	**103**		
clean up	sand	2	40,0	40,0			
	pebbles/shells	3	60,0		60,0		
		5	100				
Impact and Risk - physical sensitivity				**167**	**163**		
Low naturalistic/biological value		1	0,25	42	41		41>83
Low water quality		2	0,50	83	82		ESI I
Recreational/aesthetic value		3	0,75	125	122		
Presence of economic activities		4	1,00	167	163		
Good water quality/bathinge		5	1,25	208	204		
Clam harvesting/fishing		6	1,50	250	245		122>292
Landscape value		7	1,75	292	286		ESI II
Naturalistic value		8	2,00	333	327		327>375
Local birdlife		9	2,25	375	367		ESI III
Nature reserve		10	2,50	417	408		ESI IV

MATRIX 2 — Endolagunar intertidal elements

ESI FOR NATURAL MORPHOLOGICAL FEATURES

	CLASSES	Risk classes	Standardized weighting	low marshes	high marshes	low-lying islets	mudflats	brackish marshes
Tide height	>100 cm	1	6,7			6,7		
	50/100 cm	2	13,3		13,3			
	25/50 cm	3	20,0					
	-25/25 cm	4	26,7	26,7				
	-25/-50 cm	3	20,0				20,0	
	-50/-100 cm	2	13,3					13,3
		15	100					
Angle of slope	15°><45°	1	16,7			16,7		
	5°><15°	2	33,3					
	<5°	3	50,0	50,0	50,0		50,0	50,0
		6	100					
Wave movement	low	1	16,7				16,7	16,7
	average	2	33,3	33,3	33,3	33,3		
	high	3	50,0					
		6	100					
Risk value				**110**	**97**	**57**	**87**	**80**
clean up of the substrats	peat/mud/clay	1	20,0	33,3	33,3		33,3	33,3
	pebbles/shells	2	40,0			66,7		
		3	60					
Impact and Risk - physical sensitivity				**143**	**130**	**123**	**120**	**113**
Low naturalistic/biological value		1	0,25	36	32	31	30	28	28>72
Low water quality		2	0,50	72	65	62	60	57	ESI I
Recreational/aesthetic value		3	0,75	107	97	93	90	85	
Presence of economic activities		4	1,00	143	130	123	120	113	85>143
Good water quality/bathinge		5	1,25	179	162	154	150	142	ESI II
Clam harvesting/fishing		6	1,50	215	195	185	180	170	150>216
Landscape value		7	1,75	251	227	216	210	198	ESI III
Naturalistic value		8	2,00	287	260	247	240	227	
Local birdlife		9	2,25	322	292	278	270	255	227>358
Nature reserve		10	2,50	358	325	309	300	283	ESI IV

Table 1: the two matrices for the natural morphological features (matrix 1 for coastlines and matrix 2 for endolagunar intertidal features).

In constructing the ESI matrices we considered the following artificial features (table 2, matrix 4):
- Structures in the sea;
- jetties and breakwaters at the lagoon inlets.
- Island settlements;
- horizontal contact structures;
- vertical contact structures.

4 Risk factors and impact elements for the Venice Lagoon

The tide ebbs and flows through the three inlets and along the channels of the lagoon, registering a half-cycle high- and low-point delay of about two hours in the parts furthest from the sea. The average astronomical amplitude is about 50 cm and reaches the natural wetlands and Venice approximately one hour after entering the lagoon. When normal conditions are distorted by *scirocco* and *bora* winds - from South and North respectively - Venice and Chioggia and the smaller islands of Murano and Burano are subject to flooding. Being just 80 cm above mean sea level, St. Mark's Square is the part of Venice most frequently covered by water.

4.1 Risk assessment criteria for cases of oil spillage

Hydrodynamic conditions were taken as the guiding parameter in assessing the risk to the various natural and artificial morphological features and water stretches of the lagoon from contact with oil slicks. They have all been put in an order of increasing risk according to the potential frequency of exposure, the nature and form of the contact surface and substrata, wave movement and depth. The various classes of risk were given relative values on a scale of 1 - 100.

In constructing the ESI matrices for the risk run by shorelines and natural intertidal features, we identified three factors, each sub-divided in classes (table 1, matrices 1 and 2):
- tide height, with classes from 1 to 4 related to increasing frequency of contact;
- slope of contact surface, with classes from 1 to 3 in inverse relation to the angle of slope;
- wave movement, with classes from 1 to 3 related to high, medium and low energy action.

In constructing the ESI matrices for the risk run by stretches of water, we identified two factors, each sub-divided into classes (table 2, matrix 3):
- depth of lagoon bed, with classes from 1 to 4;

Environmental Coastal Regions 203

MATRIX 3

ESI FOR WATER STRETCHES

	CLASSES	Risk classes	Standardize d weighting	navigational channels	minor channels	mudflats	open waters	marshes	secondary channels	fish farms
RISK EXPOSURE FACTORS deepness	-0,25>-1 m	4	40			40,0		40,0		
	-1>-2 m	3	30		30,0					30,0
	-2>-10 m	2	20				20,0		20,0	
	<-10 m	1	10	10,0						
		10	100							
Wave movement	low	1	16,7		16,7			16,7		16,7
	average	2	33,3			33,3			33,3	
	high	3	50,0	50,0			50,0			
		6	100							
Risk value				60	47	73	70	57	53	47
IMPACT VALUE clean up	peat/mud/clay	1	33,3			33,3	33,3	33,3	33,3	33,3
	sand/shells	2	66,7	66,7	66,7					
		3	100							
Impact and Risk physical sensitivity				127	113	107	103	90	87	80
DAMAGE INDEX	low water quality	1	0,25	32	28	27	26	23	22	20
	barren lagoon-bed	2	0,50	63	57	53	52	45	43	40
	fishing facilities	3	0,75	95	85	80	77	68	65	60
	recreational value	4	1,00	127	113	107	103	90	87	80	20>133
	clam harvesting	5	1,25	158	142	133	129	113	108	100	ESI I
	fishing facilities	6	1,50	190	170	160	155	135	130	120
	hunting	7	1,75	222	198	187	181	158	152	140
	high water quality	8	2,00	253	227	213	207	180	173	160
	vegetation-supporting lagoon-bed	9	2,25	285	255	240	232	203	195	180	127>317
	aquaculture	10	2,50	317	284	267	258	225	217	200	ESI II

MATRIX 4

ESI FOR MORPHOLOGICAL ARTIFICIAL FEATURES

				horizontals			verticals			jetties
	CLASSES	Risk classes	Standardize d weighting	I	II	III	IV	V	VI	VII
RISK EXPOSURE FACTORS levels above m.s.l.	>115cm	4	40				40,0			
	100/115cm	3	30		30,0				30,0	
	85/100cm	2	20	20,0				20,0		
	<85cm	1	10				10,0			10,0
		10	100							
contact	vertical	1	33				33,0	33,0	33,0	
	horizontal and vertical	2	67	67,0	67,0	67,0				67,0
		3	100							
frequency	exceptional	1	17			17,0			17,0	
	rare	2	33		33,0			33,0		
	high	3	50	50,0			50,0			50,0
		6	100							
Risk value				137	130	124	93	86	80	127
IMPACT VALUE clean up monuments	medium	1	33,0				33,0	33,0	33,0	33,0
	high	2	67,0	67,0	67,0	67,0				
		3	100							
Impact and Risk for phisical sensitivity				204	197	191	126	119	113	160
			l. h.							
DAMAGE INDEX	presence of service functions	1	0,25	51	49	48	32	30	28	40
		2	0,50	102	99	96	63	60	57	80	28>153
	presence of economic activities	3	0,75	153	148	143	95	89	85	120	ESI I
		4	1,00	204	197	191	126	119	113	160
	presence of cultural functions	5	1,25	255	246	239	158	149	141	200	158>255
		6	1,50	306	296	287	189	179	170	240	ESI II
	presence of landscape values	7	1,75	357	345	334	221	208	198	280
		8	2,00	408	394	382	252	238	226	320
	features of historical/monumental/	9	2,25	459	443	430	284	268	254	360	254>510
	architectural importance	10	2,50	510	483	478	315	298	283	400	ESI III

Table 2: the two matrices for the water stretches (matrix3) and for the morphological artificial features (matrix 4)

- wave movement with classes from 1 to 3.

In constructing the ESI matrices for the risk run by artificial features, we identified three factors, each sub-divided into classes (table 2, matrix 4):
- height with respect to mean sea level, with classes from 1 to 4 related to tide height and flooding;
- contact surfaces, with classes from 1 to 2 related to the vertical or horizontal angle of the surface;
- frequency of contact, with classes from 1 to 3 in inverse relation.

A risk factor - the sum of the standardized weightings of the classes attributed - was estimated for each of the categories considered in the matrices.

4.2 Impact assessment criteria in case of oil spillage and Sensitivity Index

The impact factor was assessed in relation to the possibility and the difficulty of mounting successful Clean-up operations.

As regards the assessment of impact on natural features, the ESI matrices classify them into three groups in relation to the nature of the sub-stratum and the degree to which it will absorb and retain oil (tables 1 and 2, matrices 1, 2 and 3):
- substratum of peat/mud/clay;
- substratum of sand
- substratum of pebbles/shells.

As regards the assessment of impact on artificial features, the ESI matrices classify them into two groups according to the difficulty of Clean-up.

The capacity of these surfaces to absorb oil increases in relation to their roughness and poor state of conservation. Buildings that have not been properly maintained or restored often feature damage, cracks and serious lacunae. Almost all the buildings concerned are very old, some dating from the 9th century.

The various classes of impact were given relative values on a scale of 1 - 100. An impact and risk factor - the sum of the standardized weightings - was estimated for each of the categories considered in the matrices; this represents the physical sensitivity of the system.

5 Assessment of oil spill damage in the Venice Lagoon

In the case of pollution caused by oil spill, the term damage is used to describe the partial or total loss of value of natural, biological, anthropic and anthropogenic items in the Lagoon as identified by the summary quality descriptors illustrated below.

Damage weightings or indices have been attributed to each of these descriptors, increasing in proportion to the degree of seriousness and irreversibility of the loss of value. In the matrices the damage indices have been standardized to 100 on a base of 2,5.

For natural features reference has been made to an appropriate system of descriptors of the type, morphological and biological complexity and value of the natural systems of the Venice Lagoon, and of their values in terms of landscape/culture and use. The highest damage values have been attributed to naturalistic values and to the natural reserves (table 1, Matrices 1 and 2).

For stretches of water reference is made to a system of descriptors related to the principal activities connected with them. Important traditional economic activities include fishing, fish farming and aquaculture of mussels and shellfish, the harvesting of clams and hunting. The highest damage values have been attributed to the more important economic activities that depend on good water quality (table 2, Matrix 3).

In the case of anthropic values, the descriptors used refer to historical, monumental and architectural factors, to the degree to which the item is an integral part of the enjoyment/use of the landscape/culture and to the presence of legal protective measures. All the more important cultural events in Venetian life are closely connected with water, including sporting events and regattas commemorating historical and religious anniversaries.

As regards anthropogenic values (artificial structures), the highest damage indices relate to the loss of historical-monumental and architectural items (flood-borne oil pollution in St. Mark's Square) (table 2, Matrix 4).

6 The Environmental Sensitivity Indices (ESI)

Environmental Sensitivity Indices are the result of the analysis of numerical values obtained with two-dimensional matrices, from the meeting point of physical sensitivity indices (the sum of the standardized risk and impact values) and the damage indices.

Discrete analysis of the matrices enabled the numerical values to grouped into clusters. These define the environmental and historical-cultural sensitivity of the morphological structures analyzed. The value groupings then provide the basis for the definitions and descriptions condensed in the Environmental Sensitivity Indices (ESI) following the widely accepted procedures developed by the NOAA.

Four sensitivity matrices are presented, two for natural morphological structures, the first of which refers to the coastline (table 1, Matrix 1), the second to endolagunar intertidal elements (table1, Matrix 2), the third for stretches of water (table 2, Matrix 3) and the fourth for artificial morphological features (table 2, Matrix 4).

7 Conclusions

On the basis of the risk factors and assessment of the damage to the morphological structures characterizing the natural and anthropic components of the Venice Lagoon, the following Environmental Sensitivity Indices (ESI) were identified:

For the coast-line:
- ESI - I. beaches and seabed close to the shore with low naturalistic/biological value and low water quality
- ESI - II. beaches and seabed close to the shore with recreational/aesthetic value, with the presence of economic activities, with good water quality, also for bathing, with clam harvesting and fishing and with landscape value.
- ESI - III. beaches and seabed close to the shore with naturalistic value and local birdlife and with the presence of a nature reserve.

For natural morphological features:
- ESI - I. all intertidal elements support no or little vegetation and poor level of wild life and exchange.
- ESI - II. all intertidal elements of medium morphological and biological complexity, marshes with cultural and recreational value.
- ESI - III. all intertidal elements support vegetation, have cultural and recreational value; low-lying natural and artificial islets, mudflats and marshes all with flourishing wild life.
- ESI - IV. tidal flats and marshes with flourishing wild life and all intertidal elements with high morphological and biological complexity and with naturalistic value.

For stretches of water:
- ESI - I. all the stretches of water of low water quality and supporting no vegetation that can nevertheless support fishing facilities and have recreational value, marshes, secondary channels, fish farms and clam harvesting.
- ESI - III. all the stretches of water with high water quality, used for clam harvesting, with vegetation-supporting lagoon-beds and used for recreational fishing and hunting.

For the artificial morphological features:
- ESI - I. all the structures with mainly service functions and medium presence of economic activities. Vertical structures with high cultural and medium economic value.
- ESI - II. all the horizontal structures with high economic activity and medium level cultural function. Vertical structures with high cultural value and medium-high landscape value.

- ESI - III. all the horizontal structures with high cultural value and landscape, historical, monumental and listed building values. Listed vertical surfaces with high historical and monumental value.

The ESI identified as above were used to create Environmental Sensitivity Maps (scale 1:25,000 - website V.ESI.Maps on http://www.ivsla.unive.it/) in which lines and shading are used to identify the outlines and areas of the sensitive morphological features.

The Sensitivity Index Maps have been created as a useful contribution to the prevention of pollution from oil spills and to combating the consequences if accidents should occur along the coast or in the Lagoon.

They help to identify the more sensitive and higher priority environmental and historical morphological elements with a view to planning preventive action and emergency response in the case of accidents.

References

[1] Environment Canada, *National Sensitivity Mapping Program (NSMP) Atlantic Region, Mapping Program (ARSMP)*, 1991/1997.

[2] Groupe SILLAGE/IFREMER, *Mise en forme numérique de l'Atlas des marais maritimes atlantique*, Centre IFREMER de Brest, 1995

[3] Cinquepalmi, F., Campagnol, A. and Zitelli, A., *AdriaOil Plan: a project for multilateral spill management in the Adriatic Sea*, Proceedings of MEDCOAST '97, Qawra (Malta), Middle East Technical University, Ankara, Turkey, 1997;

[4] Krasemann, H. L., Riethmüller, R., *WATiS - The Wadden Sea Information System, Experience From an Operational System* - in Climate and Environmental Database Systems, Kluwer Academic, Boston, 1997.

[5] NOAA Technical memorandum - National Ocean Service (NOS) Office of Resources Conservation and Assessment (ORCA) 115, *Environmental Sensitivity Index Guidelines version 2.0*, Seattle (WA), 1997.

[6] Zitelli, A., Cinquepalmi, F., Benedetti, C., Campagnol, A., Bergamasco, A., *Oil spills in the Venetian Lagoon: an analysis of risk management*, in Proceedings of COASTAL ENGINEERING 97, Wessex Institute of Technology, La Coruña (Spain) 1997;

Evaluation of Sewage Outfalls by Using Tracer Techniques Combined with Oceanographics Measurements

J. Roldão, J. Pecly, E. Valentini & L. Leal
Laboratório de Traçadores - COPPE/UFRJ
Cx. Postal 68506, 21945-970
Rio de Janeiro   Brazil

Abstract

Combined techniques have been successfully used in order to evaluate the Ipanema's sewage submarine outfalls dilution patterns. The oceanographic measurements were carried out during one full year. Sea current measurements were made by an acoustic doppler current profiler (ADCP) and vertical temperature profile time series by thermistor strings deployed near the diffusers.   field campaigns using fluorescent dyes were carried out for covering stratified and non-stratified oceanographic conditions.

The acoustic doppler current profiler and thermistor strings were home programmed based on the number of profiles, sampling time, and precision constraints.

Uranine and Amidorhodamine G were simultaneously injected into the outfall's pipeline during six to eight hours by a continuous injecting device and detected at different depths in the receiving coastal water.

These combined studies for evaluating the dilution pattern of effluent in the sea water provided valuable data for calibrating and validating the hydrodynamic and water quality model.

1 Introduction

Aiming one effective evaluation of dilution capacity of the Ipanema's sewage outfall, a complete and integrated monitoring program has been developed. Such study were carried out including the following points:

1. A careful program for adequate oceanographic data gathering.
2. Monitoring of physico-chemical parameters in the near field.
3. Monitoring of the real dilution conditions by using dye tracers in specific oceanographic conditions.

2 Oceanographic data measuring program

The oceanographic measuring program included measurements of the vertical temperature profile and measurements of the sea current near the diffusers. For this study, one Acoustic Doppler Current Profiler (ADCP) and one thermistor string were used. The thermistor string is composed of 11 sensors equally distributed over the water column, at the depth of 28 m. The ADCP was home programmed in order to measure 9 layers over the water column. Both equipment were deployed for one full year, between December 1996 and November 1997, and setted up for gathering data every half an hour.

The analysis of the obtained data allowed to establish some important points, that lead to a better understanding of the submerged plume:

- The higher temperature differences between the surface and the seabed, occurred during the summer (December till March), reaching 10 °C in some situations. The lower differences were monitored in June, July and August, showing a water column practically non-stratified, with temperature gradients lower than 1 °C.

- Statistical analysis has shown a homogeneous water column, without considerable stratification due to thermal gradients in more than 80 % of the studied period of time. These results indicate that the use of bidimensional water quality models should be adequate.

Environmental Coastal Regions 211

- The current intensity, near to the diffusers, is lower than 20 cm/s in 78% of the time near the water surface and 80% near the bottom.

- The predominant current directions occurred with azimuths of 60° and 240°, parallel to the coast at the Ipanema's Beach. Figure 1 shows a summary of the data measured between December 1996 and November 1997.

- The spectral analysis shows that local circulation patterns are strongly affected by meteorological events, as cold fronts, mainly at the levels close to water surface. However, at the bottom the circulations patterns are governed by tide influence.

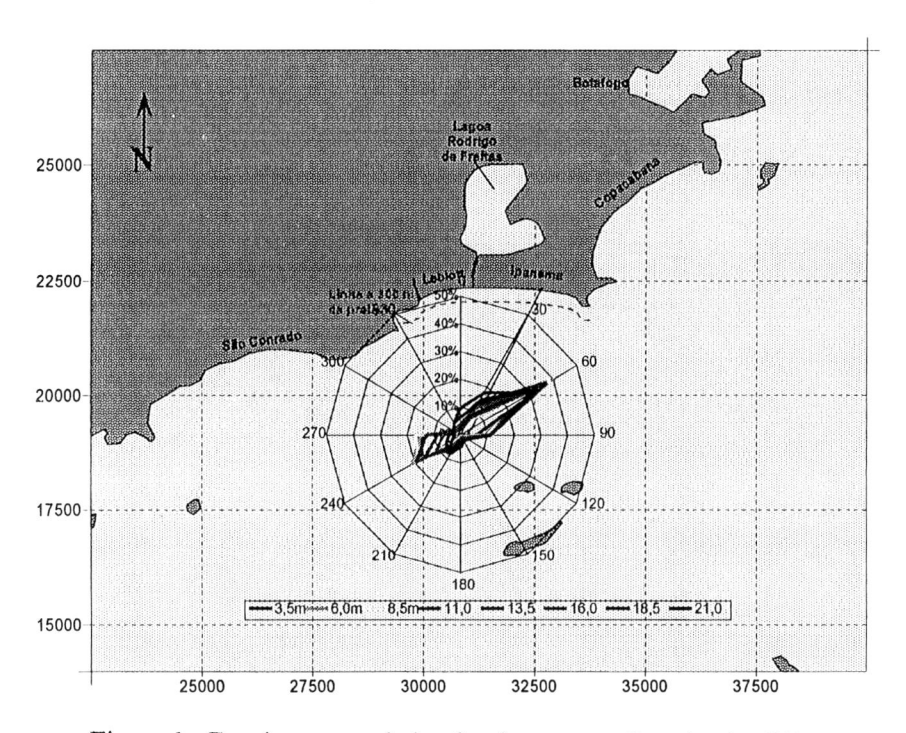

Figure 1: Dominance statistics for the current direction in different depths, from December 1997 until November 1997.

Regarding the water column temperature stratification, it is important to note that stratified and non-stratified conditions do not depend exclusively on seasonal conditions. Such conditions depend also on meteorological events. One such situation can be observed in the Figure 2. This figure represents the vertical temperature profile time series for April 1997, in a place near the diffusers.

Temperature time series

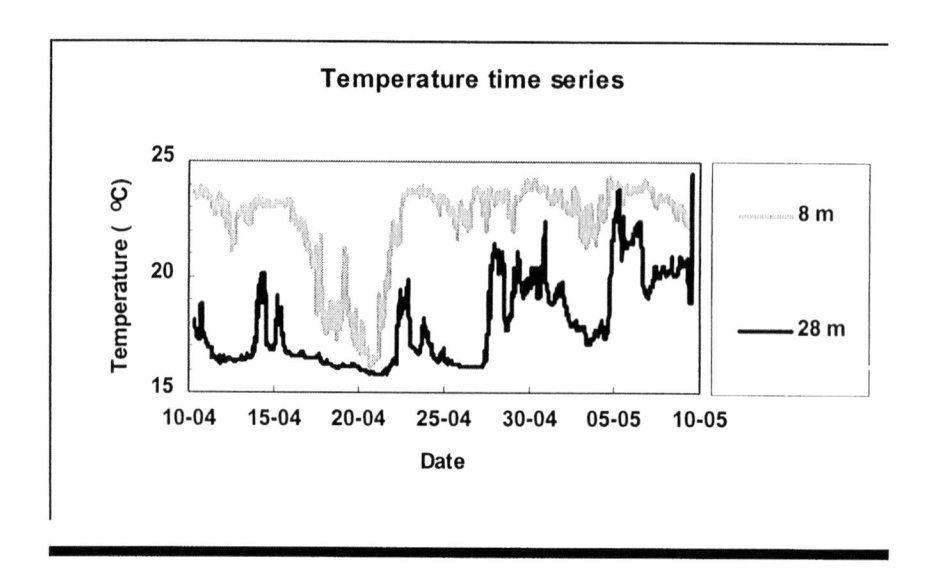

Figure 2: Time series for the upper and lower layer temperature during April 1997, showing stratified and non-stratified conditions.

3 Monitoring of physico-chemical parameters

During the dye tracers field campaigns, singular points were chosen in order to carry out vertical profiles for the parameters temperature, D.O., turbidity, conductivity, salinity and pH.

The measurement procedure includes also sample collecting activities in order to evaluate the tracer's vertical profile at the selected points.

The Figure 3 and 4 show the vertical profiles for both dye tracers, for temperature and for dissolved oxygen.

4 Monitoring of the dilution factor by using dye tracers

The field campaigns for evaluating the sewage outfall's dilution capacity, by using fluorescent dyes (Ref. 1) were carried out both in "typical summer conditions" (with stratified water column) and "typical winter conditions" (with homogeneous temperature over the water column and good mixing conditions).

Environmental Coastal Regions   213

In order to label an effluent which is continuously released into the sea, the tracer solution introduced in the pipeline should be continuously injected at a well known concentration and flow rate. The injection time was between 6 and 8 hours. A simple device that fulfills these needs uses a dilution barrel with an electro-mechanical stirrer and a peristaltic pump. The injection rate is around 1 liter/min and the tracer concentration lies between 5 and 10 %.

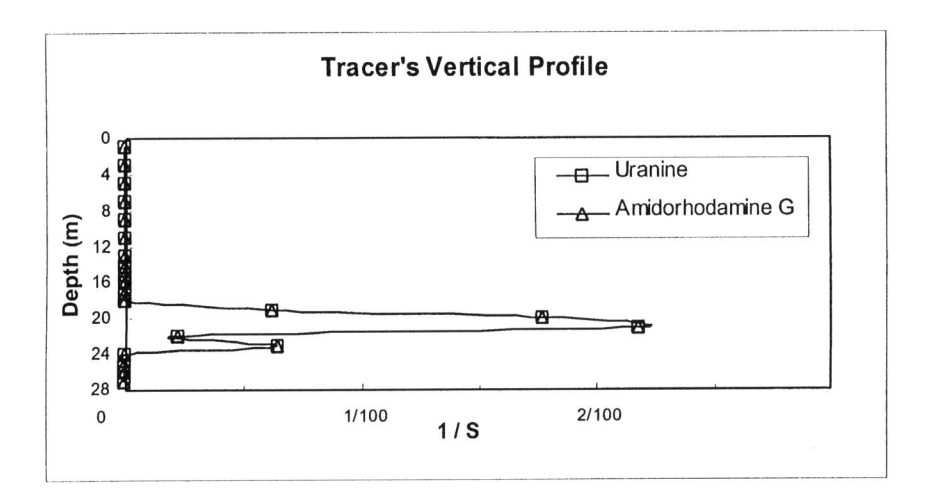

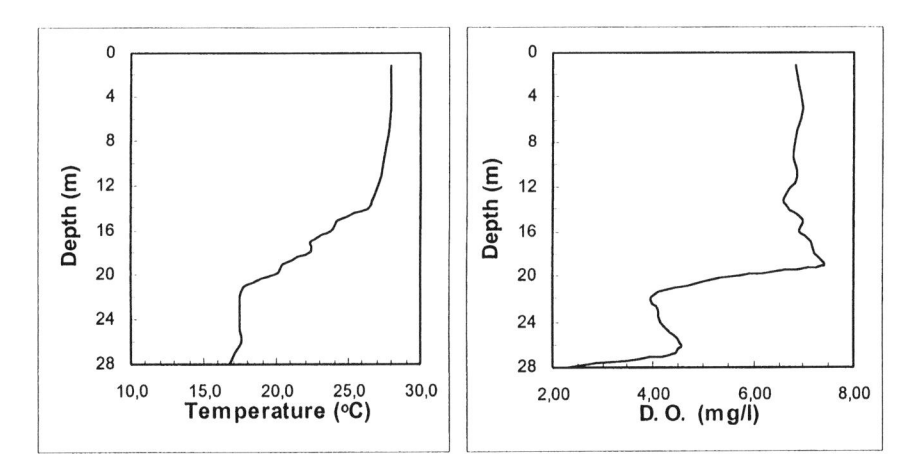

Figure 3: Vertical profile for dye tracers, temperature and dissolved oxygen during the stratified condition field campaign.

214 Environmental Coastal Regions

Tracer's Vertical Profile

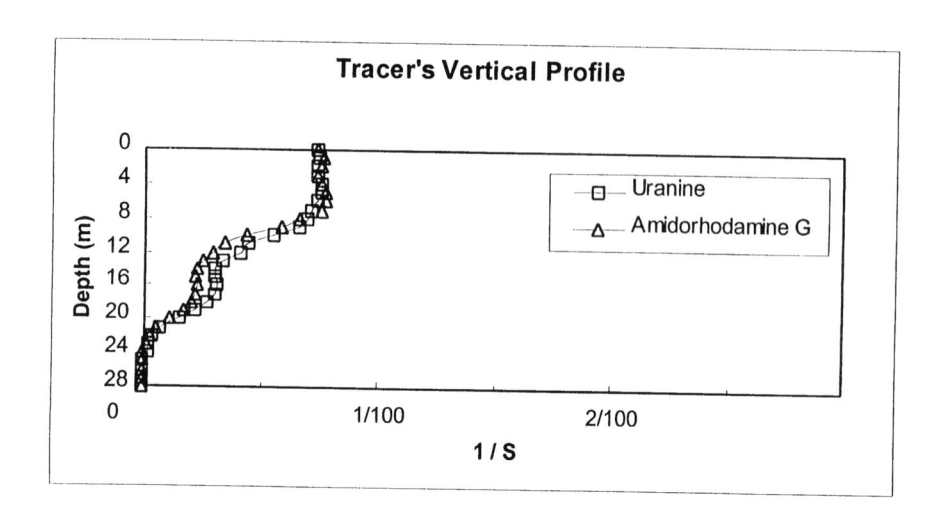

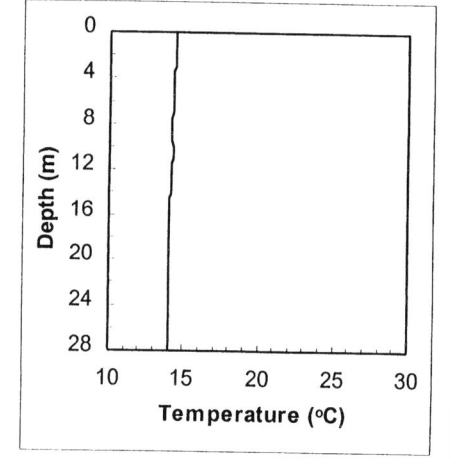

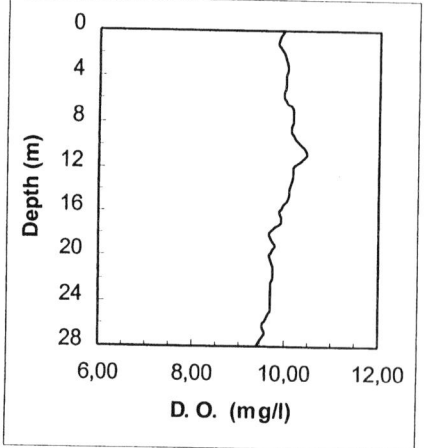

Figure 4: Vertical profile for dye tracers, temperature and dissolved oxygen during the non-stratified condition field campaign.

The mean tracer concentration of the labeled effluent was obtained when a proper sampling program is established at a point sufficiently away from the tracer injection site.

The definition of "dilution factor" is given by:

$$S = \frac{\text{mean tracer concentration in the effluent}}{\text{tracer concentration in the sea}}$$

The comparison between the tracer concentration in the pipeline and the tracer concentration for every sampling point allows the construction of a affected area map, in terms of the dilution factor (Ref. 1).

Such maps allowed to estimate the contaminated area for the 3 lower dilution factor ranges. The affected area values for the winter campaign are showed in the table below.

Table 1 : Affected area as a function of dilution factor range for the non-stratified conditions campaign with dye tracers.

Sea current direction	Depth (m)	Dilution Factor Range		
		30 a 50	51 a 100	101 a 200
West → East	4	zero	zero	0,107 km^2
	7	zero	zero	0,021 km^2
	10	zero	zero	0,001 km^2
East → West	2	zero	zero	0,215 km^2
	4,5	zero	zero	0,103 km^2
	6,5	zero	zero	0,030 km^2

During the study, four field campaigns using dye tracers were carried out. Two campaigns in typical summer conditions (stratified water column) and two campaigns in typical winter conditions (non-stratified water column). For each one of these campaigns, water samples were taken simultaneously in 3 different depths.

The Figure 6 shows the plume obtained from the tracers campaign for west-east sea current conditions, with a homogeneous water column, in "typical winter conditions".

Figure 6 shows that dye tracer flows in direction parallel to the coast, governed by the currents at the low levels of the water column. As mentioned before, the circulation patterns close to water surface, strongly influenced by local winds, do not interfere on the circulation close to the bottom, as shown in Figure 5.

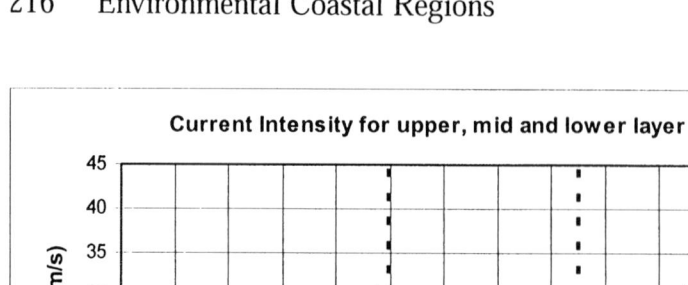

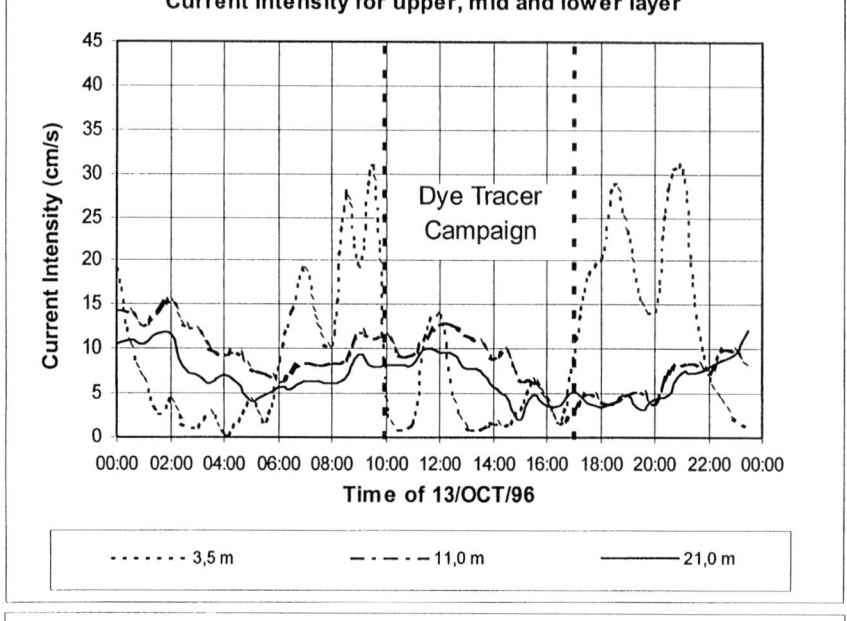

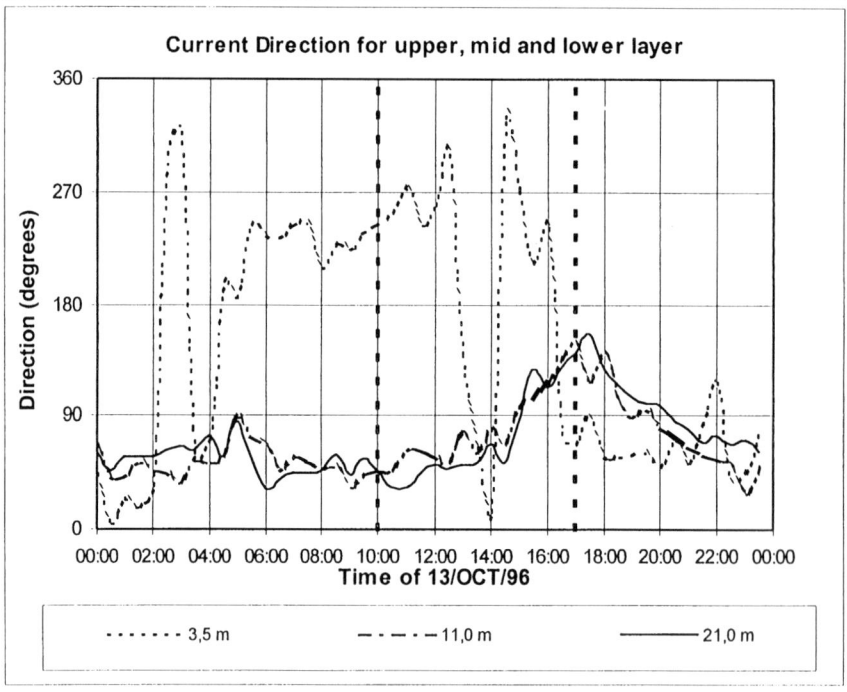

Figure 5: Current intensity and direction in the near field during the dye tracer field campaign, showing current from SE-NE.

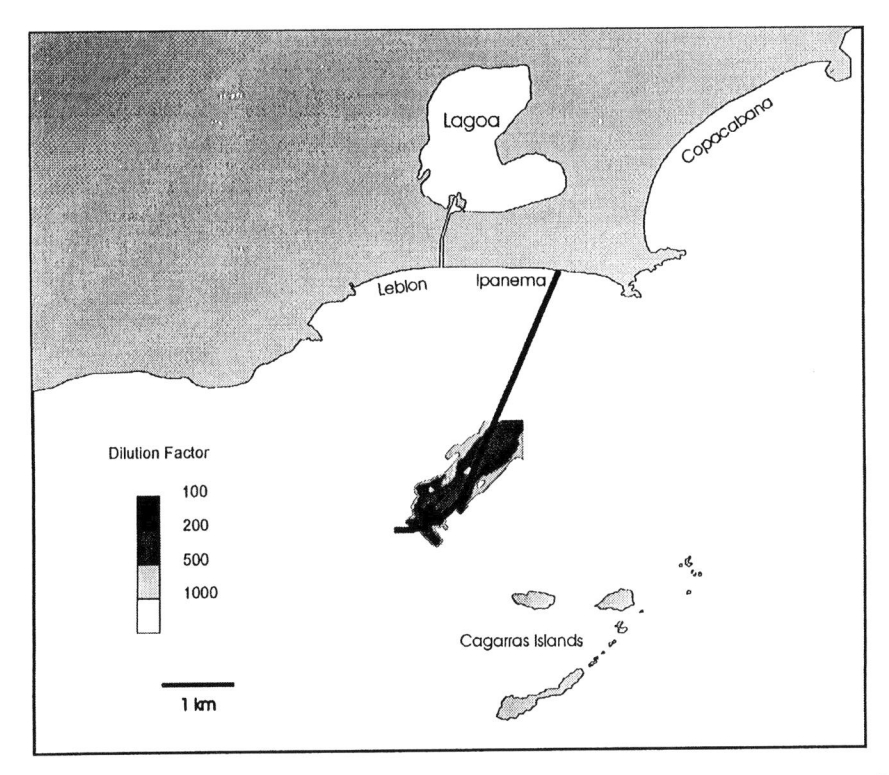

Figure 6: Contamination plume as measured with dye tracers at depth of 4 m, with *SE-NE* sea currents and non-stratified water column.

5 Conclusions

- The oceanographic program carried out between December 1996 and November 1997 showed that the predominant sea currents are parallel to the shore at Ipanema's Beach. The higher intensities are associated to meteorological events showing coherence with cold fronts.
- Sea currents in the beach direction are uncommon, with low intensity and short duration.
- Water quality models that consider the effluent mixed in a wide range of the water column (weak stratification or completely homogeneous), are adequate to simulate the effluent transport and dilution, as the field measurements has shown a water column nearly homogeneous during 80 % of studied year.

- As expected, the preferential effluent transport direction, as measured with fluorescent dyes, has perfect agreement with the sea current direction.

- The tracer vertical profile showed good agreement with the temperature and dissolved oxigen vertical profiles, presenting the same tendency. During a stratified condition, and so poor mixing conditions, the plume trapping level could be inferred from the temperature profile.

- During the winter conditions campaigns, the minimum dilution factor for the 3 monitoring depths presented approximately the same value, in agreement with the good mixing conditions expected for a homogeneous water column.

References

1. Roldão, J., Pecly, J. and Leal, L. *Tracer techniques to evaluate the dilution performance of sewage submarine outfalls*, Water Pollution 97, Bled, Slovenia, June, 1997.

2. Roldão, J. et alli, *Evaluation the dilution efficiency of the Ipanema's sewage submarine outfall*, Summer report September, 1996, Winter report December 1997, ET-150663 (in Portuguese) Rio de Janeiro,.

3. Valentini, E. et alli, *Oceanographic and Meteorological Data Analysis at the Influence Area of Ipanema's Sewage Submarine Outfall*, final report ET-150663 (in Portuguese), Rio de Janeiro, January, 1997.

4. Roberts, P. J. W. and Williams, N., *Modeling of Ocean Outfall Discharges*, Water Sci. Tech., vol. 25, No. 9, pp. 155-164, 1992.

Acknowledgements

The authors wish to thank the financial support from the Brazilian Research Council (CNPq) and CEDAE (Water and Sewage Treatment Company of Rio de Janeiro) which have made the field activities possible.

Interaction between dissolved phosphorus and suspended sediments in a tropical estuary

T.E. Payne[1], R. Szymczak[1] & T.D. Waite[2]
[1]Australian Nuclear Science and Technology Organisation, PMB 1 Menai, NSW 2234, Australia
Email: tep@ansto.gov.au, rsx@ansto.gov.au
[2]School of Civil and Environmental Engineering, University of New South Wales, Sydney, NSW 2052, Australia
Email: d.waite@unsw.edu.au

Abstract

Nutrients from the Fitzroy River (Queensland, Australia) may impact on the Great Barrier Reef, a marine area of world significance. In this study, experiments were carried out to determine the equilibrium distribution of phosphorus between Fitzroy River sediments and estuarine water under a range of conditions. The kinetics of sorption processes were also studied. Sorption reactions reached a steady state after about 2 days. The sorption of P was weaker at higher pH values, but did not decrease with an increase in ionic strength. The amount of P bound by sediment particles was limited by the surface site density of the solid phase (about 140 µmol/g). The Fe content of suspended particles increased moving down the estuary, due to the precipitation of amorphous Fe-oxides. Adsorption of P on freshly precipitated Fe-oxides may therefore be a significant mechanism of P removal from the aqueous phase. At higher pH values and salinities, solid $CaCO_3$ appears to play an increasing role in limiting the amount of dissolved P, either by surface adsorption or co-precipitation. The interaction of P with Fe-oxide surfaces was modelled using a surface complexation sorption model with the geochemical code MINTEQA2.

1 Introduction

Excessive input of nutrients to the Great Barrier Reef (located in the coastal waters of Queensland, Australia) may result in eutrophication and can have more specific impacts. In the case of phosphorus (P), increased levels may impair growth of the coral skeleton (Hunter & Rayment [11]). Understanding the processes which affect P cycling in the estuaries of the region is therefore important for the environmental management of the reef.

A number of mechanisms have been proposed as being of significance in the estuarine cycling of P (Froelich [8]). These include adsorption / desorption in response to pH or salinity gradients (Eyre & Twigg [6]), and precipitation or co-precipitation of P with Fe-oxides (Fox [7]). The aim of the present study is to develop a chemical model of the behaviour of P in the estuary of the Fitzroy River, which is the largest river on the central Queensland coast.

2 Experimental

2.1 Water and particle samples

The Fitzroy River has a pattern of summer flooding and winter drought. A barrage has been constructed across the river at Rockhampton to retain fresh water and prevent sea water intrusion into the resulting impoundment. Samples of filtered water and suspended particles were collected during the dry season at the locations shown in Figure 1. The first sample (S1) was fresh water obtained upstream of the barrage. Samples S2, S3 and S4 were obtained progressively down the estuary. Sample S5 was from near the limit of the estuary, and was similar to sea water. The major elemental composition of the water samples was determined by standard techniques, with phosphate analysed using the ascorbic acid / molybdenum blue method (APHA [2]).

A sample of bottom sediments was obtained from location S1. A slurry of fine particles (<20 μm) for laboratory sorption experiments was obtained from this sample by sedimentation in the S1 water. The mass loading in this slurry was measured by carefully drying samples of known volume. The extractable P content of the fine particles was determined after a persulfate extraction of an aliquot (APHA [2]). The surface area of the particles was measured using the BET method, which is commonly used for characterising particles in sorption studies (Davis & Kent [4]). An XRD spectrum of the particles was also obtained. The total elemental contents of sediment and suspended particles were measured after digestion with a mixture of HF and HNO_3.

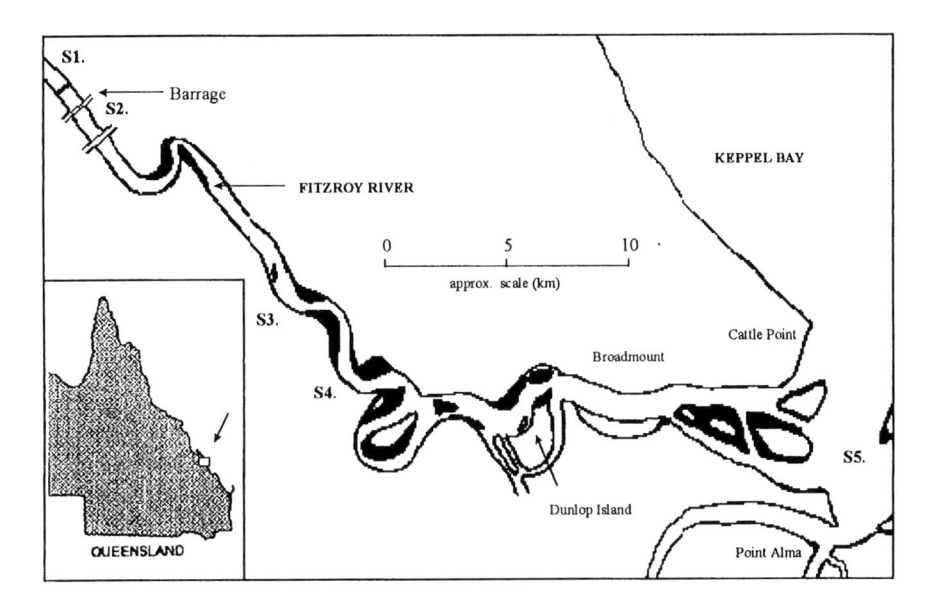

Figure 1. Sketch map of the lower section of the Fitzroy River, showing sample locations.

2.2 Sorption experiments

Sub-samples for sorption experiments were obtained from the fine particle slurry by thorough agitation followed by removal of a known volume by pipette. Sorption experiments were carried out with a mass loading of 1 g/L or 100 mg/L. The aqueous phases consisted of S1 or S5 water, containing added P of 10 or 100 µmol/L. Experiments were maintained at 25°C.

Kinetic experiments were carried out in agitated 500 mL vessels which were ventilated to ensure equilibrium with atmospheric CO_2 and to maintain aerobic conditions. The P content of the aqueous phases as a function of time was determined after filtration of sub-samples through a 0.45 µm membrane.

Batch sorption experiments were carried in ventilated 30 mL centrifuge tubes. The pH was adjusted using dilute HCl or NaOH, with $NaHCO_3$ added at pH values above 8.0 to ensure equilibrium with atmospheric CO_2. The solid and liquid phases were separated after 2 days equilibration by centrifugation (10000 rpm). The amount of dissolved P was determined by analysing the supernate. Additional experiments were undertaken with freshly precipitated amorphous Fe-oxide (ferrihydrite), prepared using the method of Waite *et al.* [15], with an aqueous phase of 0.5 mol/L NaCl.

A P sorption isotherm (at pH 8.2) was obtained using S1 water. In these experiments, high P concentrations were used to enable measurements close to surface saturation. Under these conditions, initial and final dissolved P concentrations in the experiments were similar, which meant that large errors were associated with estimating sorbed P by difference. To avoid this problem, the P sorbed on the particles was measured directly by collecting the particles on an ultrafiltration membrane and measuring their P content after extraction by the persulfate method (APHA [2]).

3 Results and Discussion

3.1 Characterisation of waters and particles

The water conductivity in the estuary showed a distinct boundary between fresh and saline water below site S1, due to the barrage (Table 1). The major ion chemistry of S1 comprised Na (0.49 mmol/L), Mg (0.38 mmol/L), Ca (0.29 mmol/L), and Cl (1.79 mmol/L). Downstream of the barrage there was an increase in dissolved P, with the P concentration in estuarine water being between about 2 and 6 µmol/L. At the time of sampling, the pH of the river water (S1) was similar to that of sea water. However, the pH of coastal rivers is affected by flow conditions, and can reach significantly lower values. For example, Eyre & Twigg [6] measured pH values of 6.5 in the Richmond River in northern NSW.

Table 1. Summary of chemical data for water and suspended particles

Sample site	pH	conductivity (mS/cm)	dissolved P (µmol/L)	Suspended sediment (mg/L)	Fe in suspended particles (mmol/g)
S1	7.92	0.189	1.2	18.9	0.45
S2	8.08	26.4	6.1	38.8	0.51
S3	7.98	43.2	2.2	29.9	0.32
S4	8.19	35.1	5.1	38.3	0.91
S5	8.19	51.3	2.6	67.5	1.77

The XRD results indicated that the S1 bottom sediments consisted predominantly of quartz and kaolinite. The Fe content of the S1 particles was 0.40 mmol/g, which was similar to the suspended particles at this location. The amount of P in the persulfate extraction of the particles was

21.8 µmol/g. This P should be considered when interpreting the results of sorption experiments with this material. The suspended solid load and the Fe content of suspended particles showed a marked increase at location S5.

The data indicate that the parameters previously suggested as being significant (pH, salinity, suspended sediment load, and Fe-oxide precipitation) could play a significant role in the cycling of P in the Fitzroy River estuary. These parameters were therefore subjected to detailed study in the sorption experiments.

3.2 Kinetic adsorption experiments

The kinetic experiments followed P sorption for time periods of up to 400 hours. In the experiments with 10 µmol/L added P, there was little difference between the sea water (S5) and river water (S1) at pH 8.0 (Figure 2a). The data suggest that about 2 days was required to reach sorption equilibrium. Adjustment of the pH to 6.0 significantly reduced the amount of dissolved P. Surprisingly, the amount of dissolved P was lower in the sea water than the river water. This indicates that, in the absence of a pH change, the salinity increase from river water to sea water should not release P from suspended particles. However, a pH increase between fresh and sea water may release adsorbed P.

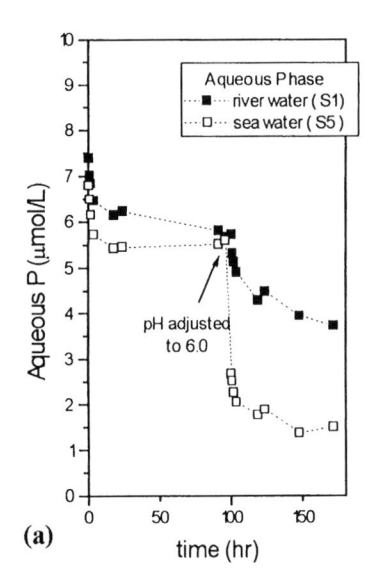

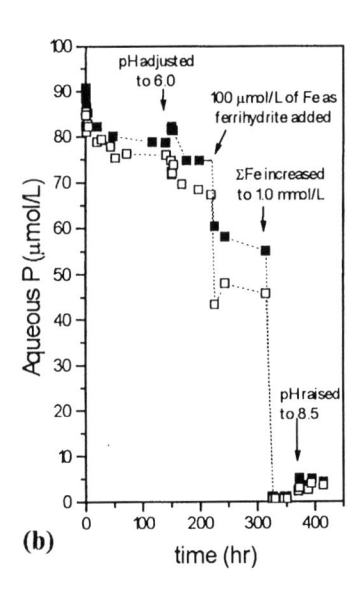

Figure 2. P in the aqueous phase in kinetic experiments with 1 g/L of S1 sediment suspended in S1 or S5 water (initial pH 8.0). Changes to the conditions are shown. Added P of a) 10 µmol/L and b) 100 µmol/L.

In the experiments with higher added P (Figure 2b), adjusting the pH to 6.0 caused a smaller relative decrease in dissolved P concentration. The data suggest that with these high levels of P, sorption sites on the particle surfaces are approaching saturation with P. During the course of the experiments, freshly precipitated ferrihydrite was added in order to study the response of the system to increases in particulate Fe-oxides. The addition of 100 µmol/L of Fe as ferrihydrite removed approximately 20 µmol/L of P from solution with both S1 and S5 waters. These data are consistent with the conclusion of Dzombak & Morel [5] that the sorption behaviour of ferrihydrite can be modelled with approximately 0.2 mol of surface sites per mole of Fe. With a total of 1.0 mmol/L of Fe as ferrihydrite, excess sorption sites were available and the P was quantitatively removed from solution (Figure 2b). The subsequent increase of the pH to 8.5 released some of the adsorbed P.

3.3 Batch adsorption experiments

The results of batch sorption experiments with S1 sediments again showed a strong pH dependence (Figure 3a). The levels of P in solution at high pH in the experiments with 1 g/L solid loading exceed the total added P (10 µmol/l) in the experiment. This was attributed to desorption of some of the existing P content of the particles.

The much higher amounts of P in solution up to pH 8.0 in the experiment with the lower mass loading (Figure 3b) show that sorption is the dominant process at low to neutral pH values, as precipitation would not be sensitive to the solid mass loading. However, the removal of P from the solution phase in experiments at high pH values with the S5 water occurred with both mass loadings, and was associated with the formation of a visible precipitate. The removal of P did not occur in the S1 river water, and appears to be related to the precipitation of $CaCO_3$ (discussed below).

3.4 Geochemical modelling

In the past 20 years, the modelling of aqueous systems has advanced rapidly due to the application of computer models for chemical speciation. The standard approach has been to define a set of components from which the stoichiometry of the system can be uniquely formulated (e.g. H^+, CO_3^{2-}, Cl^-). Each system is described by a number of linear mass balance equations and non-linear mass action equations involving these components (see Morel & Hering [13]). This type of approach permits complex chemical equilibrium problems to be solved iteratively using a computer.

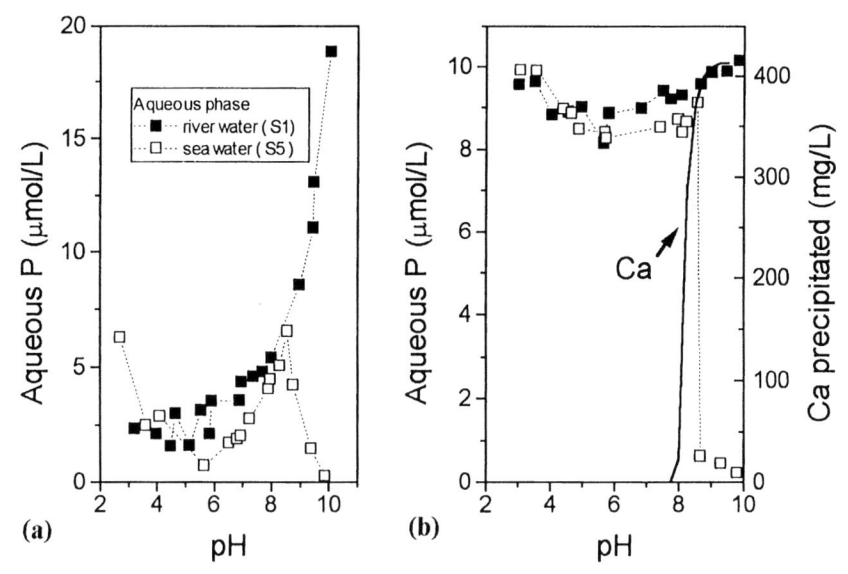

Figure 3. Dissolved P in experiments with added P of 10 µmol/L, and mass loadings of a) 1 g/L and b) 100 mg/L. The removal of P from the aqueous phase at high pH values occurred only in S5 water. This is attributed to sorption or co-precipitation with $CaCO_3$ (solid curve).

Numerous computer codes for calculating the equilibrium distribution of species in geochemical systems have been developed. The geochemical code MINTEQA2 (Allison et al. [1]) was used to examine the reason for the decrease in dissolved P in the batch experiments at high pH values in S5 water. In order to explain the data (Figure 3), a solid phase which becomes supersaturated in S5 water but not S1 water at about pH 8.5 is required. The most likely possibility is $CaCO_3$, which is predicted to precipitate from sea water (S5) but in negligible amounts from the S1 river water (which contains only low levels of Ca). A possible alternative mechanism would involve direct precipitation of a Ca-phosphate mineral such as hydroxyapatite. Although some Ca-phosphate phases were supersaturated under the experimental conditions, none was predicted to become supersaturated in the S5 water within the pH range of interest. Therefore, precipitation would not be limited to the experiments at high pH values in the S5 water. An association of P with precipitating $CaCO_3$ minerals therefore appears to be a reasonable explanation of the data.

The adsorption and co-precipitation of P on calcite has been studied by House & Donaldson [10]. These authors found that the amount of co-precipitated P increased with the precipitation rate of calcite, and did not find evidence for the control of solution compositions by a distinct calcium phosphate phase. This is consistent with the present results. The extent to which this P removal mechanism may operate in nature is uncertain, but it may be expected to become significant when the pH increases above about 8, which can result from photosynthetic activity (Richardson et al. [14]). A recent study (House & Denison [9]) has indicated that changes in sediment composition in an English river were consistent with deposition of calcite and co-precipitation of inorganic phosphate, possibly in association with algal biofilms.

3.5 Sorption modelling

To extend geochemical models to include sorption, simple isotherms (such as the Freundlich or Langmuir isotherms) have been used. However, the more mechanistic surface complexation model (SCM) treats interactions between dissolved species and surface functional groups in a similar way to complex formation with ligands in solution. Each surface reaction is expressed as a reaction involving components, with an associated equilibrium constant (log K). A significant component of the model is the inclusion of electrostatic terms to allow for changes in surface charge due to adsorption reactions. The SCM has been applied to modelling the adsorption of a wide range of inorganic cations and anions on the surface of amorphous Fe-oxide (Dzombak & Morel [5]).

In order to model the data for the pH dependence of P sorption on ferrihydrite (shown in Figure 4), it is necessary to incorporate a suite of surface reactions into the geochemical model. Dzombak & Morel [5] reviewed the available literature for P sorption on ferrihydrite, and proposed the suite of surface reactions in Table 2. These are consistent with a surface area for ferrihydrite of 600 m^2/g and a site density of 0.2 mol/mol Fe (approximately 3.8 $\mu mol/m^2$). When implemented using MINTEQA2, this model provided an acceptable fit to the sorption data in the ferrihydrite / P system (Figure 4). Although there is a slight discrepancy between the model and the experimental data, this may be due to uncertainty in the equilibrium constants. Dzombak & Morel [5] suggested that the 90% confidence interval for the log K for reaction 3 (Table 2) ranged from 17.34 to 18.09. Adjustment of this parameter within this range significantly affects the model curve (Figure 4).

Environmental Coastal Regions 227

Table 2. Surface complexation reactions used in modelling the ferrihydrite / phosphate system

	Surface reaction	Log K^{int}
1.	$>FeOH^0 + PO_4^{3-} + 3H^+ \Leftrightarrow >FeH_2PO_4^0 + H_2O$	31.29
2.	$>FeOH^0 + PO_4^{3-} + 2H^+ \Leftrightarrow >FeHPO_4^- + H_2O$	25.39
3.	$>FeOH^0 + PO_4^{3-} + H^+ \Leftrightarrow >FeH_2PO_4^{2-} + H_2O$	17.72
4.	$>FeOH_2^+ \Leftrightarrow >FeOH^0 + H^+$	-7.29
5.	$>FeOH^0 \Leftrightarrow >FeO^- + H^+$	-8.93

>Fe represents a binding site on the Fe-oxide surface

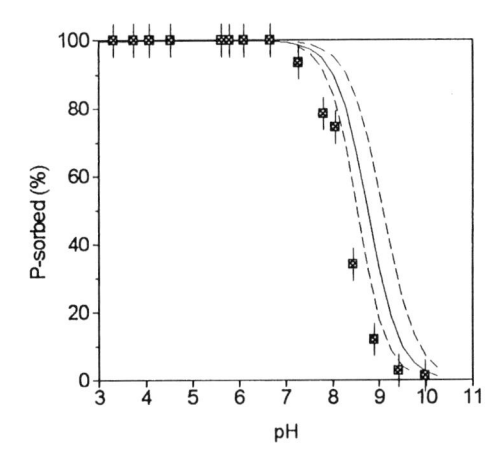

Figure 4. Adsorption of P on ferrihydrite in 0.5 M NaCl. The solid curve was calculated with the data in Table 2. The dashed lines show calculated curves with log K (for reaction 3 in table 2) of 17.34 and 18.09.

The sorption isotherm for the natural substrate in S1 water at pH 8.0 (Figure 5) levelled off at high P concentrations, indicating that site saturation was being approached. The extension of the SCM model to natural materials is a difficult problem due to the complexity of the phenomena and the multiplicity of potential sorbing sites (Davis & Kent [4]). The data for the natural substrate were modelled assuming the surface sites had the same properties as the ferrihydrite sites (using the reactions in Table 2), and using the BET surface area (81.3 m^2/g). The site density was adjusted for the best fit, which was achieved with a site density of 140 µmol/g. This corresponded to a site density of about 1.7 µmol/m^2, which is at the lower end of the range (1.7 - 11.6 µmol/m^2) for geologic substrates proposed by Davis & Kent [4].

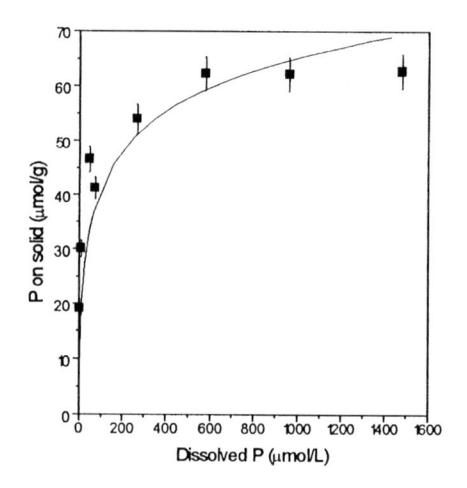

Figure 5. Isotherm for P sorption on Fitzroy River S1 sediments. The curve was calculated with the reactions in Table 2 (site density of 140 µmol/g).

The model fits the sorption isotherm on the natural substrate reasonably well, and is able to qualitatively explain the response of experimental systems to pH changes and the input of ferrihydrite (Figures 2 and 3). However, further work is required to fully understand the natural substrate. In particular, the inclusion of other site types such as those on clay minerals should improve the simulation of the experimental data.

It appears from the experimental and modelling results that that the number of sites on the S1 particle surfaces is in the range 100 - 150 µmol/g. This considerably exceeds the measured P load of the particles (21.8 µmol/g), although it is comparable to the maximum concentrations of adsorbed P reported by Lebo [12] for industrialised estuaries in the northern hemisphere (140-250 µmol/g). This would be consistent with the higher degree of pollution in the estuaries studied by Lebo [12].

Given the excess of sites relative to the P content, it would be expected that the particles may act as a sink rather than a source of P. This is shown by the net adsorption (rather than release) of P in most of the batch experiments (Figure 3a and 3b). Increases in aqueous P only occurred at high pH values, outside the natural range. Similarly, the kinetic experiments (Figure 2) showed an overall decrease in aqueous P.

While the particles tend to adsorb rather than release P there is only a modest decrease in dissolved P by adsorption with 10 μmol/L of added P and a mass loading of 100 mg/L (Figure 3b). As this exceeds the suspended particle load in the estuary (Table 1), a significant decrease in P levels by sorption on suspended particles may not occur without the precipitation of additional Fe-oxides.

In the Fitzroy River estuary, there are inputs of nutrients from sources such as sewage treatment plants and abattoirs, as well as zones of oxygen depletion (Connell *et al.* [3]). Lebo [12] suggested that phosphate may be released from sedimented Fe solids under reducing conditions. Given the high levels of dissolved phosphate at locations S2 and S4 (Table 1), and the measured suspended solid loads, it appears that suspended particles are insufficient to adsorb dissolved P in this part of the estuary.

4 Conclusions

In this study, we have shown the potential of controlled laboratory experiments and geochemical computer modelling to help understand P cycling in the Fitzroy River. The following conclusions may be drawn:

a. In the lower part of the estuary, oxygenation, a pH increase, and mixing increase the amount of suspended Fe-oxides, and $CaCO_3$ may also precipitate. In this region, sorption and co-precipitation on Fe-oxides and $CaCO_3$ are possible mechanisms of P removal from the aqueous phase.
b. The surface complexation model with parameters from Dzombak & Morel [5] adequately simulates P sorption on the Fe-oxide phase.
c. Particles flowing over the barrage from upstream probably provide a net sink rather than a source of P in the estuary. However, sorption on suspended particles appears insufficient to entirely remove dissolved P derived from local inputs within the estuary.

Acknowledgements

The financial support of the Great Barrier Reef Marine Park Authority is gratefully acknowledged.

References

[1] Allison, J.D., Brown, D.S. & Novo-Gradec, K.J., *MINTEQA2, A Geochemical Assessment Model for Environmental Systems,* USEPA, Athens, 1990.

[2] American Public Health Association, *Standard Methods for the Examination of Water and Wastewater*, APHA, Washington, 1985.
[3] Connell, D.W., Bycroft, B.M., Miller, G.J. & Lather, P., Effects of a barrage on flushing and water quality in the Fitzroy River Estuary, Queensland, *Australian Journal of Marine and Freshwater Research,* **32**, pp. 57-63, 1981.
[4] Davis, J.A. & Kent, D.B., Surface complexation modeling in aqueous geochemistry, *Reviews in Mineralogy,* **23**, pp. 177-260, 1990.
[5] Dzombak, D.A. & Morel, F.M.M., *Surface Complexation Modeling - Hydrous Ferric Oxide,* John Wiley & Sons, New York, 1990.
[6] Eyre, B. & Twigg, C., Nutrient behaviour during post-flood recovery of the Richmond River estuary, northern NSW, Australia, *Estuarine, Coastal and Shelf Science,* **44**, pp. 311-326, 1997.
[7] Fox, L.E., A model for inorganic control of phosphate concentration in river waters, *Geochimica et Cosmochimica Acta,* **53**, pp.417-428, 1989.
[8] Froelich, P.N., Kinetic control of dissolved phosphate in natural rivers and estuaries: A primer on the phosphate buffer mechanism, *Limnology and Oceanography,* **33**, pp. 649-668, 1988.
[9] House, W.A. & Denison, F.H., Nutrient dynamics in a lowland stream impacted by sewage effluent: Great Ouse, England, *The Science of the Total Environment,* **205**, pp. 25-49, 1997.
[10] House, W.A. & Donaldson, L., Adsorption and coprecipitation of phosphate on calcite, *Journal of Colloid and Interface Science,* **112**, pp. 309-324, 1986.
[11] Hunter, H.M. & Rayment, G.E., Agricultural contaminants in aquatic environments - an overview, *Proc. of Workshop on Land Use Patterns and Nutrient Loading of the Great Barrier Reef Region,* ed. D. Yellowlees, James Cook University, Queensland, pp. 53-66, 1991.
[12] Lebo, M.E., Particle-bound phosphorus along an urbanised coastal plain estuary, *Marine Chemistry,* 34, pp. 225-246, 1991.
[13] Morel, F.M.M. & Hering, J.G., *Principles and Applications of Aquatic Chemistry,* Wiley-Interscience, New York, 1993.
[14] Richardson, L.L., Aguilar, C. & Nealson, K.H., Manganese oxidation in pH and O_2 microenvironments produced by phytoplankton, *Limnology and Oceanography,* **33**, pp. 352-363, 1988.
[15] Waite, T.D., Davis, J.A., Payne, T.E., Waychunas, G.A. & Xu, N., Uranium (VI) adsorption to ferrihydrite: application of a surface complexation model, *Geochimica et Cosmochimica Acta,* **58**, pp. 5465-5478, 1994.

Acoustical response of phytoplanktonic volume scatterers at ultrasonic frequencies as an indicator of pollution in sea waters

Silvia Blanc[1], Patricia Mosto[2], Carlos Benitez[1], Ricardo Juárez[1], Marta Milou[1], Gustavo Lascalea[3]

[1] Research and Development Naval Service, Av. Libertador 327, 1638 Vte. López, Argentina; [2] Rowan University of New Jersey, NJ 08028, USA; [3] PRINSO-CITEFA, Zufriategui 4380, 1603 Vte. López, Argentina

Abstract

An experimental method to analyse the feasibility of acoustically determining numerical abundance (N) of phytoplanktonic organisms is presented. When unialgae cultures of diatoms and dinoflagellates are insonified, they intercept and reirradiate a fraction of the incident energy producing Acoustic Volume Scattering. An attempt to use this phenomenon as an alternative method for getting N determination of phytoplanktonic sound scatterers, is being carried out. The final objective of the proposed technique is to correlate the algae acoustic scattering parameters to sea waters pollution levels. Its theoretical foundations are based on the assumption that the total scattered intensity from a volume with a random distribution of scatterers is equal to the sum of the scattered intensities from each individual. From this hypothesis it follows a simple equations system linearly connecting the Volume Backscattering Coefficient (σ_V) and the Volume Scattering Cross Section (σ), which can be solved for N determination. In order to obtain experimental σ_V values, piezoelectric transducers with 2.6 MHz emission frequency, wide band mini-hydrophones for signals reception and the associated electronic circuits, were developed. Significant scattering effects were detected. Acoustic signals were acquired by a digital oscilloscope and transferred to a PC for further processing. From theoretical models, σ values could be computed for these insonified species. From the obtained experimental σ_V values and the predicted σ ones, N determination is being carried out.

1 Introduction

Whenever an acoustical signal propagates within the sea, it meets a great variety of inhomogeneities that imply discontinuities in the physical properties of the medium. These inhomogeneities intercept and reradiate a fraction of the acoustical incident energy producing sound volume scattering. This phenomenon has long been used as a relatively reliable procedure for the detection of zooplankton groups as well as for studying their spatial distribution and classification according to their different size and physical properties [1,2,3,4]. However, as far as we are aware no attempt has been done with the aim of monitoring sound volume scattering parameters from insonified phytoplankton as an indicator mechanism of pollution levels at coastal sea waters. For this final purpose a research programme has been initiated in the sense of accomplishing a rigorous theoretical-experimental approach to the determination of numerical abundance of phytoplankton sound volume scatterers through indirect acoustical measurements.

At the first stage there is a great need to conduct controlled at laboratory measurements as a first step to provide an experimental data frame for further models predictions validations. Experiments with unialgae cultures of diatoms and dinoflagellates have been planned due to their well known correlation with marine ecosystem integrity.

In this paper it is summarised the progress performed in the development of an alternative technique to correlate the acoustical response of phytoplankton, when insonified by piezoelectric transducers with ultrasonic emission frequencies, to coastal sea water pollution.

2 Experimental technique

2.1 Biological cultures

Phytoplankton organisms (algae) are used as biological indicators of coastal marine pollution due to their rapid response to water quality changes and also because they are at the bottom of the food web. In coastal waters, dinoflagellates, in particular, certain species of _Gymnodinium, Gonyaulaux, Amphidinium and Peridinium_ play an important role as ecosystem indicators of polluted waters. On the other hand, diatoms, such as _Skeletonema costatum_ may be associated to clean waters.

Environmental Coastal Regions   233

Traditional estimates of phytoplankton abundance held on the basis of quantitative counts in collected samples, using planktonic nets or pumps, lead to tedious and time consuming processes with a non-neglectable time lag after the sampling is performed.

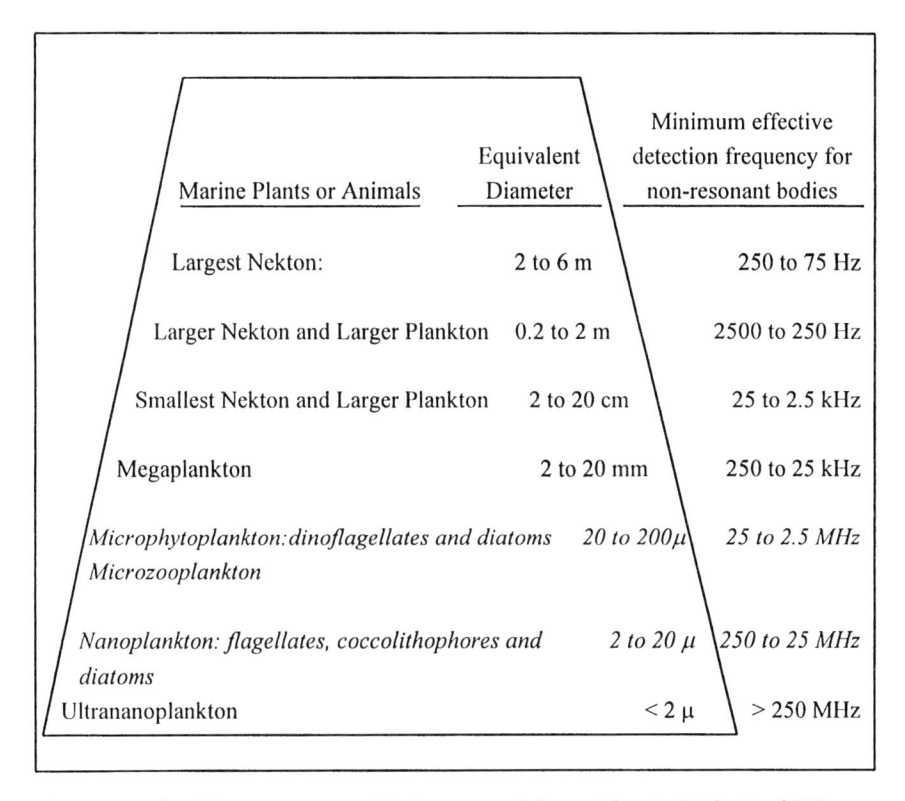

Marine Plants or Animals	Equivalent Diameter	Minimum effective detection frequency for non-resonant bodies
Largest Nekton:	2 to 6 m	250 to 75 Hz
Larger Nekton and Larger Plankton	0.2 to 2 m	2500 to 250 Hz
Smallest Nekton and Larger Plankton	2 to 20 cm	25 to 2.5 kHz
Megaplankton	2 to 20 mm	250 to 25 kHz
Microphytoplankton: dinoflagellates and diatoms *Microzooplankton*	20 to 200 μ	25 to 2.5 MHz
Nanoplankton: flagellates, coccolithophores and diatoms Ultrananoplankton	2 to 20 μ < 2 μ	250 to 25 MHz > 250 MHz

Figure 1. The biomass pyramid (extracted from Fig. 1.6.1 in Ref.[5])

(1) The equivalent diameter for detectability of a single non-resonant body is the diameter of the equivalent spherical volume of the organism. (2) The minimum effective detection frequency for non-resonant bodies is the frequency below which sound diffraction makes the body appear smaller than it is. Substantially reduced backscatter (Rayleigh scatter) occurs when the sound wavelengths in water is greater than the equivalent circumference. For this column we used the criterion $f > c / (\pi \text{ diameter})$ for Rayleigh scattering.

234   Environmental Coastal Regions

In view of our interest in acoustically determining phytoplankton numerical abundance (N), it was apparent that ultrasonic frequencies in the 2.5 - 250 MHz range would be necessary (Figure 1).

Figure 2 shows an image taken from an unialgae culture of *Skeletonema costatum* used during the first attempts to detect scattering effects that might be expected specially due to their siliceous cell walls when insonified with a 2.6 MHz emission frequency. In order to have available unialgae cultures of this specie in log-to-stationary phase for being acoustically monitored, when necessary, through the measurements, they were at first obtained from The Culture Collection of Algae at the University of Texas at Austin. Culturing methodology has mainly consisted of growing the algae in the ES-Enrichment sea water media[6] in 250 ml erlermeyers, on a shaker at 16:8 hours light : darkness cycle, under cool-white fluorescent light (400 feet candles) at 20 °C.

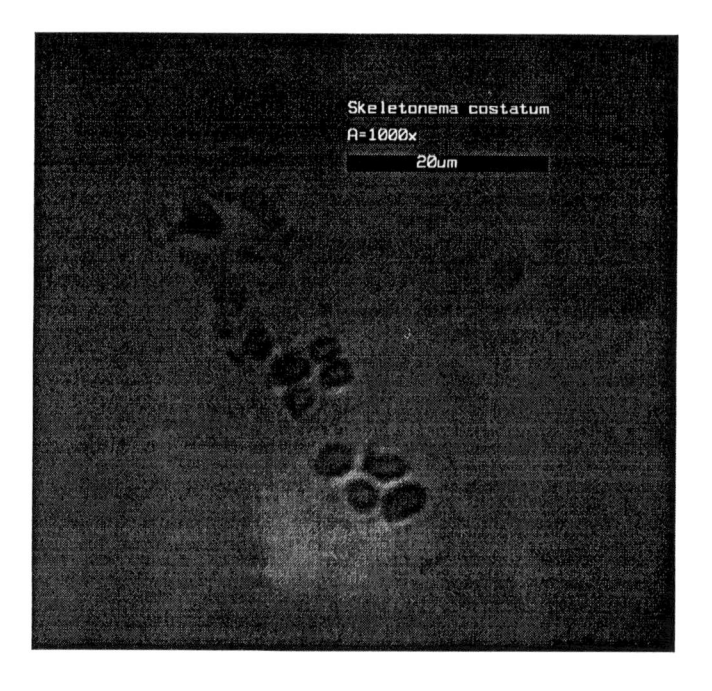

Figure 2. An image from the culture of the selected diatom specie (*Skeletonema costatum*) for the at laboratory acoustical measurements.

2.2 Experiment design

Because of the limited availability of culture media, their sonorization was conducted in a half full 1000 ml precipitate glass. Non-commercial, but "ad-hoc" piezoelectric transducers were developed for the experiment. The highest emission frequency that could be obtained with relatively simple manufacture techniques of piezoelectric transducers was 2.6 MHz, with the corresponding narrow beam of sonorization for the projector, 5° beamwidth at -3 dB. It was tuned with the objective of getting the maximum power since the expected backscattered signals in the receiver were very weak. Two closely spaced transducers were used (nearly monostatic geometry), one as a transmitter and one as a receiver, to get backscatter signals.

For our purposes two transmitter-receiver pairs were constructed. Each transmitter transducer contained a 15-mm-diam. flat ceramic with a resonance frequency of 2.6 MHz. Backing materials were added so that the ceramic would be damped to avoid undesirable ringing effects. With the objective of allowing backscatter signals reception, less directional (20° at -3 dB) mini-hydrophones containing 3 mm-diam flat ceramic material were also constructed.

A pulse-echo electronic equipment was designed and constructed for the acoustic measurements. Bursts of sound of 20 cycles at a 1 Hz repetition rate with 7.69 µs pulse-length, generated by a Hewlett Packard 8165 A waveform generator, were injected to the transmitting transducer through the driver circuit. Since some spurious low frequency emissions modes could not be completely avoided, the detected echoes at the receiver were amplified and filtered with a 2 MHz high-pass filter. Data acquisition was held with a Tektronix THS 710 digital oscilloscope and transferred to a PC for further processing. Figure 3 schematically shows the whole experiment design.

Calibration of each transmitter-receiver pair was performed dipping them into a vessel containing distillate water aiming them toward the water-air interface and measuring emitting electric power (18 watts) and the corresponding rms echo voltage. Since there were no available high frequency calibrated hydrophones, at this first stage of the programme measurements, a 30-50 % efficiency range for the transducers was assumed, which is a reasonable estimation in the case of backed tuned transducers. It was obtained an approximate sensitivity of 72.8 µv/Pa with an uncertainty of about 12 %.

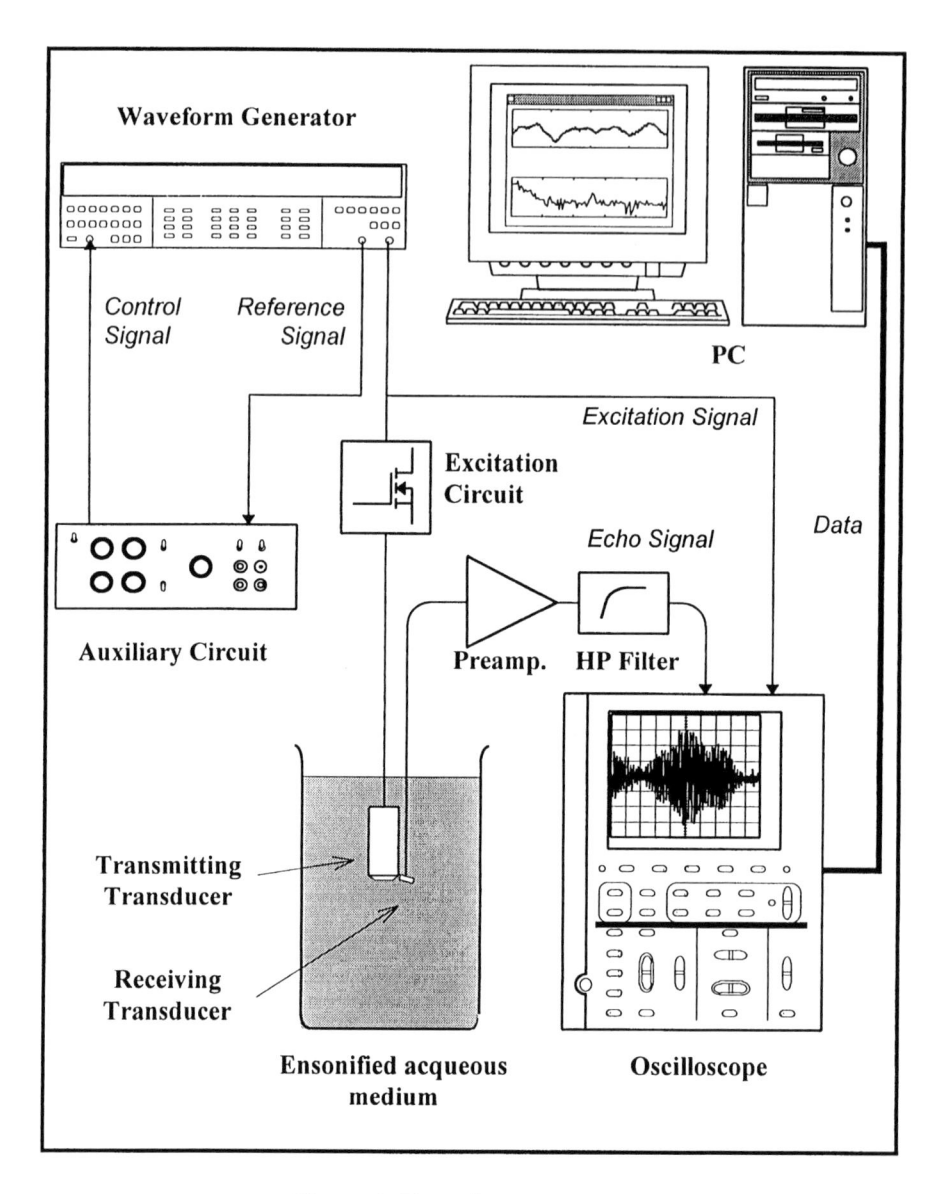

Figure 3. Experiment geometry.

3 Theoretical foundations revision

One of the usual parameters used to describe the ability of a medium to scatter acoustic waves in the direction back toward the source is the volume backscattering strength defined as the ratio between the scattered intensity from a unit volume containing scatterers, referred to a 1 m distance from the volume, I_s; and the incident plane wave intensity on the volume, I_i. In symbols,

$$s_V = I_s / I_i \qquad (1)$$

Another parameter also used to describe sound volume backscattering is the volume backscattering coefficient σ_V, defined as,

$$\sigma_V = P_s / I_i \qquad (2)$$

where P_s is the acoustic scattered power per unit volume and I_i is the incident plane wave intensity. The units of σ_V are m^{-1}. Under the hypothesis of isotropic sound scattering,

$$s_V = \sigma_V / 4\pi \qquad (3)$$

Since the well known superposition principle is assumed, the total scattered intensity from a volume containing a random distribution of scatterers is, on average, equal to the sum of the scattered intensities from each individual. Therefore, a linear equation can be expressed in terms of the volume backscattering coefficient, σ_V, and the backscattering cross-section, σ,

$$<\sigma_V (f,r)> = N(r)\ \sigma(f, a_0) \qquad (4)$$

for any distance r from the sound source, when sonorization is held at an unique f frequency and a single size class scatterer with a known size a_0, dominates the acoustic scattering. N(r) is the numerical abundance per unit volume. The symbol < > denotes average values that account for an aggregation of phytoplankton moving scatterers in our case.

As it has been already reported in the literature[5], acoustically small, non-spherical bodies whose dimensions are less than the wavelength have a sound scattering behaviour equivalent to actual spheres of the same volume and same average physical properties. Since Rayleigh first derived the backscattering cross-section for a non-resonant sphere in 1896, some models have been developed[3,7,8] to solve this problem with different orders of approximation and with consequent different degree of complexity in the involved mathematics. However, in the range k a << 1 (k, is the wavenumber and a, is the sphere radius) all the presented algorithms can be reduced to Rayleigh first-order approximation,

$$\sigma / \pi a^2 = 4\ (ka)^4\ [(1-gh^2)/(3\ g\ h^2) + (1-g)/(1+2g)]^2 \qquad (5)$$

where g and h are the density ratio ρ'/ρ and sound speed ratio c'/c of the sphere to the medium. Values of g have been reported in the literature for this organism[9]; compatible values of h were estimated from the wave compressional velocity-density relationship[5].

Experimental values of $<s_V>$ (or $<\sigma_V>$) have been obtained from the average echo rms acoustic pressure measured at the receiving transducer (p_h), taking into account range-dependent losses due to the spreading along the propagation path, equivalent two-way beam width of the transducer pair (Ψ), pulse length (τ), and acoustical rms pressure in the centre of the measuring window (p_w).

$$<s_V> = 2\, p_h^2 / (p_w^2\, c\, \tau\, \Psi) \qquad (6)$$

Accordingly, from Equations (3), (4), (5) and (6) it follows that the numerical abundance N can be acoustically estimated from computed experimental values of $<s_V>$ and predicted values of σ.

$$N = 4\pi <s_V> / \sigma \qquad (7)$$

4 Results and conclusions

Experimental technique as described in Section 2 allow us to report the first achievements in the study of the sound scattering behaviour of phytoplankton individuals. Some illustrative results are shown in Fig. 4.

Computed acoustic pressure from measured echo voltage at the receiving transducer, are plotted versus time in Fig. 4 (a). The recorded signal shows the direct emitted burst with its corresponding ringing effects, along the first approximate 20 µs. Its echo from the water-glass vessel bottom interface is located around the 60 µs. Signal portion recorded between 30 µs and 50 µs has been selected for sound volume scattering phenomenon analysis. Figure 4 (b) and (d) show a zoom image of the selected time window for distillate water and for an unialgae culture of *Skeletonema costatum* insonified at 2.6 MHz, respectively. Figure 4 (c) and (e) show the corresponding spectrum levels (averaged over ten spectrums) for the signals showed in (b) and (d). It can be noted that the acoustic power value obtained for the algae culture (Fig. 4 (e)) is 50 times greater than for distillate water (Fig. 4 (c)) for the emission frequency spectral component. Scattering effects due to phytoplankton presence could be clearly observed.

Direct optical determination of numerical abundance of the insonified unialgae culture was held with a Leco-300 microscope. Equation (7) was used for indirect acoustical determination of the same parameter. Both results lead to values of N in the range of (1.6 - 7) 10^{12} cells/m^3. The degree of agreement between the estimations

Environmental Coastal Regions 239

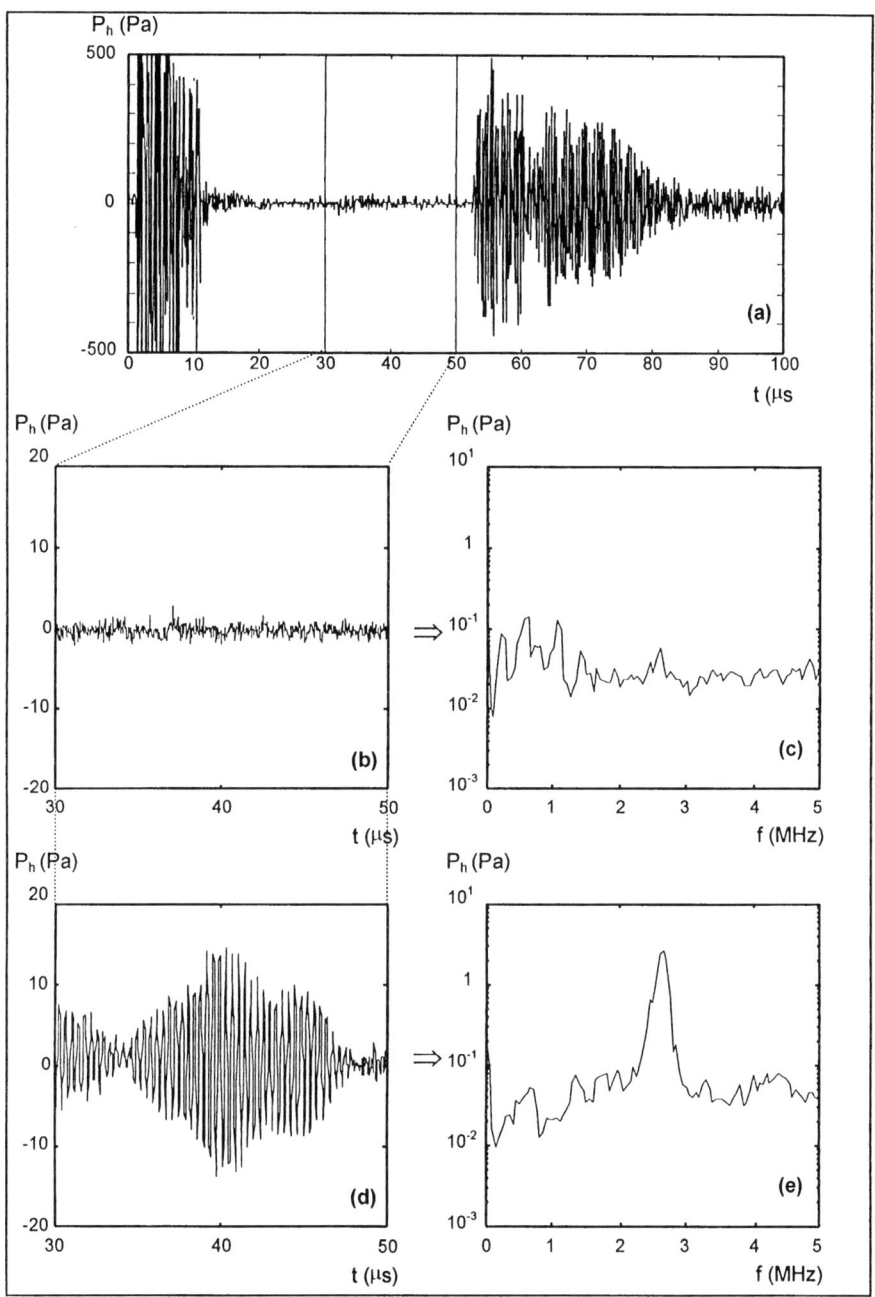

Figure 4. (a) Acoustic pressure at the receiving transducer vs. time; (b) Zoom image of the selected time window for distillate water; (c) Average spectrum level of (b); (d) Analogous to (b) for the *Skeletonema costatum* culture; (e) Average spectrum level of (d).

obtained by both methods suggests the feasibility of the proposed alternative acoustic method to evaluate the numerical abundance.

Much work remains to be done in this field as regards to the involved experimental measuring procedures, the signal processing techniques, and the theoretical approach. However, this is a foundation for proceeding to the next step in a further more rigorous association of phytoplankton acoustic signatures with coastal sea water pollution levels.

References

[1] Greenlaw, Ch. F., Acoustical estimation of zooplankton populations. Limn. Oceanogr. 24 (2), pp. 226-242, 1979.

[2] Holliday, D. V., Volume scattering strengths and zooplankton distributions at acoustic frequencies between 0.5 and 3 MHz. J. Acoust. Soc. Am. 67 (1), pp. 135-146, 1980.

[3] Kristensen, A. and Dalen, J., Acoustic estimation of size distribution and abundance of zooplankton. J. Acoust. Soc. Am. 80 (2), pp. 601-611, 1986.

[4] Stanton, T. K. et al. Sound scattering by several zooplankton groups. I. Experimental determination of dominant scattering mechanisms. J. Acoust. Soc. Am. 103 (1), pp. 225-235, 1998.

[5] Clay, C. C. and Medwin, H. *Acoustical oceanography: Principles and Applications.* Ed. by John Wiley & Sons, pp. 24, 1977.

[6] Provasolli, L., Mc Laughin, J. and Droop, M. The development of acquisition media for marine algae. Archives Mikrobiology, v25, 392-428, 1957.

[7] Johnson, R. Sound scattering from a fluid sphere revisited. J. Acoust. Soc. Am. 61, 375-377, 1977.

[8] Stanton, T. K. and Chu, D. Sound scattering by several zooplankton groups. II. Scattering models. J. Acoust. Soc. Am. 103 (1), pp. 236-253, 1998.

[9] Lee, R. *Phycology*. University of Cambridge Pub. pp. 645, 1995.

Assessment of natural radioactivity in the marine environment in Croatia

G. Marovic, Z. Franic and J. Sencar

Institute for Medical Research and Occupational Health, Radiation Protection Unit
HR-10000 Zagreb, Ksaverska cesta 2, PO Box 291,
CROATIA
E-mail: marovic@imi.hr

Abstract

Investigations of natural radioactivity, particularly radium and uranium, in the Adriatic Sea water and biota have already been performed on selected locations along the Croatian coast, as a part of an extensive monitoring program of the Croatian environment, conducted by the Radiation Protection Unit, Institute for Medical Research and Occupational Health, Zagreb. The paper deals with increased levels of natural radioactivity in a bay at the Croatian cost, which is due to geographic characteristics exposed to any kind of pollution including radioactivity originating from the coal fired power plant situated about 5 km from the seaside. Slag and ash are accumulating continuously, consequently to regular operation of the power plant. Previous investigations of used coal resulting slag, showed increased activity concentrations of natural radioactivity. In order to assess the sensitivity of the bay to the radioactive pollution, natural radioactivity concentrations were studied in the zone between the plant and the seaside, as well as in sea water. ^{226}Ra concentrations in the sea water were found to be increased in comparison to ^{226}Ra levels measured at the selected open sea locations.

1 Introduction

The central role of energy in economic and social development has long been recognized and a great deal of effort has been devoted to developing technologies for the extraction, production, and use of all types of energy. The aim has always been to reduce costs, make systems more efficient, and provide access to previously untapped energy sources. Since electrical energy production is always associated with the emission of several pollutants into the atmosphere, reducing the adverse environmental and health effects have become the questions of public and scientific priority in recent decades. [1,2]

The natural radioactive series of uranium and thorium with their decay products in secular equilibrium are always present in coal. By burning coal, the activities have been redistributed, taking them from underground where the impact on living environment was none and liberating them into air and over the surface of the ground where it can modify ambient radiation fields and population radiation exposure. Several coal mines in Croatia have more or less elevated concentrations of natural radioactivity. The presence of uranium varies from mine to mine and even from one coal-layer to another in the same mine.[3]

The paper deals with increased levels of radioactivity in the bay at the Croatian cost of the Adriatic Sea, which is due to its geographic location exposed to any kind of pollution including the radioactivity emitted from the coal fired power plant situated about 5 km from the seaside. The coal used for the regular plant operation originates from a domestic coal mine, which in addition to higher levels of natural radioactivity (uranium) has an elevated concentration of sulphur, up to 10%.[4] Coal has been mined mostly at adjacent coal mine area, about 10 km from the plant site. Over the past few years the coal used in the plant includes about one half of imported coal characterized by different radioactivity characteristics. The aim of this study was to determine radioactivity level at the ash and slag deposits and to assess the risk to the Adriatic Sea water and biota. Investigations of natural radioactivity have already been performed at selected locations along the Croatian coast, as a part of an extensive monitoring program of the Croatian environment conducted by the Radiation Protection Unit, Institute for Medical Research and Occupational Health, Zagreb.

2 Description of the Area

The plant is situated in the coastal area, about 5 km from the seaside, and 10 km from the more populated urban area in the continental part of the region. The plant is surrounded by the hills with sparsely dissipated villages. A more populated urban center is situated at the seaside. The direct combustion of coal, or its conversion to other fuels, results in the concentration of the noncombustible mineral matter, including most of the radionuclides, in the ash or gaseous residues. Besides the immediate surrounding area of the plant, radioactively polluted material affects also the zone between the plant and the seaside, as well as submarine area of the bay. In view of this, the location of the deposit sites, i.e. the storage of large quantities of ash and slag, presents a remarkable environmental problem.

The coal ash and slag deposit site is situated close to the power plant site. Ash and slag are accumulating continuously, consequently to regular operation of the power plant. There have been performed some measurements of natural radiation in marine in vicinity of coal mine and power plant, as well as in ashes generated at power plant. After incineration the initial radioactivity becomes 7-10 times more concentrated than in coal. The particulates and the solid incombustible ash containing uranium and ^{226}Ra together with sulphur bound to the mineral substance, are here the main problem.

The slag and ash pile is situated on low permeable Eocene flysh. However, this sediment could be very easy weathered and affected by proluvial process. Geological setting of the broader area is characterized by Mesozoic carbonates (limestone, dolomites) which are due to irregular circulation of groundwater environmentally very sensitive and, thus, the monitoring of radionuclides which would possibly migrate in the groundwater from this pile, is not easy.

3 Material and Methods

Investigations of radioactive contamination were based on measurements performed in the field and in the laboratory.

In situ gammaspectrometrical measurements were carried out using HPGe Ortec detector (resolution 1.74 keV on 1.33 MeV ^{60}Co, relative efficiency 21.6%), and included several measuring points at the protected and still operating part of the deposit and the seaside control points.

Determination of radioactive contamination required analyses of different types of samples: the samples of sea water, waste water from the plant at the point of mixing with sea water, and sea bioindicators The collected samples of sea bioindicators were gammaspectrometrically analysed in the laboratory using Ge(Li) Ortec detector (resolution 1.78 keV on 1.33 MeV ^{60}Co, relative efficiency 16.8%) with a 4,096 channel analyser and a personal computer. All samples were measured in Marinelli beaker, volume 1 L. Measurement time was 80,000 sec.[5]

The liquid samples were radiochemically separated, after which ^{226}Ra was determined by alpha spectrometric measurements using Ortec Si(Li) surface barrier detector. The counting time for each measurement was 60,000 sec or longer.

4 Results and Discussion

In situ gammaspectrometrical measurements involved several measuring points at the deposit site (still operating part and covered part of the deposit site) and at several control points at the seaside. At the still operating part of the deposit site the measurements showed the presence of natural radionuclides of uranium and thorium decay series, ^{40}K and ^{137}Cs. Corresponding contribution of measured radionuclides to the absorbed dose rate was calculated for each measuring point. The results of measurements at 20 measuring sites indicated a great dispersion of the obtained absorbed dose rates. Over the recent years was used the mixture of domestic and imported types of coal, characterized by the different radioactivity levels that resulted in varied radioactivity values obtained at the operating part of the ash and slag deposit site.

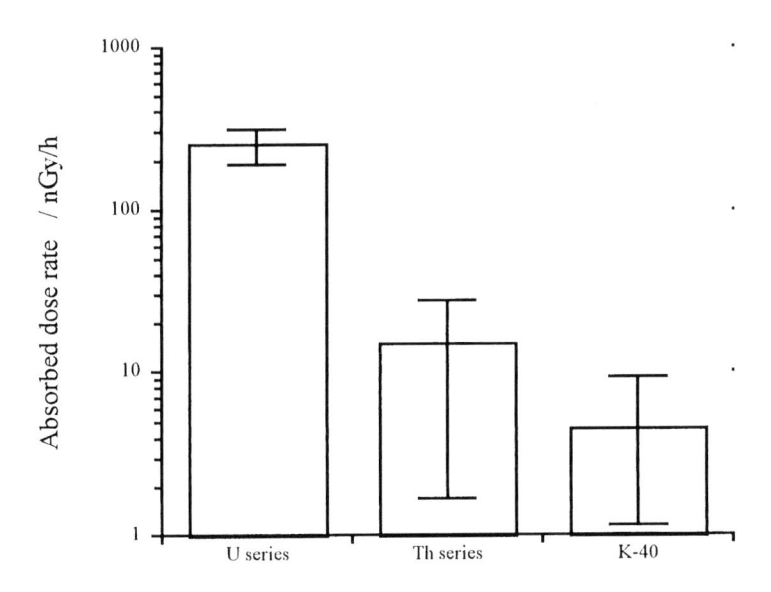

Figure 1: Contributions of radionuclides of uranium and thorium decay series and ^{40}K to the absorbed dose rate at the operating part of the ash and slag site

Figure 1 shows contributions of radionuclides of uranium and thorium decay series and ^{40}K to the absorbed dose rate of 20 measuring points at the operating-deposited ash and slag site.

In order to determine the efficacy of the protective layer at the soil-covered part of the deposit site, in situ gammaspectrometrical measurements were carried out at several measuring points. For each measurement was calculated corresponding contribution of measured radionuclides to the absorbed dose rate. The calculated absorbed dose rates at some measuring points of the operating ash and slag deposit were up to five times higher than the average calculated absorbed dose rate at the covered part of the deposit site (177±40 nGyh^{-1}). The statistical analysis of the difference between the absorbed dose rate at the operating pile and protected part of the deposit site showed statistically significant difference of P(t)<0,01 which indicates the satisfying efficacy of the protective covering of the deposit site.

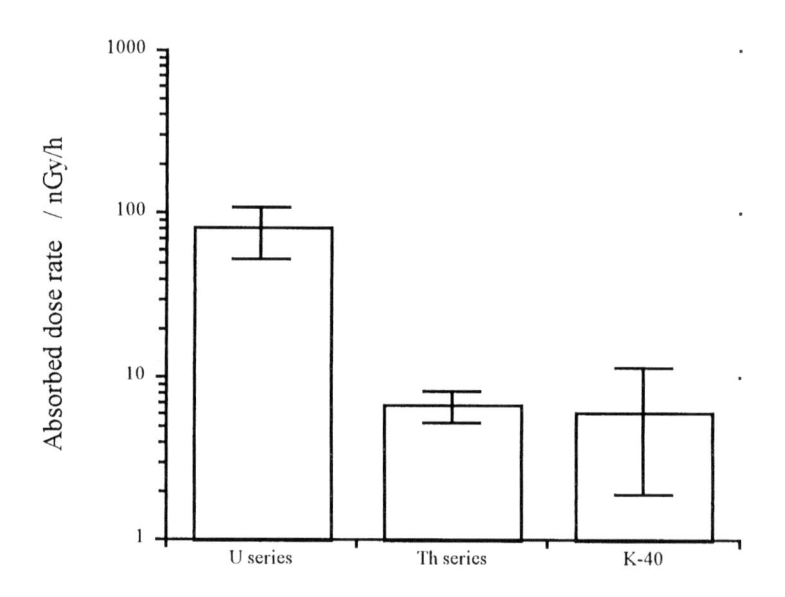

Figure 2: Contributions of radionuclides of uranium and thorium decay series and ^{40}K to the absorbed dose rate for several location at the seaside

Figure 2 shows the average absorbed dose rates for several locations at the seaside. There is a noticeable difference between the still active part of the deposit site and the results obtained at the seaside which correspond to the absorbed dose rate measured in other parts of Croatia.

In the samples of bay water, the mean ^{226}Ra concentration of 5.2±0.9 Bqm^{-3} was within the same order of magnitude with ^{226}Ra concentration in the Mediterranean Sea (3.7 Bqm^{-3})[6], and as well as in the open Adriatic Sea far from the plant deposit bay.[7]

Figure 3 shows ^{226}Ra specific activities determined in the samples of sea water (in the bay) and waste water from the plant at the point of mixing with the bay water. ^{226}Ra activity concentrations decrease with increasing distance from the mixing point.

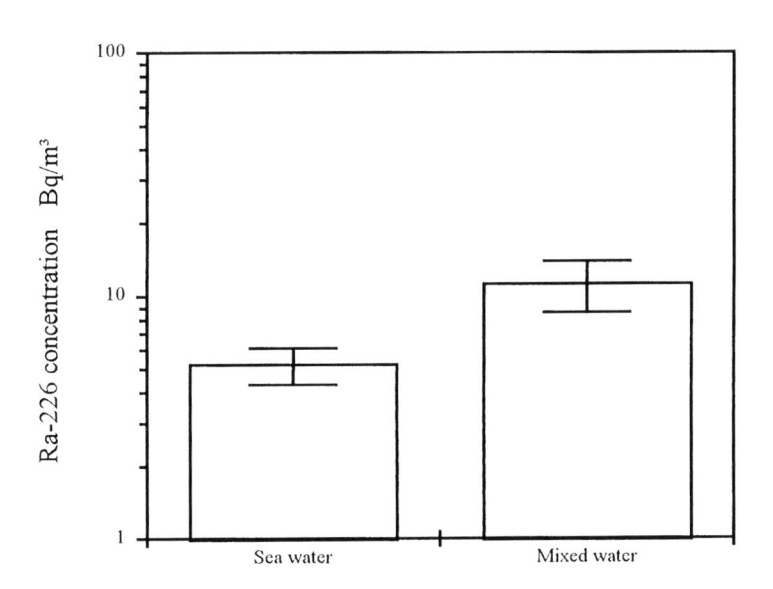

Figure 3: ^{226}Ra activity concentrations in the samples of sea water and waste water at the mixing point with the sea water

The samples of mussels were collected at the seaside as most suitable indicators of a sea biota. Gammaspectrometrical analyses carried out to detect possible accumulation of natural radionuclides in sea organisms showed that all values except ^{40}K were below the detection limit of the instrument.

5 Conclusion

On basis of our study and the obtained data it can be concluded that the investigated power plant and its deposit site present no significant risk to the inhabitants and the environment of the region. This is due to the geographical location of the plant in a sparsely inhabited area, relatively distant from more populated and a tourist region of the seaside.

The results of the measurements also confirm that the ash and slag deposit site of both the protected and still operating part of the deposit site are well monitored and involve all the necessary protective measures. All obtained data can be used as a valuable database for future estimations and modeling of the impact of radioactive pollution to the marine environment and developmental prospects of the region.

References

[1] Bauman, A. & Horvat, Dj., The impact of Natural radioactivity from a coal fired power plant, *The Science of the Total Environment*, **17**, pp.75-81, 1981.

[2] Bauman, A., Horvat, Đ., Kovač, J. & Marović, G., Risk from exposure to radiation and chemical agents at a coal fired power plant, *Proc: Comparison of risk resulting from major human activities, Xth Regional Congress of International Radiation Protection*. Avignon, pp. 173-180, 1982.

[3] UNSCEAR (United Nations Scientific Committee on the Effects of Atomic Radiation). *Sources and Effects of Ionizing Radiation.* UNSCEAR, United Nations, New York, 1993.

[4] Marović, G., Maračić, M., Lokobauer, N. & Bauman, A., Technologically enhanced natural radiation in coal fired power plants, *Proc. of the Italian-Yugoslav Radiation Protection Association Symposium*, Udine 1988, ENEA, Rome, pp. 349-352, 1989.

[5] IAEA, *Measurements of radionuclides in food and the environment*, Tech. Rep. Ser. No. 295, Vienna, 1989.

[6] UNEP. *Assessment of the State of Pollution in the Mediterranean Sea by Radioactive Substances.* UNEP (OCA) MED WG. 25 Athens, 1991.

[7] Marović, G., Kovač, J. & Franić, Z., Impact of technologically enhanced natural radioactivity on marine environment in Croatia. *Proc. of The Second Regional Mediterranean Congress on Radiation Protection*, Tel-Aviv, pp. 291-294, 1997.

Eutrophication modelling of a tidally influenced mangrove area in Bali subject to major dredging and reclamation activities

Hanne K. Bach, Erik Kock Rasmussen & Tom Foster
EMC, Agern Allé 5, DK-2970 Hørsholm, Denmark
Email: hkb@vki.dk; ekr@vki.dk;tmf@dhi.dk;

Abstract

A large fashionable tourist resort area is under construction at Benoa Bay, Bali. The location is an island called Turtle Island, which is an enlargement of the existing Serangan Island. Extensive dredging operations of surrounding sea bed are needed for land reclamation, resulting in a general change in current pattern and water exchange for the Benoa Bay and the strait between the coast and Turtle Island. Prediction of the water quality before and after construction was accomplished in order to find the preferable of three layouts from an environmental point of view.

The existing Serangan Island is relatively small, with no tourism and inhabited by a small number of local people. The shallow waters around the island are covered by benthic vegetation. A coral reef to the east of Serangan Island forms a border to the ocean.

The water quality in terms of benthic vegetation, concentrations of phytoplankton, nutrients, dissolved oxygen and water turbidity was modelled using the MIKE 21 eutrophication model.

The water quality was predicted to improve in the strait between the coast and the island due to increased flushing of the dredged channels. On the other hand, the improved flushing resulted in a larger spreading of the nutrients and deterioration of the water quality north and south of the proposed construction site. Lagoons proposed for leisure crafts also showed a deterioration of the water quality. For a lagoon close to a conservation area of the reef a large decrease in submerged vegetation was predicted. One layout was finally recommended as the optimal combining the various modelling results and results of other related studies.

1 Introduction

Tourism is seen as a sector with a high potential for development in many places around the world. This includes South-East Asia, where warm climate, beautiful beaches and an exotic and varied underwater flora and fauna e.g. coral reefs make the coastal area particularly attractive to tourism. Exploitation of these resources may involve large reclamation operations providing the land for resorts etc. This creates immediately a potential conflict between the expectations of the tourists for undisturbed nature and the detrimental effects of the dredging and reclamation.

Environmental impact assessments are required for such activities in most cases. One way of overcoming the conflict between disturbance and preservation of the environment is to design a reclamation with a minimum of environmental impact by e.g. using models for prediction of the impact (Bach et.al. [1], Brøker et.al. [2]); and to mitigate impacts of the dredging operations by planning and applying feedback monitoring to control the effects during the construction phase (Gray & Jensen [3]).

This paper describes the methods and outcome of a specific part of the environmental impact assessment for a reclamation project at Bali namely the impact on water quality, phytoplankton and the benthic vegetation. Considerations concerning hydraulic impacts, morphology and sediment spreading in the construction phase are described elsewhere (Driscoll et.al. [4]).

2 Materials and Methods

2.1 Study Site

The tourism development area will be located at an island called Turtle Island, which is an enlargement of an existing Serangan island situated at the entrance to Benoa Bay, Bali, Indonesia (Figure 1). The existing Island is relatively small, with no tourism and inhabited by small number of local people. The shallow waters around the Island are inhabited by macrophytes. A coral reef to the east of Serangan Island forms a border to the ocean. The water on the coral reef flats and slopes is clear, has low nutrient levels, and is well saturated with respect to oxygen. This area seems largely unaffected by plumes from Benoa Bay. From aerial photos it appears that the benthic vegetation dominated by seagrass (Hunting [5]) is concentrated in many small areas with high density, surrounded by areas with almost no benthic vegetation.

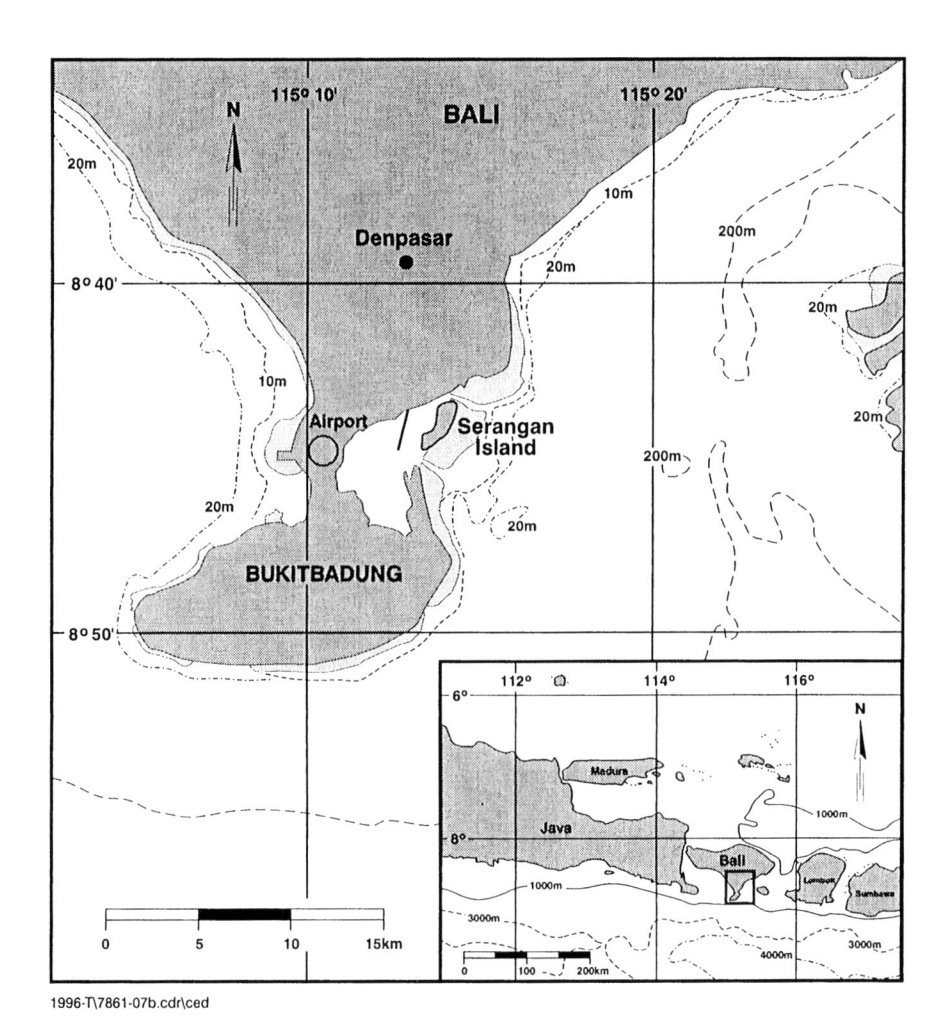

Figure 1. The study site including the southern part of Bali and the Serangan Island.

Benoa Bay is a tidally influenced shallow lagoon (25-30 km^2) covered by mangroves in the innermost parts and with some patchy distributed submerged rooted vegetation in the gullies. The tidal falts dry out at low water. Channels that never dry traverse the bay. On tidal flats random distributed spots of macroalgae occur. The water in the bay is slightly turbid and under-saturated with respect to oxygen (Hunting [5]). Total nitrogen (N) and phosphorus (P) concentrations are in the order of 1.3 mg N/l and 0.3-0.6 mg P/l, and 5-27 mg/l BOD$_5$.

Extensive dredging operations of surrounding sea bed is needed for the planned land reclamation, which is expected to result in a general

254 Environmental Coastal Regions

change in current pattern and water exchange for the Benoa Bay and the strait between the coast and Turtle island. Three different layouts were investigated. The size and shape of Turtle Island is the same for all three layouts (Figure 2). Layout #1 has borrow areas north of Turtle Island (area A) and a short channel half of the maximum length in the area between the island and the mainland until the bridge crossing over to Turtle Island. Layout #2 is as Layout #1, but with an additional borrow area between the Turtle Island and the mainland (area B1) and a channel at full length down to the south channel and Benoa Port. Layout #3 is as Layout #2, but with construction of a large harbour in the area between the Turtle Island and the mainland (extension of Benoa Port). Turtle Island is planned to have three artificial lagoons for leisure crafts, (lagoon B & C) and beaches (lagoon A). Due to tourism and natural preservation interests the area with a varied fauna and flora at the reef east of the present Serangan Island has to be preserved.

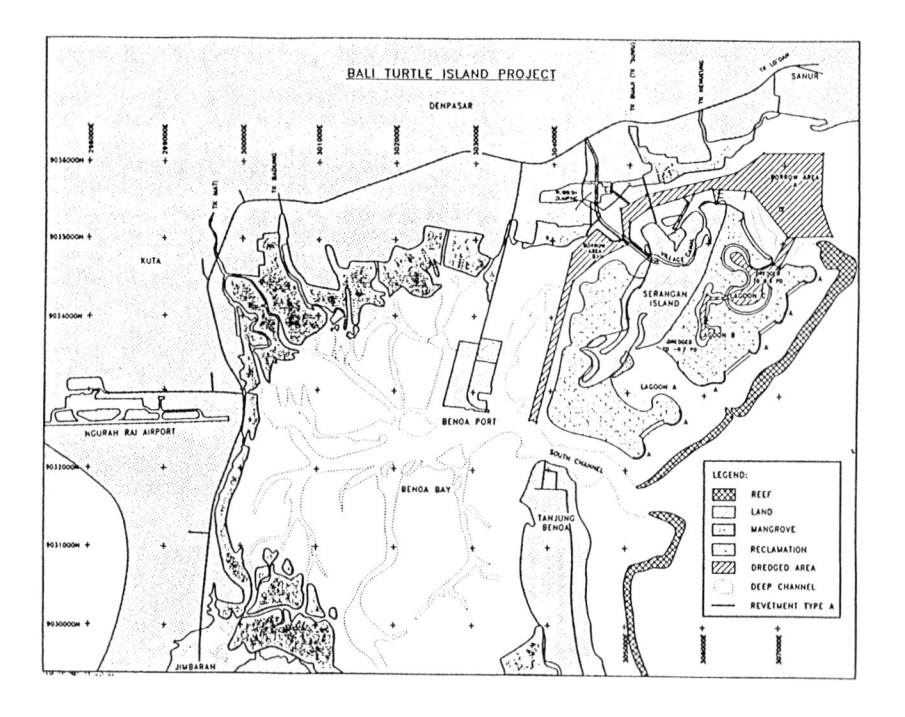

Figure 2. A sketch showing the original Serangan Island and the Turtle Island development, Layout #2.

Environmental Coastal Regions 255

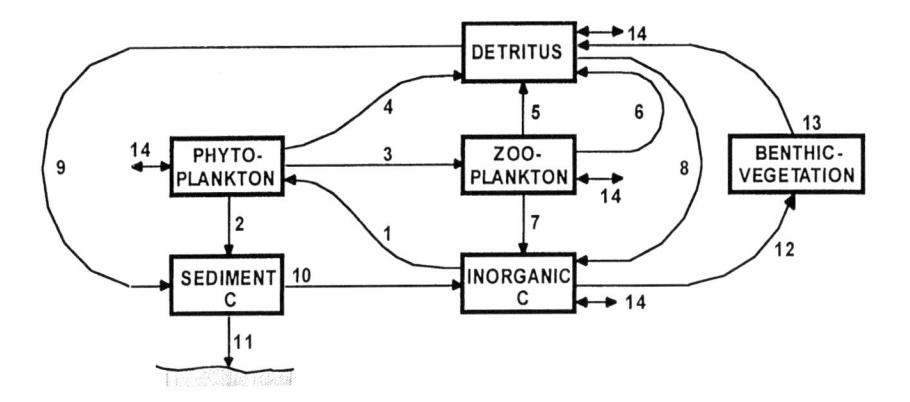

[Figure: diagram showing state variables DETRITUS, PHYTOPLANKTON, ZOOPLANKTON, BENTHIC-VEGETATION, SEDIMENT C, INORGANIC C with numbered process arrows]

1. production, phytoplankton
2. sedimentation, phytoplankton
3. grazing
4. extinction, phytoplankton
5. excretion, zooplankton
6. extinction, zooplankton
7. respiration, zooplankton
8. mineralisation of detritus
9. sedimentation of detritus
10. mineralisation of sediment
11. accumulation in sediment
12. production, benthic vegetation
13. extinction, benthic vegetation
14. exchange with surrounding waters

Figure 3 State variables and processes in the eutrophication model.

2.2 The Ecological Model

Quantification of the effects on water quality in terms of rooted bentic vegetation, macroalgae, concentrations of phytoplankton, nutrients and oxygen was made modelling the important components and processes using the MIKE 21 EU (Eutrophication) model.

The MIKE21 model is a comprehensive 2-dimensional modelling system solving the Navier Stokes equations and the mass transport equation using the finite difference method (Warren & Bach [6]). The eutrophication module describes the condition in an area by using a number of state variables including carbon, nitrogen and phosphorus in phytoplankton, zooplankton, detritus and benthic vegetation and, with regard to nitrogen and phosphorus, also the dissolved form in the water. Dissolved carbon in the water, i.e. carbon dioxide, is not included explicitly in the model because dissolved carbon is normally present in excessive amounts. The model describes the seasonal and spatial variations of the state variables. The seasonal variations depend on a number of forcing functions: hydraulic conditions, influx of light, water temperature, nutrient loadings, and the conditions in the surrounding areas (boundary conditions).

A total of 16 state variables are used, 11 of which comprise the pelagiale and the remaining 5 the benthic vegetation. The state variables and processes in the biological system are illustrated in Figure 3 and described in detail elsewhere (Bach [7], Bach et.al. [8]). The advective-diffusive transport is calculated using the MIKE21 AD model by an interactive coupling of the two models (Vested et.al. [9]). The MIKE21 AD model is on the other hand fed currents and water levels by the MIKE21 HD model from which the transport is calculated. The sediment is modelled implicitly as a pool to which fluxes of nutrients arrive due to settling of organic matter and from which fluxes of nutrients reach the water due to mineralisation.

2.3 Model set-up and forcing

The model area covers Benoa Bay, the Turtle Island/Serangan Island, Sanur Beach to the north and the reef and a part of the ocean to the east (Figure 4). The grid spacing of the finite difference model is 120 m.

Two different seasons of the hydrodynamic conditions exist mainly due to variation in rainfall. The model includes the wet season and the dry season by applying differences in rainfall and land based run-off (Table 1), but using the same oceanic current regime as boundary conditions. The hydrodynamic model was calibrated using data from April-May 1996 (Driscoll et.al. [4]). These data were also used for the eutrophication model simulations. The model simulations covered one month for each of the seasons.

Table 1. Land based run-off: the measurements obtained from a field campaign in April 1996 used for the hydrodynamic model calibration, the annual average and the values applied for the wet and dry seasons.

April meas. m^3/sec	Mean annual m^3/sec	Mean wet season m^3/sec	Mean dry season m^3/sec
3.90	7.81	19.49	2.41

The data available for model set-up and calibration was very limited. Arial photos showing the distribution of vegetation were used to a.o. determine the initial conditions for the benthic vegetation. Some older data on nutrients and oxygen concentrations exist and these were used also as indications of the concentration levels (Hunting [5]). The boundary conditions could not be derived from measurements directly. The available data were used to determine realistic oceanic values.

At present, sewage from about 320.000 people is discharged into an area covering Benoa Bay and an area just north of the bay. In year 2010 this population is expected to increase to 369.000 inhabitants. The load estimates were made using standard contribution per capita since no measurements or detailed estimates could be found. The load was distributed between the dry and wet season according to the rainfall/land based run-off as indicated in Table 1. The annual load for the present situation was estimated at 1526 tonnes N/year and 333 tonnes P/year.

Typical values for temperature (27°C) and light irradiation (40 E/m^2/day) were derived from local sources.

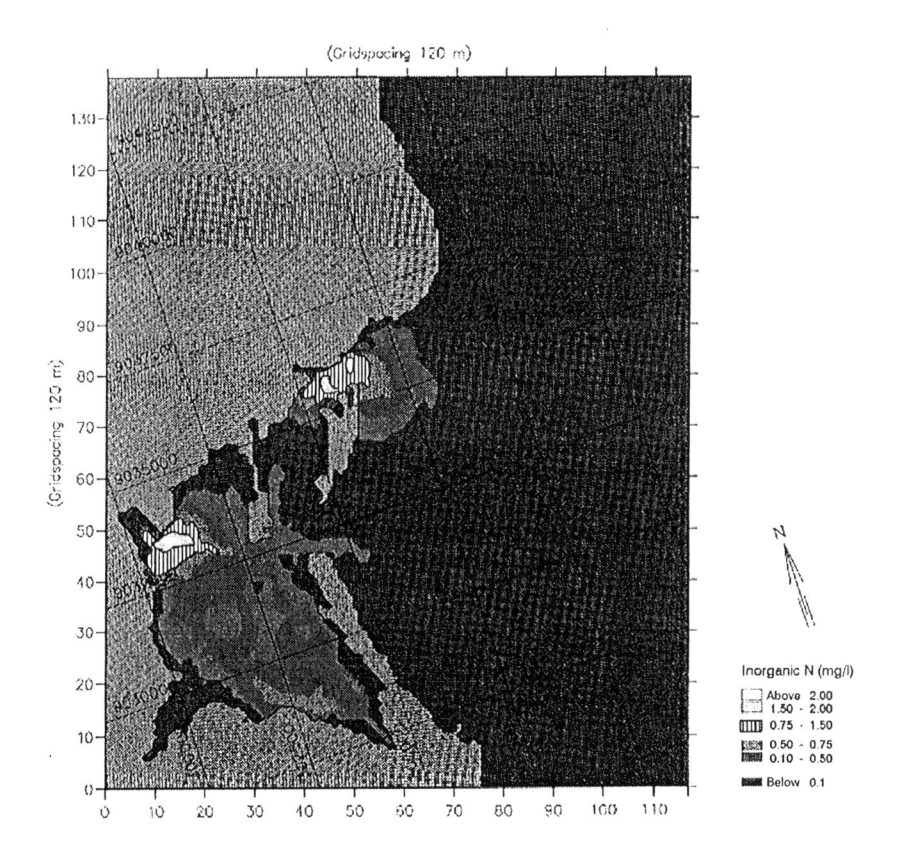

Figure 4 Average inorganic nitrogen concentration (mg/l) during the wet season for the existing situation.

258 Environmental Coastal Regions

3 Results

It was decided that the impact assessment should be made comparing the situation with and without the reclamation at the point in time where the development would be ready i.e. around the year 2010. Five different situations were investigated.
1. Future situation, no tourist resort, year 2010 and nutrient load
2. Future situation, tourist resort with layout #1, year 2010 nutrient load
3. Future situation, tourist resort with layout #2, year 2010 nutrient load
4. Future situation, tourist resort with layout #3, year 2010 nutrient load

As an example of the simulation results for the existing situation the nitrogen concentration during the wet season is shown in Figure 4.
Comparison of simulated water quality between year 1996 and year 2010 with no tourist resort shows increased concentrations of nutrients, decreased water transparency and increased concentration levels for phytoplankton (2-5%).

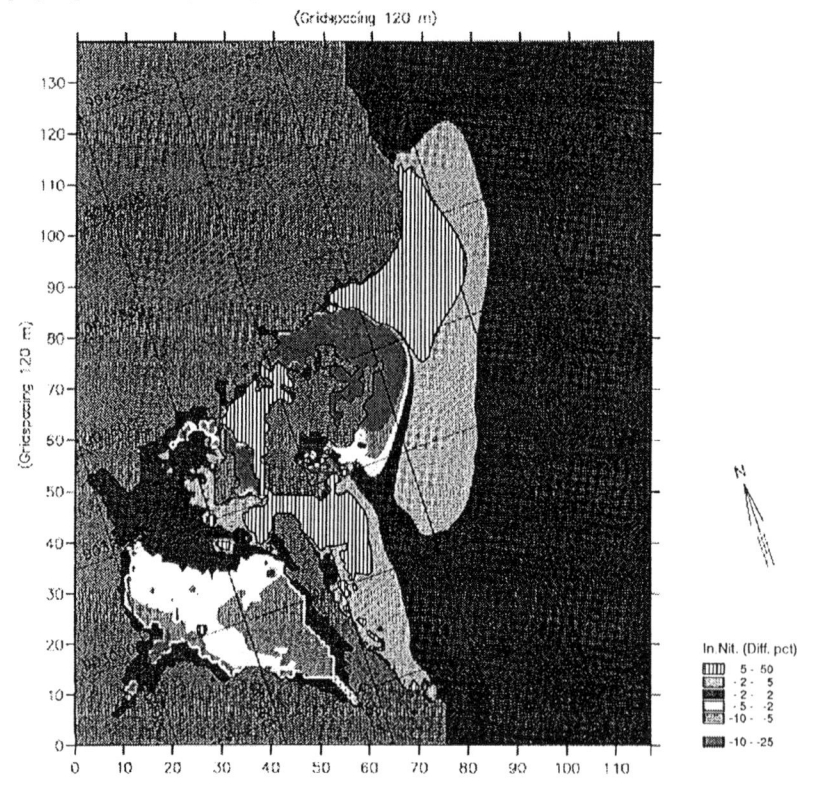

Figure 5 Percent change year 2010, inorganic nitrogen concentration. Layout #2, wet season.

Deterioration (5-10% decrease) is predicted for rooted submerged vegetation in Benoa bay and an area north of the Serangan Island, where two major sources enter the coast. The impact is highest during the wet season due to a higher load of BOD and nutrients from land.

The 3 layouts change the currents around the Turtle Island resulting in an increased transport of pollutants towards the sea caused by deepening of the channels and borrow areas (Driscoll et.al. [4]). According to the model predictions, the water quality improves in the borrow areas north of Turtle Island in all 3 layouts due to increased water exchange (Figure 5, Figure 6). Decrease in nutrient and chlorophyll-a concentrations and increase in secchi disc depth are predicted. The benthic vegetation coverage decreases due to the removal of parts of the seabed. The water quality in the strait between Turtle Island and the mainland is worsened concerning the nutrient concentrations, but the plankton biomass decreases due to the dilution with water from the borrow areas.

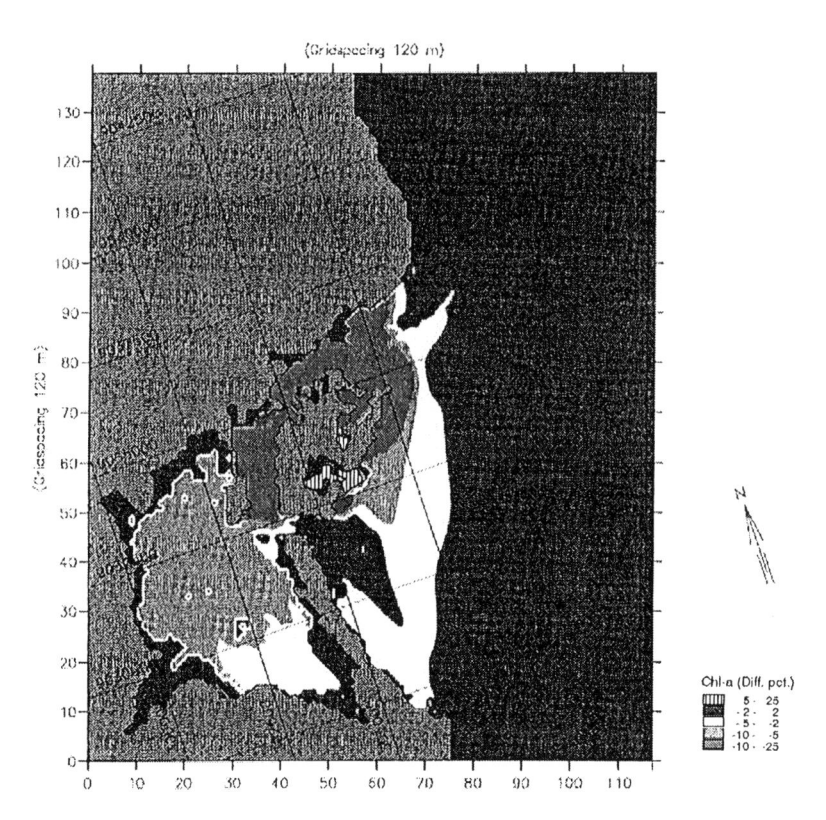

Figure 6 Percent change year 2010, chlorophyll-a concentration, Layout #2, dry season

Improvements are predicted on the reef outside the tourist beaches on the east side of the Island e.g. the secchi disc depth increases. The water quality becomes worse in layout #1 in Benoa Bay, whereas it improves in layout #2 and #3.

The water quality off Sanur beach on the main land north of the Island is affected negatively for the wet season (Figure 5), but not significantly affected in the dry season. The reason for this impact is the increased water exchange that will lead to and increased transport and spreading of pollutants in the area in general. The lower concentrations of pollutants in the borrow areas are opposite to the increased concentrations in the area north of the island.

The water quality in lagoon B and C to be used for leisure crafts is predicted to be poor for all layouts. The water quality in lagoon A close to the conservation area of the reef is predicted to deteriorate in all layouts, with decrease in submerged vegetation biomass and coverage.

4 Discussion and Conclusion

Ranking the three layouts, the best solution in terms of water quality is achieved by Layout #2. The conclusion was similar for the other parts of the investigation even though the results of e.g. flushing calculations did not show exactly the same results in terms of changes as the eutrophication model. The flushing calculations were aimed at calculating retention times in the area of a pollutant (conservative tracer) and water exchange rates. An initial concentration of 100 was applied and the dilution simulated assuming a zero concentration at the boundaries.

The differences between this approach and the approach using the eutrophication model are caused by the fact that the interactions between the components are included explicitly in the eutrophication model.

Furthermore, the correct concentration gradients are obtained in the area because the sources are positioned according to the reality. The results have shown that it is important to model the relevant components specifically and that results of flushing calculations are not necessarily representative for what is actually going to happen.

The study elucidated the fact that the protection of the seagrass areas and the coral reefs is very important. A feedback monitoring programme (Gray & Jensen, [3]) was consequently designed and implemented for the construction phase using eelgrass and some specific corals as feedback variables. During the first period of construction sediment plume modelling was accomplished to facilitate planning of the dredging work and the biological monitoring programme. For the rest of the construction period the monitoring programme includes only the biological variables.

References

[1] Bach, H.K., Jensen, K. & Lyngby, J.E., Management of Marine Construction Works Using Ecological Modelling, *Estuarine, Coastal and Shelf Science,* **44** (Supplement A), pp. 3-14, 1997.

[2] Brøker, I.; Foster, T.; Mangor, K.; Hasløv, D., Towards integrated design of marinas and beaches, *AIPCN; PIANC, Ravenna, Italy, 12-14 Oct.,* 1995.

[3] Gray, J.S. & Jensen, K., Feedback Monitoring: A New Way of Protecting the Environment. *Trends in Ecology and Evolution,* **8**, pp. 267-268, 1993.

[4] Driscoll, A.M., Foster, T., Rand, P., Tateishi, Y. Environmental modelling and management of marine construction works in a tropical environment . *2nd Asian and Australian Ports and Harbour Conf., Ho Chi Minh City, Vietnam, 16-18 April* 1997.

[5] Hunting Aquatic Resources, Environmental Baseline Studies for the Bali turtle Island Development Zone II – dredging and Reclamation Baseline Studies, *Report No. 390/293/02,* 1996.

[6] Warren, I.R. & Bach, H.K., MIKE21: a modelling system for estuaries, coastal waters and seas. *Environmental Software,* **7**, pp. 229-240, 1992.

[7] Bach, H.K., A Dynamic Model Describing the Seasonal Variation in Growth and Distribution of Eelgras (*Zostera marina* L.). I Model Theory. *Ecological Modelling,* **65**, pp. 31-50, 1993.

[8] Bach, H.K., Rasmussen, E.K.& H.H. Riber: Use of an Ecological Model in Assessing the Impact from Sediment Spill on Benthic Vegetation. *Proceedings of the 12'th Baltic Marine Biologists Symposium.Elsinore, Denmark, 25-30 August.* pp. 7-17, 1991.

[9] Vested, H.J., Jensen, O.K., Ellegaard, A.C., Bach, H.K.& Rasmussen, E.K., Circulation Modelling and Water Quality Prediction, *Proccedings of the 2nd. Int. Conf. On Estuarine and Coastal Modelling, 13-15 November, USA,* 1991.

Section 6: Pollutant Transport and Dispersion

Climate change and coastal zone: the importance of atmospheric pollutant transport

C. Borrego & M. Lopes
Department of Environment and Planning, University of Aveiro, 3810 Aveiro, Portugal
Email: borrego@ua.pt

Abstract

The importance of atmospheric circulation, particularly the typical phenomena associated with the discontinuity between sea and land, as well as the analysis of their contribution to air pollutant transport and dispersion is presented. The influence of climate on transport/deposition processes and the impact of climate change on coastal zones will be discussed, using a climate scenario induced by the increase of greenhouse gases. Additionally a case study have been included, based on work done about coastal zone vulnerability in Portugal, particularly the numerical simulation of air pollutants acidic deposition and ozone atmospheric concentrations resulting from pollutants dispersion over the littoral area.

1 Introduction

Coastal zones are very complex regions and very sensitive to stress caused by natural forces and human activities. On coastal zones live the most significant part of population and are located the most important cities and anthropogenic activities, such as industries and road traffic, with large emission rates to the atmosphere. For example, half of the humanity and 80% of Portuguese population lives on or near the coast and of the 25 largest cities of the world, 14 are located on a coast and support port facilities[1]. Figure 1 shows the location of the subset of coastal cities catalogued as polluted cities based on measured levels of sulphur dioxide[2]. This list is undoubtedly incomplete, since some polluted cities may be excluded simply because pollution monitoring in those

266 Environmental Coastal Regions

cities has not been established or because other compounds related with pollution scenarios has not been considered.

Figure 1: World map showing very polluted coastal cities[2].

Human activities (primarily the burning of fossil fuels and changes in land use and land cover) are increasing the atmospheric concentrations of greenhouse gases, which alter radiative balances and tend to warm atmosphere. However, in some regions, aerosols have an opposite effect on radiative balances and tend to cool atmosphere. The evidence of the human influence on the climate change is currently accepted, but most acknowledge uncertainties as to the speed and extent of this changing process, due to the complexity of involved phenomena.

If climate warming occurs, it will rise sea-level, heat shallow waters and modify precipitation, wind and water circulation patterns, especially in coastal zones and estuaries. Since the coastal zone is a major component in global budgets and global resources availability and utilisation, effective sustainable management strategies in such areas require a special ability to think beyond traditional sectoral divisions between different types of resources and human activities.

Other important atmospheric pollution phenomena influenced by anthropogenic activities are the photochemical pollution, the depletion of ozone on stratospheric layer, pollution due to particulate matter and aerosols. Photochemical pollution episodes, characterised by the increase of tropospheric levels of photochemical compound, particularly ozone, is directly related with emissions of nitrogen oxides and organic compounds principally due to the traffic and combustion processes. In spite of the preferential distribution of human activities on the littoral this photochemical pollution phenomena have significant importance on coastal urban areas.

2 Effect of climate change in coastal zones

Climate change is a reality on present days. In fact, it took an important role on the evolution process on the earth, during thousands of millions of years. Nevertheless, predicted changes on climate over the next century are larger than any since the appearance of civilisation and scientific community concerns due to the potential rapidity of the changes. For many scientists there are no doubts about the humanity contribution on this process.

Coastal zones are particularly at risk due to the potentially predicted effects of climate change and the ecological and socio-economic vulnerability of these areas. Coasts have been modified and intensively developed in recent decades and thus made even more vulnerable to higher sea levels, flooding and coastal erosion.

2.1 Greenhouse gases emissions and temperature increase

The increase on atmospheric levels of greenhouse gases (figure 2) as a direct result of human activity is changing the energy balance on the earth causing an "enhanced greenhouse effect". To keep the global "energy budged" in balance, the climate system must change somehow to adjust to rising greenhouse gas levels and the most probable result is a global warming of the earth's surface and lower atmosphere.

An analysis of temperature records shows that the Earth has warmed an average of 0.5°C over the past 100 years. Climate models predict that the global temperature will rise by about 1-3.5°C by the year 2100[3]. This projected change is larger than any climate change experienced over the past 10,000 years. It is based on current emissions trends and assumes that no efforts are made to limit greenhouses gas emissions.

Changes on global temperature have important effects. An example of the potential effect could be found from the *El-Niño* Southern Oscillation (ENSO) events and its consequences on inter-annual climate variability. ENSO is a short-term cycle in the ocean-atmosphere system with a mean recurrence interval of 3-8 years, characterised by an increase on ocean temperature, which produce a local atmospheric warming[4]. These atmospheric anomalies propagate horizontally such that the effects of the ENSO event are present in distant regions, with different types of affectation phenomena, characterised by extreme whether events (figure 3) like storms, floods and droughts.

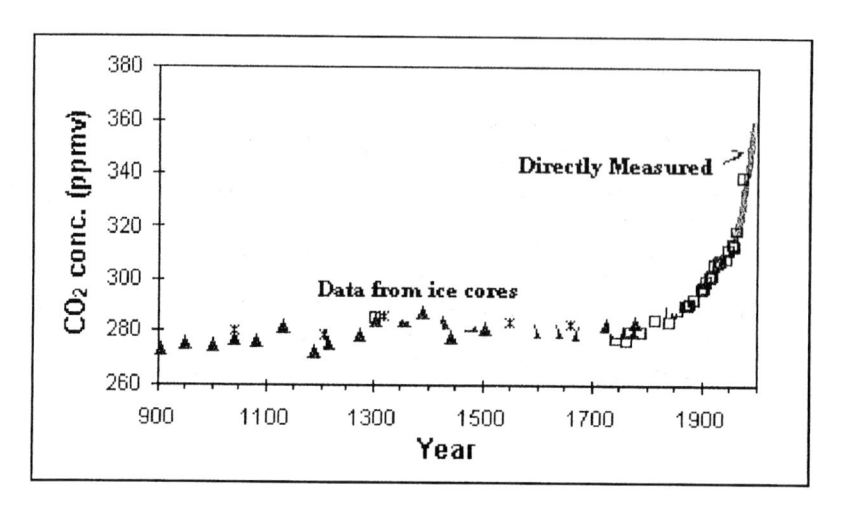

Figure 2: Trends in atmospheric concentrations of CO_2 on past 1000 years (Environment Canada's www site).

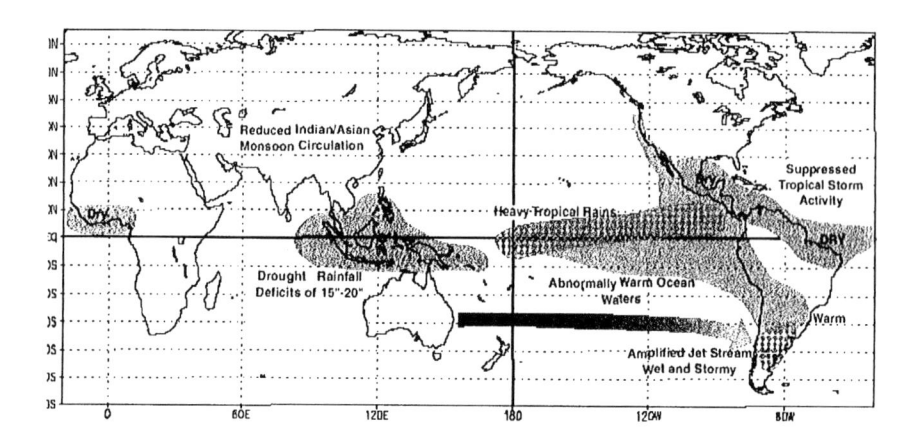

Figure 3: Sketch of the consequences of the 1997 *El-Niño* during the period of June-September 1997 (from Climate Prediction Centre, USA).

Some evidences could be taken from the 1982-83 ENSO. Disastrous effects and meteorological changes took place around the world during this event, caused total damages estimated at over $8 billion: Australia - drought and bush fires; Indonesia, Philippines - crops fail, starvation follows; India, Sri Lanka - drought, freshwater shortages; Tahiti - 6 tropical cyclones; South America - fish industry devastated; Pacific ocean - coral reefs die; Colorado river basin - flooding, mud slides; Gulf States - downpours cause death, property damage; Peru, Ecuador - floods, landslides; Southern Africa - drought, disease, malnutrition. The 1997-1998 ENSO is expected to have an even more disastrous effect!

2.2 Sea-level rise

During the last hundred years the sea-level has risen by 10 to 25 cm. It is likely that much of this rise is related to an increase of 0.3-0.6°C in the lower atmosphere's global average temperature since 1860. Models project that sea levels will rise another 15 to 95 cm by the year 2100, due to the thermal expansion of ocean water and an influx of freshwater from melting glaciers and ice[3].

Sea-level rise will affect the coastal zones of some 180 nations and territories. Sea-level rise of one metre would cause estimated land losses of 0.05% in Uruguay, 1% in Egypt, 6% in the Netherlands and 17% in Bangladesh. Other consequences of sea-level rise are flooding, coastal erosion and salt-water intrusion which will reduce quality and quantity of freshwater supplies. Higher sea levels could also cause extreme events such as high tides, storm surges and seismic sea waves. As a result of those damages, important economical sectors will be affected directly fisheries, aquaculture and agriculture. Indirectly other human activities, such as tourism, human settlements and insurance could be at risk.

The displacement of flooded communities, particularly those with limited resources, would increase the risk of various infections, psychological and other illnesses. Insects and other transmitters of diseases could spread to new areas. The disruption of systems for sanitation, storm-water drainage and sewage disposal would also have health implication.

2.3 Precipitation and water circulation changes

Compared with predictions about rising sea-level or temperature, there is less certainty involved in prediction either changes in precipitation, water circulation or the ecological effects of such changes. Some potential impacts could be taken from the ENSO episodes (see fig. 3). Two types of changes on precipitation patterns could be expected: increase rainfall in some areas result on more floods and runoff; decline precipitation in other areas, which drier regional climate.

270 Environmental Coastal Regions

On coastal areas, changes on precipitation patterns combined with sea-level rise could result on shrink estuarine habitat or, on other hand, on salinity encroachment into the tidal freshwater. Changes on water circulations could have also secondary effects reducing soil infiltration and ground/surface water supplies. Arid and semi-arid areas could also become more sensitive and some typical habitats could be loss.

2.4 Wind circulation changes

Changes on wind speed and direction could have important effects on air pollution transport (thus on air quality). Coastal areas, with significant atmospheric emissions and where wind circulations took an important role on pollutants dispersion, are particularly at risk. Climate changes are also expected with more intense and frequent extreme weather events.

The topology of coastal zones is often very complex resulting from the presence of typical features, such as the sea/land interface, the irregularity of coastline, the presence of estuaries, the complex topography, with mountains, hills and valleys, etc.. This topology in conjugation with synoptic or mesoscale meteorological conditions, contributes to the development of a characteristic atmospheric circulation that plays an important role on the wind flow over coastal areas.

The most relevant coastal circulations are the sea and land breezes, due to the differential heating between sea and land. More complex flow patterns can be produced, when breezes are combined with orographic flows (for example, mountain and channel winds). Sea/land breezes are developed under slack synoptic gradients and moderated or stronger insolation. This conditions diurnal reversing produces a thermal driven circulation on the land/water interface[5] (figure 4).

The effects of coastal breezes on the dispersion patterns of the air pollutants are diversified, being their interaction with the flow on a synoptic scale very complex. However, it is possible to subdivide the impact of these mesoscale atmospheric circulations in two different effects: accrued mixture due to ventilation associated with the sea air intrusion and potential recirculation of air pollutants. The importance and combination of these aspects depend on the morphological characteristics of the region and on the meteorological conditions under consideration[6].

A recent work[7] assessed the impact of ENSO in the atmospheric physical properties over Portugal, by means of a downscale approach, ranging from global scale to regional scale. Simulation results shows that large-scale atmospheric circulation induced by ENSO generates some changes, both in wind direction and in potential temperature field over the studied region. However, more developments are needed in order to study the relation between possible storm and drought condition in Portugal related to ENSO events.

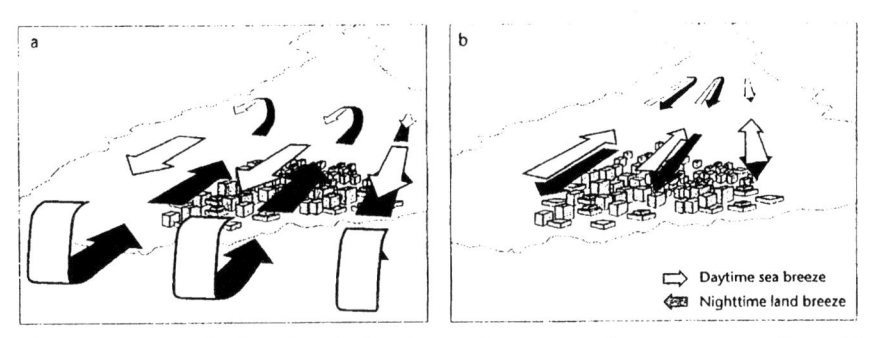

Figure 4: a) Vertical recirculation in sea breezes and upper return flow, b) horizontal recirculation in daytime sea breezes and night-time land breeze[2].

3 Study case: Portuguese coastal zone

Portugal has a quite extensive coastline (850 km) associated to significant terrain features and sea-land breezes circulation, which result in complex wind field with strong implications on the production and transport patterns of atmospheric pollutants, particularly photochemical species such as ozone. Like in other countries, Portuguese littoral is the region with higher population density and economic development. On a narrow band of 50 km from the coastline (figure 5) live about 80% of the population and are settled almost all the main industries and power plants. The resulting pollutant emissions carry out a high pressure on the Portuguese coastal areas environment and the natural resources[8].

3.1 Atmospheric emissions on Portuguese littoral

A significant part of the Portuguese anthropogenic emissions - 85% of total NO_x and 95% of NMVOC - are emitted on the narrow band of 50 km nearby the coastline. Table 1 shows that "per capita emissions" and "emissions per km^2" for Portugal mainland are significantly smaller than EU similar parameters. On other hand, considering "emissions per GDP" the figures are larger than the EU, what might be means that the industrial production in Portugal mainland is still less efficient than the average industrial production on EU. All parameters for the coastal areas are closer to the ones of EU, particularly the "emissions per km^2" parameter, which is the same order of magnitude[7].

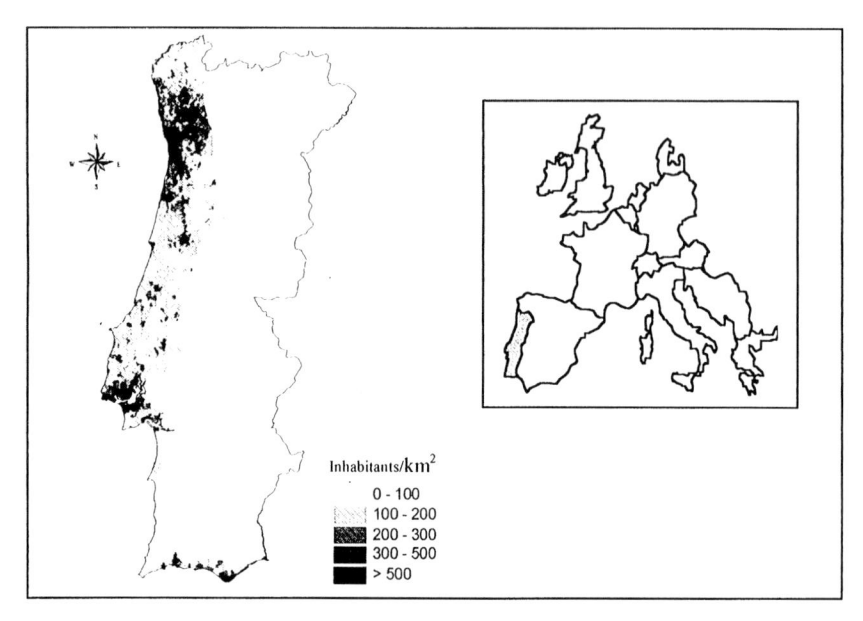

Figure 5: Population density of Portuguese littoral (50 km).

Table 1: Current emissions situation in Portugal mainland, Portuguese coastal areas and EU (12).

	Portugal Mainland NOx	Portugal Mainland NMVOC	Coastal Area NOx	Coastal Area NMVOC	European Union (12) NOx	European Union (12) NMVOC
Total emissions (kton/year)*	220.8	201.8	183.4	180.6	12145.0	12455.0
Per capita (kg)	23.5	21.5	24.1	23.7	34.8	35.7
Per GDP (kg)	3.1	2.8	na	na	2.2	2.3
Per km2 (ton)	2.5	2.3	4.8	4.7	5.1	5.3

source : Eurostat and Portuguese Institute of Statistic
* : excluding biogenic emissions; na : not available; GPD: Gross Domestic Product

3.2 Simulation of pollutants transport, deposition and transformation

In order to evaluate the vulnerability of the Portuguese coastal zones, an extended study have been carried out, focused on different environmental parameters, including the atmospheric system[9]. To analyse critical load capacity of the coastal ecosystems, numeric simulations (MAR system) have been made to calculate the spatial distribution pattern associated to air pollution at a national level. A specific episode (typical summer day) has been considered and simulation domain includes all Portuguese continental territory.

Globally, the obtained hourly atmospheric levels, both in ozone concentrations and in NO_x deposition levels (figure 6), are below the critical loads almost over the entire domain. However, the maps shows also that higher pollution levels are located on a twin band near the littoral.

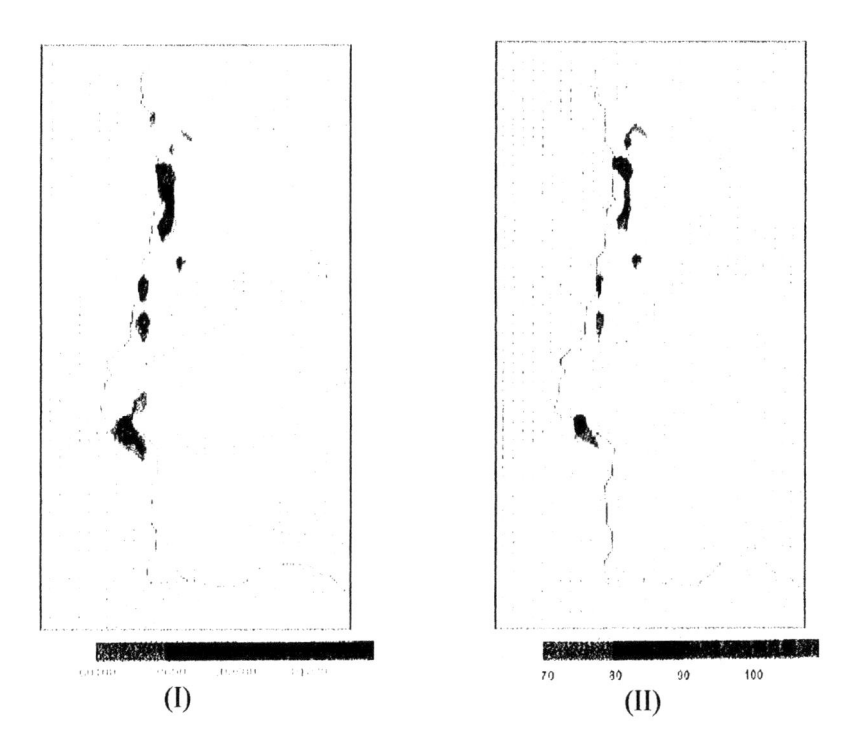

(I)   (II)

Figure 6 - Hourly wind field and distribution pattern of atmospheric pollutants levels over Portugal for a typical summer conditions: (I) acidic deposition of NOx (keq H^+/ha) at sunrise; (II) ozone concentration ($\mu g/m^3$).

If local meteorological conditions contribute to stagnation or recirculation processes, pollutants transport and dispersion could be strongly affected, and thus air pollutants concentration and deposition levels could be higher than those predicted. Taken into account the spatial distribution of Portuguese population along the coast, probably an important fraction could be occasionally exposed to a significant air pollution level.

4 Conclusions

The difficulties of predicting the direction and scale of change in the climate, whether globally or regionally, are mirrored by the difficulties in predicting the responses of ecosystems and atmospheric patterns. The limited data allow some general predictions, but the complexity of biological systems and climate behaviour prevents more detailed forecastings at present. It would seem prudent to scale correctly the release of greenhouse gases, to begin practicing sustainable economic approaches to resource use and to develop the data base needed to mitigate challenges to the coastal zone without damaging the diversity of ecosystems and allowing the real quality of human life.

Acknowledgements

This work is supported by PRAXIS XXI grants and funded by the projects PRAXIS/3/3.2/EMG/1949/95, PRAXIS/3/3.2/AMB/38/94, ENV4-CT98-0700 and included in GLOREAM (EUROTRAC–2).

References

[1] Borrego, C., Climate change and coastal zone management: towards the sustainable development, Proc. *Global change, Environmental integrity, and Sustainable development of coastal areas*, Carvoeiro 26-28 February, 1998, pp. 1-13.

[2] Steyn, D.G., Air pollution in coastal cities, *Proc. of the 21ST NATO/CCMS International Technical Meeting on Air Pollution*

Modelling and Its Application, eds. American Meteorological Society, Baltimore, USA, pp. 350-362, 1995.

[3] UNFCCC, Climate change information kit, UNEP's Information Unit for Conventions, Geneva, 1997.

[4] Haston, L. & Michaelsen, J., Long-term central coastal california precipitation variability and its relationship to El-Ninõ Southern Oscillation, *Journal of Climate*, 7, 9, 1994, pp. 1373-1387.

[5] Hsu, S.A., *Coastal Meteorology*, Academic Press, 260p, 1988.

[6] Borrego, C., Atmospheric pollution in coastal zones: mesoscale modelling as applied to air quality studies, *Air Pollution IV – monitoring, simulation and control*, Ed. Caussade, B., Power, H., Brebbia, C.A., Computational Mechanics Publications, Southampton, 1996, pp.59-69.

[7] Carvalho, A. C., A. I. Miranda, C. Borrego, Ferreira, C. and Rocha, A.: Effects of climate change in regional weather patterns, 1st GLOREAM Workshop, 10-12 September, Aachen, Germany, 1997, pp. 13-19.

[8] Borrego, C., Barros, N., Tchepel, O., Valinhas, M.J., Ozone production over coastal areas in Southern Europe: the Portuguese case, submitted to *Journal of Applied Meteorology*.

[9] Borrego, C. & Barros, N., Estudo de avaliação da vulnerabilidade da capacidade de recepção das águas costeiras em Portugal – Efluentes e cargas poluentes não associadas a bacias drenantes resultantes da poluição atmosférica (Relatório R4.1), IDAD – Projecto nº13/94, 1996.

A high-accuracy approach for modeling flow-dominated transport

Jung L. Lee
Department of Civil Engineering, Sung Kyun Kwan University,
Suwon Science Campus, Suwon 440-746, Korea
EMail: jllee@yurim.skku.ac.kr

Abstract

In this study a new hybrid method is proposed for solving flow-dominated transport problems accurately and effectively. The method takes the forward-tracking particle method for advection. However, differently from the random-walk Lagrangian approach, it solves the diffusion process on the fixed Eulerian grids. Therefore, neither any interpolating algorithm nor a large enough number of particles is required. The method was successfully examined for both cases of instantaneous and continuous sources discharged at a point. Comparison with a surrounding 5-point Hermite polynomial method (Eulerian-Lagrangian method) and the random-walk pure Lagrangian method shows that the present method is superior in result accuracy and time-saving ability.

1 Introduction

Since the most serious environmental problems facing us today are encountered in industrialized and urbanized estuaries due to the growing incidence of the contamination, several distinct numerical improvements have been made for predicting the environmental pollution transport in flow-dominated area. Application of Eulerian numerical models to the solution of sharp-front problems often results in oscillations, phase errors, peak depression, and/or numerical dispersion, unless very fine temporal and spatial steps are adopted. The representative Eulerian scheme is Quadratic Upstream Interpolation for Convective Kinematic with Estimated Streaming Terms

(QUICKEST) scheme first presented by Leonard [1]. As the second generation, the mixing Eulerian-Lagrangian method has been proven to provide high accuracy with reduction of oscillations and numerical dispersion. However, its accuracy depends on interpolating algorithm used. This type of method uses a split operator approach in which the advection term is treated by a Lagrangian approach along characteristic paths and the other diffusion term is solvd on Eulerian grids. The Lagrangian approach to advection usually takes either Forward Particle Tracking Method (FPTM) (Garder et al. [2]; Dimou and Adams [3]; Huang et al. [4]) or a single-step Reverse Particle Tracking Method (RPTM) (Holly and Preissmann [5], Komatsu et al. [6]): The RPTM method requires the interpolation to evaluate the unknown value between grid points by using the known values of surrounding grid points; and the FPTM requires the four consecutive steps which are somewhat complicated; tracking the concentration front, single-step forward tracking, single-step reverse tracking and finite difference/element approximation.

In this study a new hybrid method is developed for solving flow-dominated transport problems accurately and effectively.The method takes a Lagrangian viewpoint for advection step, by means of moving particles to track their assigned concentration continuously forward. At each time, the concentration of particles is re-assigned through the diffusion step and a new particle is released at the center of each cell where the concentration is newly diffused. Any interpolating algorithm is not required here because the diffused value obtained at the Eulerian cell center is assigned back to its own position where the particle posed. If there is no diffusion effect, the conventional Lagrangian random-walk model requires the numerical implementation of a small number of particles so that it becomes quite economic and effective. However, with diffusion effect, the method requires a large enough number of particles to simulate the diffusion process by random walk (Lee [7]; Lee and Wang [8]). That makes the Lagrangian method quite time-consuming and labourous. Therefore, the present method solves the diffusion step on a fixed Eulerian grid but particles move continuously forward. Differently from the FPTM which requires the four consecutive steps, the present method requires only the two steps; forward tracking for advection and finite difference approximation for diffusion.

2 Formulation

The depth-averaged partial differential equation describing advective and dispersive processes for two dimensional problems is written in the conservative form as

$$\frac{\partial C}{\partial t} + u\frac{\partial C}{\partial x} + v\frac{\partial C}{\partial y} = K\frac{\partial^2 C}{\partial x^2} + K\frac{\partial^2 C}{\partial y^2} \tag{1}$$

where K is the dispersion coefficient and (u, v) are the x, y components of depth-averaged velocity vector. The advection-dispersion equation is solved using split-operator approach which is based on the recognition that the physical phenomena of pollutant transport are represented by superimposing two individual operations, advection and diffusion. Therefore, Eq. (1) is decoupled into the two fractional steps at each small time increment and solved; (i) by the transport of concentration elements due to advection, and (ii) by the ADI (Alternating Direction Implicit) factorization techniques due to dispersion.

Advection Step:

$$\frac{D\hat{C}}{Dt} = 0 \tag{2}$$

The above equation states that $\hat{C}$ of a particle moving with flow must be constant along its characteristic path predicted by using a first-order difference formula given below.

$$x_p^{n+1} = x_p^n + u_p^n \Delta t, \quad y_p^{n+1} = y_p^n + v_p^n \Delta t \tag{3}$$

Diffusion Step:
Once the concentration, $\hat{C}$ by the pure advective motion is obtained, the following pure dispersion operator is solved for the coupled effect.

$$\frac{\partial \tilde{C}}{\partial t} = K\frac{\partial^2 \tilde{C}}{\partial x^2} + K\frac{\partial^2 \tilde{C}}{\partial y^2} \tag{4}$$

which is discretized by using ADI scheme as follows.

$$C_{ij}^{*n+1} - \Delta t \alpha K D_{xx} C_{ij}^{*n+1} =$$
$$\hat{C}_{ij}^{n+1} + \Delta t K[(1-\alpha)D_{xx}\hat{C}_{ij}^{n+1} + D_{yy}\hat{C}_{ij}^{n+1}] \tag{5}$$

$$C_{ij}^{n+1} - \Delta t \alpha K D_{yy} C_{ij}^{n+1} = C_{ij}^{*n+1} - \Delta t \alpha K D_{yy} \hat{C}_{ij}^{n+1} \qquad (6)$$

where α is an implicit weighting factor and D_{xx} and D_{yy} are the second-order difference operators.

3 Solution Strategy

A new hybrid method proposed here is useful in particular for flow-dominated transport problems. This method takes the switching approach between advection and diffusion prosesses, keeping the concentration particles continuously moving forward rather than a sigle-reverse particle tracking. The discrete position of each particle by the advective motion is not always posed at a cell center. Therefore, this study lets its diffusion process accomplished by proxy at the center of Eulerian cell posed. Here any interpolating algorithm is not required because the diffused value obtained at the Eulerian cell center is assigned back to its original position where it posed. Since the equation of dispersive transport is linear in the concentration variable, such diffusion process for all polluting particles is done at a time by the Eulerian ADI scheme. The following is the list of the computing procedure:

1) Move particles forward according to pure advection
2) Find the grid cell where each particle poses
3) Assign the concentration of particle to the cell center posed
4) Do diffusion step by using Implicit FDM
5) Re-assign the values to corresponding particles
6) Release a new particle at the cell where concentration is newly diffused.

Without diffusion, step 1 is only required and without advection, step 4 is only required. The more details are shown in Figure 1.

4 Numerical Examples

4.1 Impulse Loading

The numerical model is tested for both one- and two-dimensional transports under a uniform flow. The one-dimensional analytic solution of Eq. (1) is given by

$$C(x,t) = \frac{M}{\sqrt{4\pi K t}} \exp\left(-\frac{(x-ut)^2}{4Kt}\right) \qquad (7)$$

A point source of $M = 3{,}000 kg/m^2$ having a deviation of zero is initially given and transported downward for a time of $12{,}800 sec$

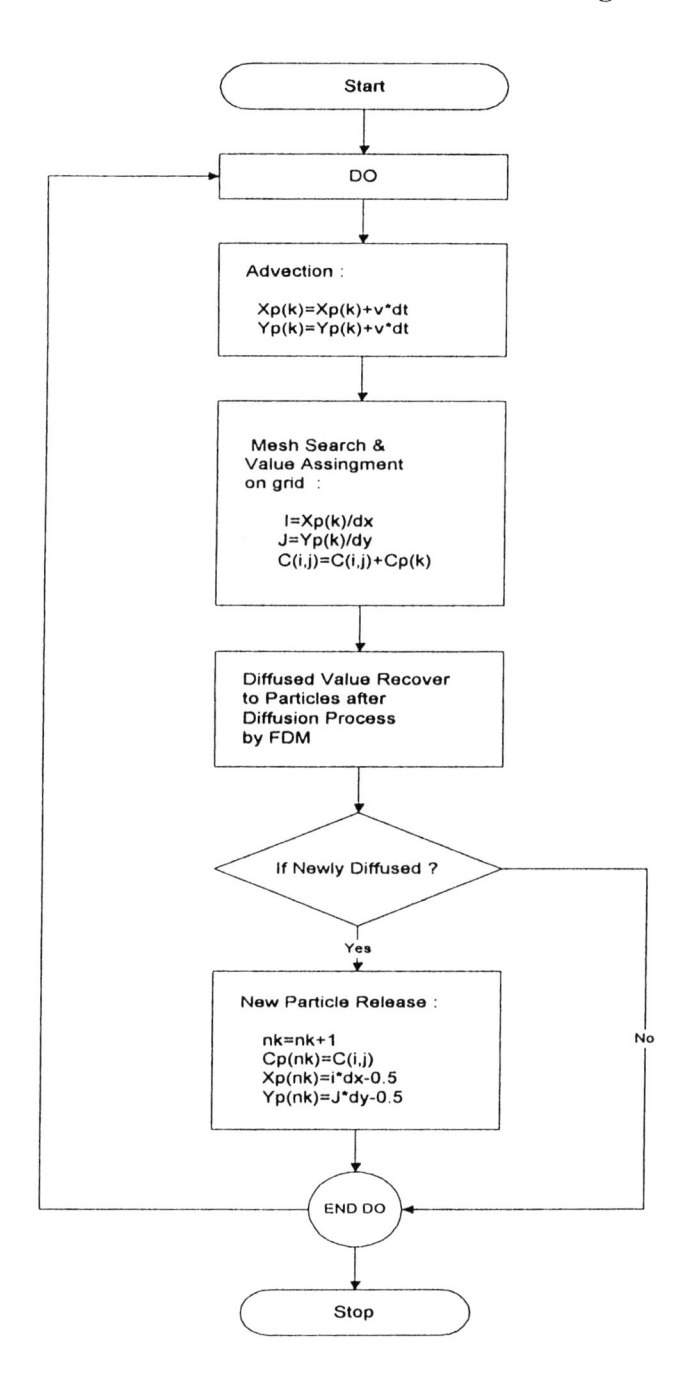

Figure 1. Flow chart of computing procedure.

by a uniform current of 0.5m/s. The node spacing and time step chosen are 50m and 100sec, respectively. Figures 2 and 3 show the numerical solutions by the present method is very close to the analytic solution. The present method is also examined to the two dimensional problem. The two-dimensional analytic solution is given by

$$C(x,y,t) = \frac{M}{4\pi Kt} \exp\left(-\frac{(x-ut)^2 + y^2}{4Kt}\right) \tag{8}$$

The initial profile has a point source of $M = 1,000,000 kg/m^3$ and is transported in the x direction for 20,000sec by a uniform current of 0.1m/s. The node spacing and time step chosen are 100m and 100sec, respectively. The computational times required by the employed schemes for 2D example are shown on Table 1 in which the values indicate ratios of computational times to the present method for the case $K = 2m^2/s$. The computational effort required by five-point Hermite polynomial method is more than four times than that required by the present method. The ratio for random-walk method was resulted from 100,000 particles discharged into flow fields and a 50×50 concentration grid was used. Figure 4 shows the results by the five-point method, random-walk method (100,000 particles) and the present method, respectively. It can be seen from Figure 4 the present method is very consistent with analytical solutions.

Table 1. Comparison of computational time

Method Type	Present Model	5-Point Model	Random-Walk Model
$K = 2m^2/s$	1.0	4.4	5.1
$K = 10m^2/s$	1.05	7.5	5.1

4.2 Continuous Loading

Lastly the method is examined for the continuous source loading at a point. The analytic solution for such case is as follows.

$$C(x,y,t) = \frac{\dot{M}}{4\pi K} \exp\left[\frac{xu}{2K}\right] \int_0^t \frac{1}{\tau} \exp\left[-\frac{x^2+y^2}{4K\tau} - \frac{u^2\tau}{4K}\right] d\tau \tag{9}$$

The given values of $\dot{M}$, u, Δt and grid spacing are $10000 kg/m^3/s$, 1m/s, 10sec, and 20m, respectively. The solution obtained after 1000sec is compared as shown in Figure 5. Only numerical results by the present method closely approximate the analytic solution regardless of the Peclet number. Totally 1,000,000 particles were discharged for the random-walk method.

5 Conclusion

The hybrid method proposed here requires only the two steps without need to use the interpolation; forward tracking for advection and finite difference approximation for diffusion. The present method has a high accuracy and fast computation compared with other schemes such as five-point Hermite polynomial method of mixed Eulerian-Lagrangian type and random-walk Lagrangian method. Based on the results, it is concluded that the present method is easily extended to various flow-dominated fields and very effective in calculating advectiive diffusion not only in one-dimensional cases but also in two-dimensional ones. For both instantaneous and continuous point sources, the results show that the present method works very well.

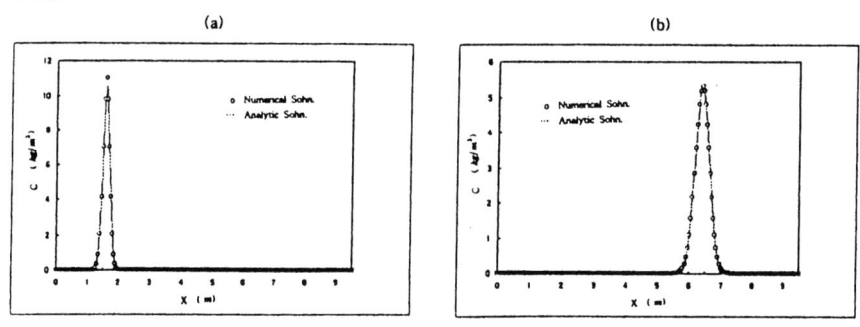

Figure 2. Comparison between analytic solutions and results by the present study for $K=2m^2/s$; (a) t=3200sec and (b) t=12800sec).

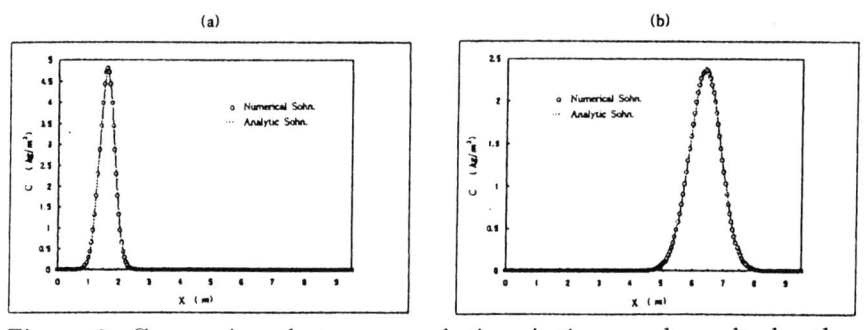

Figure 3. Comparison between analytic solutions and results by the present study for $K=10m^2/s$; (a) t=3200sec and (b) t=12800sec).

284 Environmental Coastal Regions

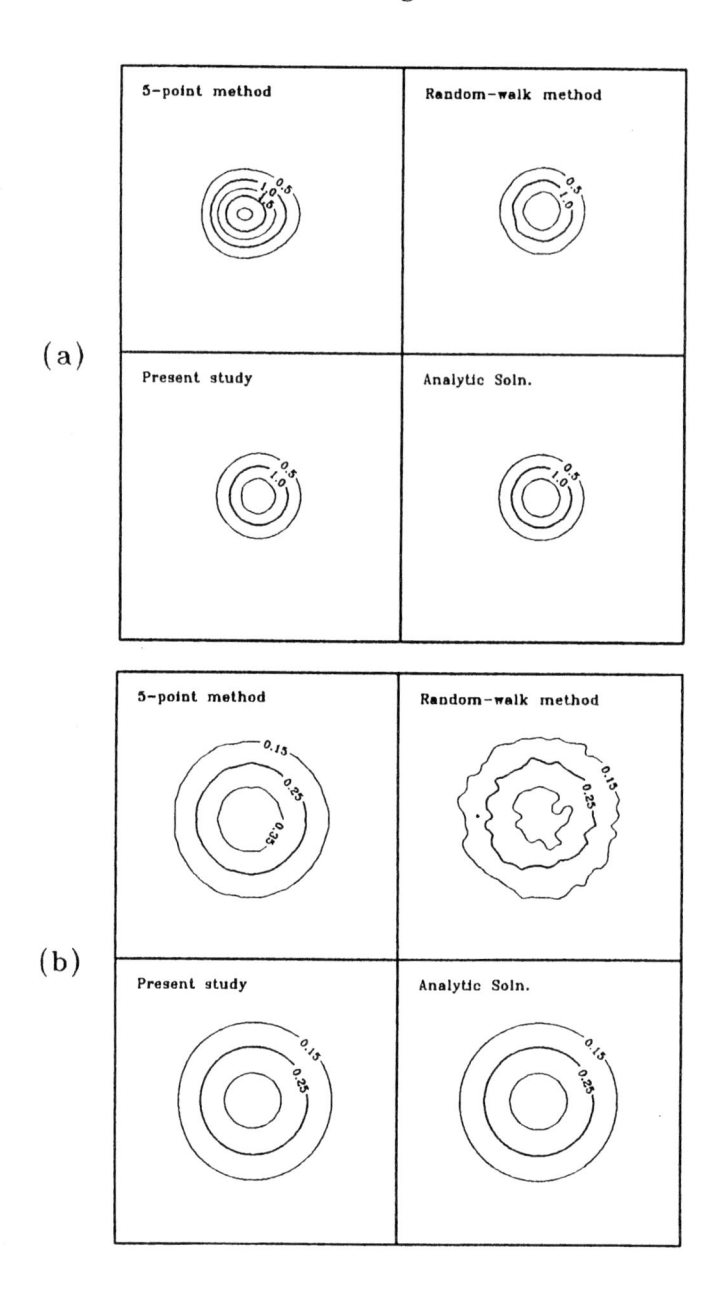

Figure 4. Comparison for an impulse point loading after t=20000sec (a) $K = 2m^2/s$, (b) $K = 10m^2/s$ (contour unit: kg/m^3).

Environmental Coastal Regions 285

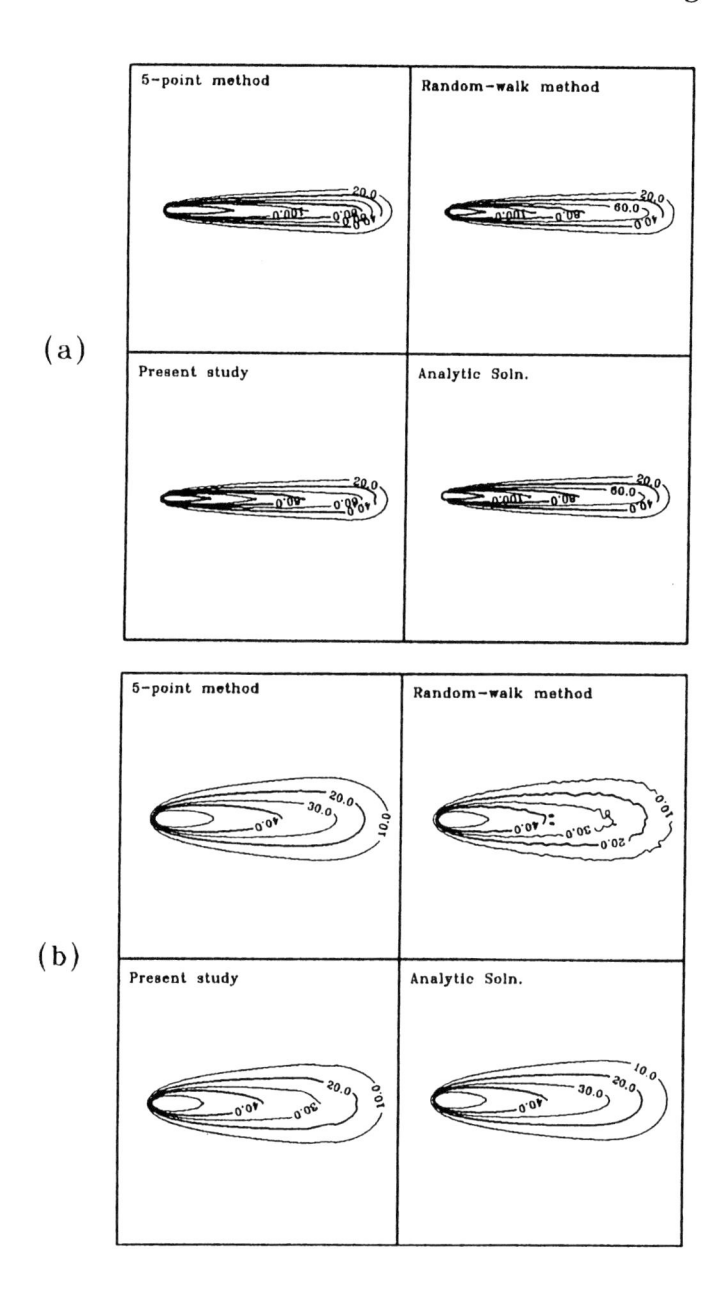

Figure 5. Comparison for a continuous point loading after t=1000sec;
(a) $K = 2m^2/s$, (b) $K = 10m^2/s$ (contour unit: kg/m^3).

References

[1] Leonard, B.P., 1979. A stable and accurate convective modelling procedure based on quadratic upstream interpolation, *Computer Method in Applied Mechanics and Engineering*, 19, 59-98.

[2] Garder, A.O., Peaceman, D.W., and Pozzi, A.L., 1964. Numerical calculation of multidimensional miscible displacement by the method of characteristics, *Soc. Pet. Eng. J.*, 4, 26-36.

[3] Dimou, K.N., and Adams, E.E, 1991. Representation of sources in a 3-D Eulerian-Lagrangian mass transport model, in Water Pollution: Modelling, Measuring and Prediction (Ed. Wrobel, L.C. and Brebbia, C.A.), 251-264, *Proceedings of the 1st Int. Conf. on Water Pollution*, Elsevier Applied Science, New York.

[4] Huang, K., Zhang, R., and van Genuchten, M.Th., 1992. A simple particle tracking technique for solving the convection-dispersion equation, in Vol. 1: Numerical Methods in Water Resources (Ed. Russell, T.F., Ewing, R.E., Brebbia, C.A., Gray, W.G., and Pinder, G.F.), 87-96, *Proceedings of the 9th Int. Conf. on Computational Methods in Water Resources*, Elsevier Applied Science, New York.

[5] Holly, F.M.Jr., and Preissmann, A., 1977. Accurate calculation of transport in two dimensions, *J. Hyd. Div.*, ASCE, 103, 1259-1277.

[6] Komatsu, T., Holly, F.M.Jr., Nakashiki, N., and Ohgushi, K., 1985. Numerical calculation of pollutant in one and two dimensions, *J. Hydroscience and Hydraulic Engineering*, 3(2), 15-30.

[7] Lee, J.L., 1994. Boundary flow under a sluice gate, *J. Korean Association of Hydrologic Society*, 27(3) (in Korean).

[8] Lee, J.L., and Wang, H., 1994. One-D model prediction of pollutant transport at a canal network, *J. Korean Society of Coastal and Ocean Engineers*, 6(1), 51-60.

Section 7: Hydrodynamic Pollutant Transport Modelling

Two- and three-dimensional modelling of mercury transport in the Gulf of Trieste

R. Rajar[1], D. Žagar[1], A. Širca[1], and M. Horvat[2]
[1] University of Ljubljana, Faculty of Civil and Geodetic Engrg., Hajdrihova 28, Ljubljana, Slovenia
[2] Josef Stefan Institute, Jamova 2, Ljubljana, Slovenia
EMail: rrajar@fagg.uni-lj.si

Abstract

The Gulf of Trieste is subject to mercury pollution from the Soča river which drains contaminated sediments from the region of a former mercury mine in Idrija, Slovenia. This results in elevated mercury levels in some marine organisms. Due to concern for human health, a study has been undertaken to predict mercury contamination trends through the use of a field program and mathematical modelling. In previous studies, a two-dimensional (2D) steady-state advection-dispersion model for the simulation of mercury cycling in the Gulf was developed. As the 2D model has several important limitations for proper simulation of the phenomenon, a three-dimensional (3D) unsteady state model was developed. Several measurements have also shown that the inflow of total dissolved Hg into the Gulf is almost negligible in comparison with the inflow of Hg adsorbed to suspended particles. Because of the importance of this phenomenon, a 3D sediment-transport module was developed, which is included in the integrated model PCFLOW3D. The present contribution deals mainly with simulations of transport and dispersion of suspended sediment and Hg adsorbed to it. Some simulated cases show that the sediments from the river settle to the bottom relatively near the river mouth (mostly in the distance of some kilometers from it). But simulations with the strong Burja wind (the most frequent in this region) show resuspension of the bottom sediments and their displacement along the northern shore towards the west. This is in agreement with measured distribution of mercury concentration in the bottom sediments of the Gulf.

1 Introduction

The Gulf of Trieste is situated in the northern part of the Adriatic Sea and covers about 25 x 30 km^2 (Fig. 1). The depth reaches 25 m in the central part and the average depth is about 16 m. The Gulf is subject to natural and anthropogenic mercury load from the Idrija region where mercury mining has been active for almost 500 years. The former mine, which is closing down, is situated on the Idrijca River, a tributary of the Soča river, flowing into the Gulf. Recent measurements by different authors have shown increased mercury contents in water, sediment and biota (Horvat et al. [4], Horvat et al. [5]). Concentrations in sediments and suspended matter were as much as two orders of magnitude higher than the corresponding natural background values from the Central and Southern Adriatic. The developing eutrophication of the Gulf and connected occasional anoxias may lead to methylation of mercury which, in its organic form, is bio-accumulative and highly toxic, particularly for humans at the end of the food-chain.

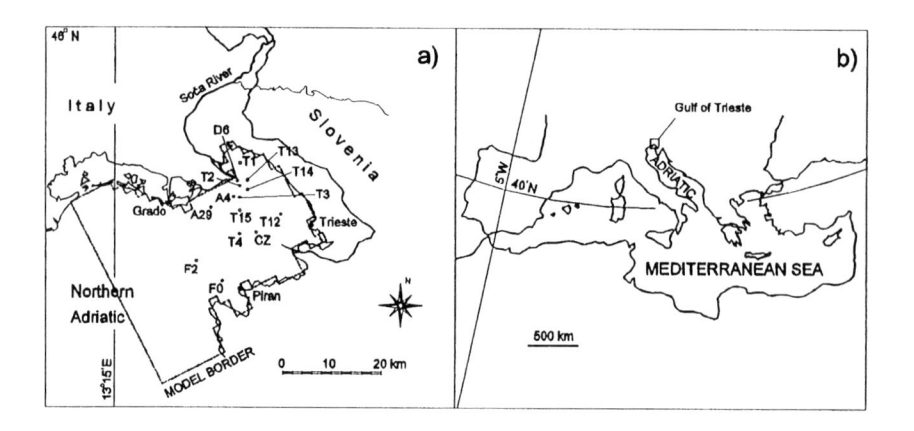

Fig. 1 The Gulf of Trieste showing the extent of the computational domain and measuring points.

Even 10 years after closure of the mercury mine, Hg concentrations in river sediments and water are still very high and do not show the expected decrease in the Gulf of Trieste. Therefore extensive research on mercury cycling in the Gulf is in progress. Many measurements of physical, chemical and biological parameters have been carried out during recent years. A two-dimensional (2D) steady-state model was first developed for the simulation of hydrodynamic (HD) circulation and mercury transport and fate. Some of the previous results have been described in Rajar, Četina and Širca [6] and in Širca and Rajar [9]. A brief description of these results is given in chapter 2.

As the 2D model has several important limitations for proper simulation of the phenomenon, a three-dimensional (3D) unsteady state model PCFLOW3D was developed. Several measurements have shown that the inflow of total dissolved Hg into the Gulf is almost negligible in comparison with the inflow of Hg adsorbed to suspended particles. E.g. Širca and Rajar [9] determined that during one year, the Soča river brings only 3.3 kg of dissolved Hg, while the mass of particulate Hg is 1773 kg. Therefore the present contribution deals mainly with simulations of transport and dispersion of suspended sediment and Hg adsorbed to it.

2 Two-Dimensional Simulations

A two-dimensional hydrodynamic (HD) and steady-state mercury transport and fate model, STATRIM, of the Gulf has been developed to evaluate possible trends in mercury pollution in the future. In this model, the hydrodynamics of the Gulf is represented by an average velocity field due to the average annual wind and the discharge of the river. The STATRIM advection-dispersion module shows the transport of non-methylated (HgII) and methylated (MeHg) mercury. In each control volume, concentrations of HgII and MeHg are represented by single values which include dissolved, particulate and plankton fractions. Mercury processes include the input of atmospheric mercury, sedimentation, reduction, methylation and demethylation. A sediment transport model MIKE 21 MT, developed by the Danish Hydraulic Institute was used to determine the spatial distribution of suspended sediment in the Gulf.

The first results of the described model (Rajar et al [6]) have shown excessive concentrations of all forms of mercury in the northern part of the Gulf, and too low concentrations in its southern part, compared to some measurements. A detailed analysis has shown four main reasons for the disagreement: (a) unrepresentativeness of measurements performed at too low discharges of the Soča river, when almost no sediment transport is present; (b) depth averaging due to the limitation of the two-dimensional model; (c) too coarse numerical grid near the river mouth and (d) too low dispersion of mercury forms which, in nature, is greater as a result of changing winds and an oscillating tide.

In the next step of the research our interest was directed towards item (d). In previous research, calibrations have shown the horizontal diffusion coefficient in 2D modelling to be of the order of 5 m^2/s. However, this was connected to steady-state measurements. In the case of 2D mercury modelling we tried to simulate the cycling as a steady state phenomenon. As it was found that the main forcing factor is wind, an average annual wind was determined, while the influence of the tide was neglected, as its maximum velocities are of the order of 7 cm/s only.

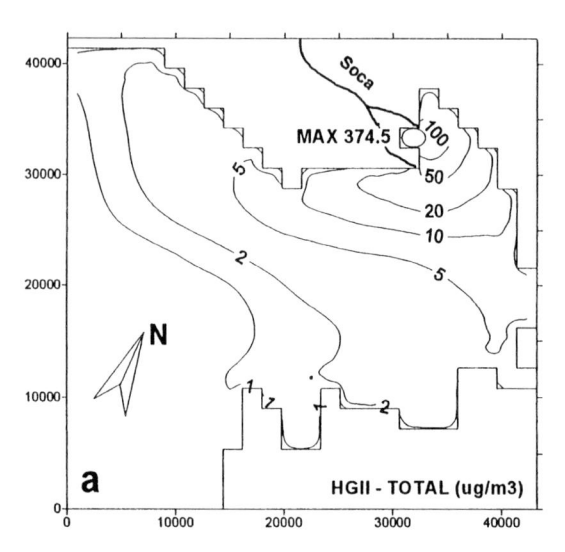

Fig. 2 Simulated concentration of total HgII in the Gulf water (dissolved, particulate and plankton HgII) with annual average wind and river inflow conditions.

It is well known that unsteady phenomena cause greater dispersion than steady-state simulations with average parameters. Therefore we tried to find a more appropriate value for the dispersion coefficient. Best-fit value of the dispersion coefficient was found to be K = 20 m^2/s.

A comparison between measured and computed values with both dispersion coefficients, the starting 5 m^2/s and best-fit 20 m^2/s, of dissolved, particulate, plankton and total inorganic mercury was done. (Širca and Rajar [9]). At most points, the calibrated results are of the same order of magnitude as the measured data, or at least the proper trend of concentration increase/decrease is noted. A considerable improvement was achieved at points in the central and southern part of the Gulf. Near the river mouth there still was a significant disagreement between the measured and the computed results, due to a too coarse numerical grid.

The results of the 2D simulation are presented in Fig. 2. The concentration distribution of the total HgII is presented. The simulation was done with the average annual wind (with wind speed of 1.61 m/s and direction of 73 ^{0}) and the Soča river discharge which was 150 m^3/s, with a suspended sediment concentration of 50 g/m^3. The input of Hg from the river, taken into account in the simulations, was obtained previously from a rough mass balance of mercury in the Gulf and was set to 1780 kg/year. The atmospheric Hg input was estimated to be 5 kg/year.

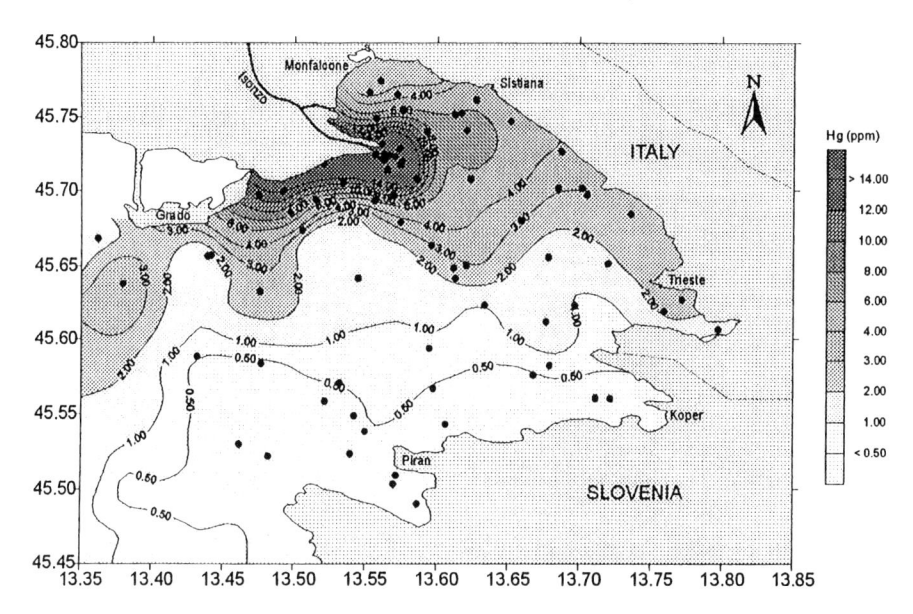

Fig. 3 Areal distribution of mercury in bottom sediments of the Gulf of Trieste (from Covelli et al. [1]).

An interesting qualitative comparison with measured results in Fig. 3 can be made. The concentration of total mercury in the bottom sediments are presented. The quantitative values of the computed and measured results cannot be directly compared as the computed results (Fig. 2) are for concentration of total HgII *in water*. However, the distribution in water (and suspended particles) is quite similar to the distribution of mercury in sediments. These two phenomena should really have similar trends as the majority of mercury inflow is from the Soča river and it is and deposited (on the particles) to the bottom sediments.

3 Description of the Three-Dimensional Model

In further research, measurements have shown that several parameters are far from being uniformly distributed over the depth. Therefore a three-dimensional (3D) model was needed. We have completed the existing 3D hydrodynamic and transport-dispersion model PCFLOW3D developed at the University of Ljubljana. Also, measurements and experience on river flow and suspended sediment transport show that most of a total annual sediment transport can be carried out during a single high flood. This requires an *unsteady* model, with which we can simulate such extreme events. Besides, it was found that most mercury transport is related to suspended sediment particles, consequently a new 3D sediment transport module was developed and included into the PCFLOW3D model. A short description of both sub-models is given below.

3.1 Hydrodynamic and transport-dispersion modules

In the 1980s, in the Faculty of Civil and Geodetic Engineering of the University of Ljubljana, we developed a 3D numerical model called *PCFLOW3D*, which is continuously being improved, calibrated and applied to practical hydrodynamic and pollutant dispersion problems. This is a non-linear baroclinic model, composed of three modules: a hydrodynamic (HD) module, a transport-dispersion (TD) module, and a recently developed sediment-transport (ST) module. The model is based on the finite volume method, the system of differential equations is solved using a hybrid (central-upwind) implicit scheme. In the horizontal plane the turbulent (eddy) viscosity coefficients are constant, while in the vertical direction the simplified one-equation turbulence model of Koutitas (Rajar et al [6]) is included. Recently a simulation of some biochemical processes was included. The model is described in detail in Rajar and Četina [3] and in Rajar et al. [6].

The TD module, which is solved coupled with the HD module, simulates temperature, salinity or any contaminant which can influence the water density and with it also the velocity field. The simulation of transport and dispersion of heat from heat sources and from the atmosphere was also included recently (Rajar and Širca [7]) to enable the simulation of thermal pollution in rivers, lakes and coastal seas.

3.2 The sediment-transport module

The sediment transport module is based on the equations of Van Rinj [11] and is similar to the 3D sediment-transport module described in Lin and Falconer [2]. Non-cohesive sediment material can be simulated. The module basically resolves the advection-diffusion equation for suspended sediment concentration, where the empirical equation for the sedimentation velocity (Van Rijn [11]) of the particles is accounted for. Resuspension of sediments from the bottom, which is important in the described case, is calculated first and depends on the bottom shear stress caused by current velocities (these are the result of the HD module). The boundary condition at the bottom is sedimentation of the particles when the shear velocities at the bottom are smaller than critical, or resuspension of the bottom particles when the shear velocities are greater than critical. The continuity equation of the sediment mass is used to calculate erosion/deposition at the bottom.

The module was completed with the calculation of wave parameters and their influence on the bottom shear stresses. This phenomenon is especially important with strong winds in shallow northern regions of the Gulf.

4 Simulated Cases and Results

This part of the research was directed especially towards the simulations of transport, dispersion and deposition of suspended sediments from the river, as well as resuspension and relocation of the sediments in the vicinity of the Soča's mouth. We no longer took into account average annual conditions, but instead seasonal average conditions with some peak flow and wind inserts.

4.1 Case 1 - Average winter conditions

Input Data. The seasonally averaged wind and river inflow for winter conditions were taken into account: wind 2.2 m/s from ENE (66° 40'), inflow of the Soča at 144 m³/s. As measurements showed that at this low inflow discharge the concentration of suspended sediments is extremely low, no sediment inflow with the river was taken into accont. The Gulf is supposed to be well mixed during the winter months, with a water temperature of 8° C and salinity of 37.8 ‰ constant along the depth. Other input data for the simualtion were: sediment density 2500 kg/m³, porosity 0.6, and mean diameter of sediment particles D_{50}=15 μm.

Results. As expected there was *no* resuspension of the bottom sediments, since the critical value was not exceeded. Within the Gulf, bed shear stress nowhere exceeded 30 % of the critical bed shear stress which is $\tau_{b,cr}$ = 0.179 N/m².

4.2 Case 2 - NE wind - 13 m/s

Input Data. To simulate realistic conditions when the bottom sediment resuspension should be important, we simulated strong winds of 13 m/s

296 Environmental Coastal Regions

from ENE approx. (66° 40'), for which statistical data show to occur several times a year, mostly during the winter season. A two-day duration was simulated and seasonally averaged river inflow (144 m³/s) was accounted for. Other data are the same as in Case 1.

Results. Fig. 4 presents the velocity field, which is computed by the 3D model PCFLOW3D, but is depth-averaged. Maximum flow velocities of up to about 35 cm/s are attained near the northern shore, where the depth is shallow (at a distance of 5 km from the northern shore the depth is under 10 meters). Since the influence of wind waves on the bottom shear stress is greatest in shallow regions, the resuspension is greatest along the N shore.

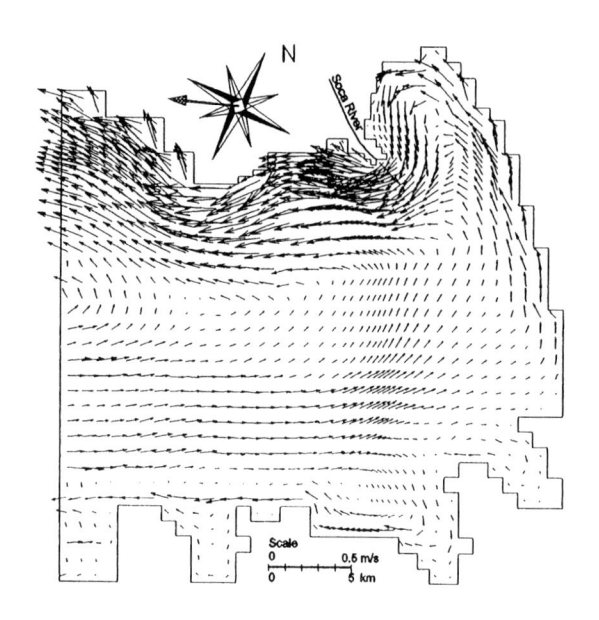

Fig. 4 Velocity field for Case 2: Wind speed 13 m/s, from ENE. Velocities are depth averaged from 3D simulation. Soča river inflow is 144 m³/s (average winter discharge).

Fig. 5 shows the erosion/deposition height of the bottom sediments. Resuspension occurs in the northern part of the Gulf, along the Italian coast, where water depth does not exceed 7 to 10 m. The maximum bed shear stresses were approx. 10 times greater than the critical value. It is interesting to note that obviously strong winds cause resuspension of mercury-rich sediments near the river's mouth and displace them along the northern coast where they are deposited near Grado. This distribution is again similar to the measured distribution of mercury in the bottom sediments in Fig. 3.

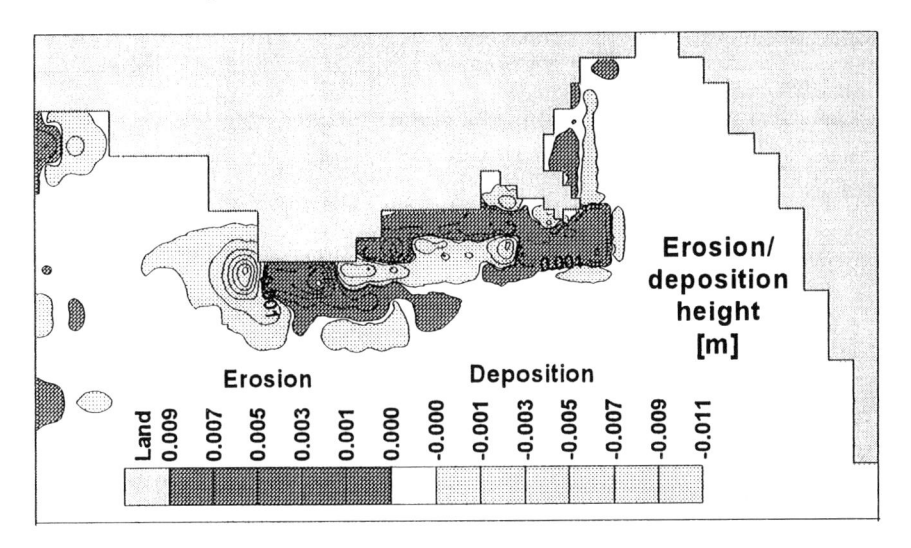

Fig. 5 Simulated erosion/deposition height of the sediments for Case 2: Wind ENE of 13 m/s, Soča river inflow is 144 m^3/s (average winter discharge).

4.3 Case 3, sediment inflow from the Soča river

Input Data. As it is supposed that most of the sediment is being transported into the Gulf during one or two annual floods that usually occur in fall, a simulation has been done with a real discharge curve of the Soča between November 6 and 14 1997. It was a two-peak flood, with the first peak at about 2200 m^3/s and the second at 1600 m^3/s. A total amount of about 470 million m^3 of water has been discharged into the Gulf. As the concentration of suspended matter in the water was 365 g/m3, the total suspended matter inflow during the described flood was about 180.000 tons. A previous rough estimation of annual inflow (Širca [8]) was 240.000 tons.

The concentration of Hg on suspended matter was measured at the discharge of 1600 m^3/s and was 49 µg/g. Another comment about the mercury budget: with these data the inflow of particulate Hg into the Gulf during a single flood is more than 8000 kg. This is much more than the previously estimated value of 1780 kg (Širca [8]). However, as the described flood is statistically with a 5 year return period, the mass of incoming Hg is over the annual average.

Other data are the same as in cases 1 and 2.

298   Environmental Coastal Regions

Results. Fig. 6 shows the peak concentrations of suspended matter, where the conclusion can be made how far from the river mouth suspended matter can be transported.

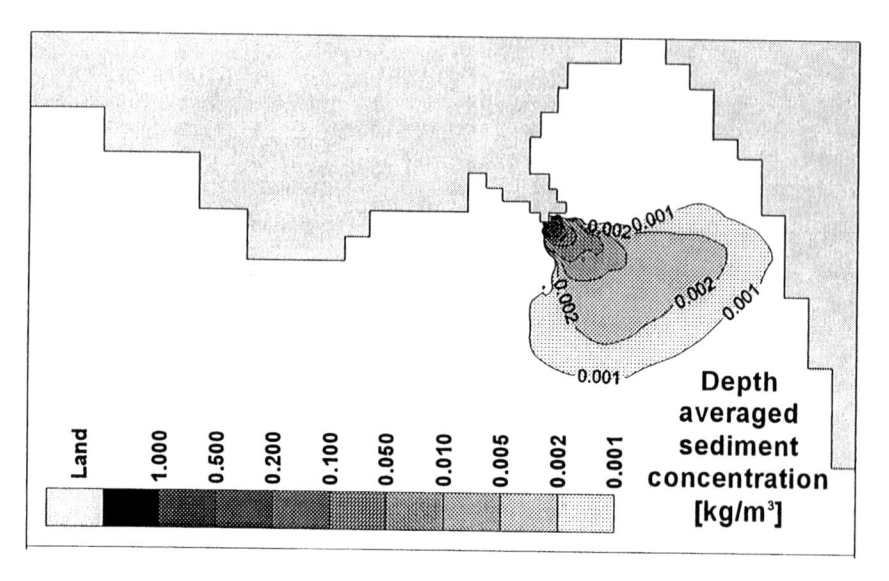

Fig. 6 Simulated depth-averaged suspended sediment concentration. Results are presented 4 hrs after the maximum peak river discharge, when max. spreading of suspended sediments is attained.

5 Conclusions

As the previous research showed that most of the mercury is transported with sediment particles, this part of the research was dedicated to simulations of the sediment transport, both from the Soča river and from the resuspension of the bottom sediments.

The results of the simulation in Case 3 (peak inflow from the Soča river, see Fig. 6) show that, due to settling, the sediments are deposited relatively near the river mouth. But the results of Case 2 show (Fig. 5) that during periods of strong ENE wind, which is statistically the most frequent wind in the Gulf, cause resuspension of these sediments which are further transported towards the west and deposited mostly near Grado. So the combination of the two phenomena can be a plausible explanation of the measured mercury distribution in the bottom sediments presented in Fig. 3.

What is the connection between the mercury concentration in the suspended sediment of the river, in the suspended sediment of the Gulf itself and in the bottom sediments in the Gulf?

Measurements of mercury concentration in the suspended sediments of the river were made during a high water discharge in November 1997. At the discharge of 1600 m^3/s, mercury concentration in the suspended matter was 49.6 µg/g. A measurement in 1996 (Širca and Rajar [9]) showed that the concentration of particulate HgII (which is very near to total Hg) on suspended matter *inside the Gulf* was between 0.6 and 41 µg/g. Measurements (Fig. 3) show that in the bottom sediments of the Gulf, concentrations of total Hg are over 14 µg/g near the river mouth, but decrease to about 3 µg/g near Grado and are below 0.5 µg/g in the southern part of the Gulf.

Thus in the bottom sediments of the Gulf the concentration is lower than in the suspended mater in the river. There are several possible causes for this: (a) The Hg concentration on suspended matter is much lower with low river discharge, when there is almost no erosion of flood plains along both rivers. At a low discharge (about 100 m^3/s) in October 1997, the measured concentration of Hg on suspended matter in the river was below 3 µg/g. Although the mass of these sediments is much smaller than during high waters, it contributes to the diminished average Hg concentration in the bottom sediments. (b) The suspended sediments in the river, when they arrive into the sea water, are diluted by coagulation and when settled they are mixed with detritus, which has much lower Hg concentration (c) As was shown by the results of Case 2, strong winds and currents cause important displacements, mixing and mutual covering of the bottom sediment layers with a different Hg concentration. (d) Due to bio-geochemical processes of mercury cycling, mercury in the bottom sediments is permanently leached and transformed into other forms.

In the previous step of the research (2D modelling) some of these reactions were included, more or less successfully. In the next step we shall concentrate our efforts on item (d). Simulation of bio-geochemical processes, revised and based on more numerous measurements for calibration will be included in the 3D model.

References

[1] Covelli, S., Faganeli, J., Horvat, M., Brambati, A. (1998) Benthic fluxes of mercury and methylmercury in the Gulf of Trieste. *Geochemistry*, Submitted.

[2] Lin, B., and Falconer, R.A. 1996: Numerical Modelling of 3D Suspended Sediment for Estuarine and Coastal Waters, *Journal of Hydraulic Research*, Vol. 4, pp. 435-455, 1996.

[3] Rajar, R., Četina, M.,.: Hydrodynamic and Water Quality Modelling: An Experience. *Ecological Modelling*, 101, pp. 195-207, 1997.

[4] Horvat, M., Faganelli, J., Planinc, R., Prosenc, N., Azemard, S., Coquery, M., Širca, A., Rajar, R., Byrne, A.R., Covelli, S.: Mercury Pollution in Trieste Bay . *Int. Conf. Mercury as Global Pollutant*, Hamburg, 4-8. August 1996, (poster).

[5] Horvat, M., Faganelli, J., Planinc, R., Logar, M., Mandić, V., Rajar, R., Širca, A., Žagar, D., Covelli, S.: The impact of mercury mining on the Gulf of Trieste. 2nd Int. Conf. Coastal Environment, Cancun, 8-10 Sept. 1998.

[6] Rajar, R., Četina, M., Širca, A.,.: Hydrodynamic and Water Quality Modelling: Case Studies. *Ecological Modelling*, 101, pp. 209-228., 1997.

[7] Rajar, R. and Širca, A.: Three-Dimensional Modelling of Thermal Pollution of the River Sava and its Reservoirs. 3rd *int. Conf. on Hydroscience and Engineering*, Cottbus, 1998 (submitted).

[8] Širca, A. (1996). Modelling of the hydrodynamics and of the transport of mercury compounds in Trieste Bay. Ph.D. *Thesis, University of Ljubljana*, Slovenia. (In Slovene, extended abstract in English).

[9] Širca, A., Rajar, R.: Calibration of a 2D mercury transport and fate model of the Gulf of Trieste. *Proc. of the 4th Int. Conf. Water Pollution 97*, Eds. Rajar, R. and Brebbia, M., Computational Mechanics Publication, Southampton. pp. 503-512. 1997.

[10] Širca, A., Horvat, M., Covelli, S., Faganelli, J., Rajar, R.: Mercury Fluxes and mass balance in the Gulf of Trieste, *Proc. of the 5th Intern. Conf. Mercury as a Global Pollutant*, Rio de Janeiro, 1999. (submitted).

[11] Van Rijn,: *Principles of Sediment Transport in Rivers, Estuaries and Coastal Sea*, Aqua publications, Amsterdam, 490 pp., 1993.

Nonlinear model of random wave load on pipelines

Živko Vuković & Neven Kuspilić
*Faculty of Civil Engineering, University of Zagreb,
Kačićeva 26, 10 000 Zagreb, Croatia
EMail: kuspa@master.grad.hr*

Abstract

In this paper the authors present a developed analytical procedure for predicting the wave load on submarine pipelines to the action of random waves, representing the nonlinear drag force by cubic approximation. Using Morison's equation, the spectra of the horizontal and vertical component of wave force on pipelines are defined by the spectral analysis method. A numerical example is included to demonstrate the potential of the analytical model and to quantify the effects of nonlinear drag force on the structural load.

1 Introduction

The in-line wave forces acting on a slender member of marine structure consist of two parts; drag and inertia. In the case when the wave length is much larger than the characteristic structural diameters, the drag forces can be significant. Since the drag forces, as normally defined through Morison's equation, are nonlinear, it is difficult to obtain an analytical solution, particularly when sea state is random.

With some assumptions and additional statistical approximations, one can easily study the above problem. The same is applied in this paper, which deals with the nonlinear drag force along the lines given by Borgman.[1,2]

In fact, an attempt is made to exemplify and quantify the effects of nonlinear drag force on the second-order statistics of pipeline wave loading.

2 Assumptions

(a) In the following analysis the linear wave theory is employed to relate surface wave parameters and wave motions. The sea surface elevation is assumed to be an ergodic, normally (Gaussian) distributed random process of zero mean.

(b) It is assumed that on the analysed section the submarine pipeline is laid horizontally; that it is of a constant outside diameter; rigid; fixed; slender (in the sense that its characteristic dimension - diameter is relatively small in relation to wave length); and that it is located in the horizontal co-ordinate $x = 0$.

(c) For the evaluation of wave force, Morison's equation[3] is used in this study. That is, wave force, $F(z,t)$, is considered to be consisting of two parts; the drag component, nonlinearly related to water particle velocity; and the inertia component, linearly proportional to water particle acceleration. Horizontal component, $F_x(z,t)$, and vertical component, $F_z(z,t)$, of this force per unit length on the submarine pipeline, Fig. 1, can be expressed as:

$$F_x(z,t) = F_{Dx}(z,t) + F_{Ix}(z,t) = u_D v_x(z,t)|v_x(z,t)| + u_I a_x(z,t) \quad (1)$$

$$F_z(z,t) = F_{Dz}(z,t) + F_{Iz}(z,t) = u_D v_z(z,t)|v_z(z,t)| + u_I a_z(z,t) \quad (2)$$

where:

$$u_D = 0.50 C_D \rho D \quad (3)$$

$$u_I = 0.25 C_I \rho \pi D^2 \quad (4)$$

In the above equations $F_{Dx}(z,t)$ and $F_{Ix}(z,t)$ = horizontal components of drag and inertia force, respectively; $F_{Dz}(z,t)$ and $F_{Iz}(z,t)$ = vertical components of drag and inertia force, respectively; C_D and C_I = drag and inertia coefficients, respectively; ρ = water density; D = outside pipeline diameter; x and z =

Environmental Coastal Regions 303

horizontal and vertical co-ordinates, i.e. horizontal and vertical axes, respectively; $v_x(z,t)$ and $v_z(z,t)$ = horizontal and vertical components of water particle velocity, respectively; $a_x(z,t)$ and $a_z(z,t)$ = horizontal and vertical components of water particle acceleration, respectively; and t = time.

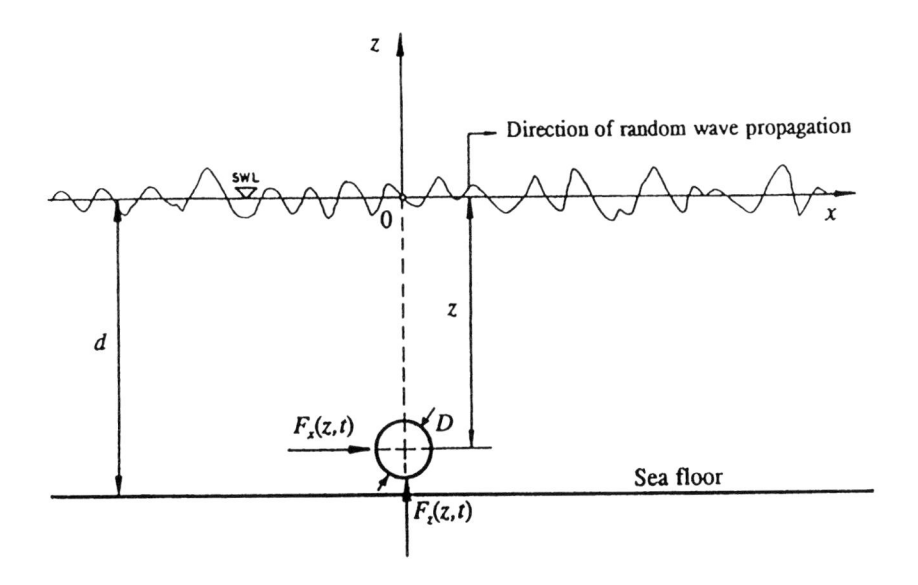

Figure 1: Definition drawing of submarine pipeline wave load

3 Review of wave load analysis

For the case of deep water random waves action, Borgman[1,2] obtained the theoretical autocorrelation function, $R_{FF}(z,\tau)$, for the horizontal force, $F(z,t)=F_x(z,t)$, on a rigid and vertical cylinder per unit length, as:

$$R_{FF}(z,\tau) = u_D{}^2(\sigma_{v_x}^0)^4(z)G\left[\frac{R_{v_x v_x}(z,\tau)}{(\sigma_{v_x}^0)^2(z)}\right] + u_I{}^2 R_{a_x a_x}(z,\tau) \qquad (5)$$

where:

$$G(r) = \frac{(4r^2+2)\arcsin r + 6r\sqrt{1-r^2}}{\pi} \qquad (6)$$

$$r = \frac{R_{v_x v_x}(z,\tau)}{(\sigma_{v_x}^0)^2(z)} \tag{7}$$

$$\sigma_{v_x}^0(z) = \left[\int_0^\infty S_{v_x v_x}^0(z,\omega)\,d\omega\right]^{1/2} \tag{8}$$

$$S_{v_x v_x}^0(z,\omega) = \left(\omega\,\frac{\cosh[k(z+d)]}{\sinh kd}\right)^2 S_{\eta\eta}^0(\omega) \tag{9}$$

In the above equations $\sigma_{v_x}^0(z)$ = standard deviation of $v_x(z,t)$; $R_{v_x v_x}(z,\tau)$ and $R_{a_x a_x}(z,\tau)$ = autocorrelation functions of $v_x(z,t)$ and $a_x(z,t)$, respectively; τ = time lag in autocorrelation functions; $S_{v_x v_x}^0(z,\omega)$ = deep water spectrum of $v_x(z,t)$; $S_{\eta\eta}^0(\omega)$ = deep vater (uni-directional) wave spectrum; ω = wave frequency; k = wave number; and d = water depth.

The wave number is related to wave frequency by:

$$\omega^2 = gk\tanh kd \tag{10}$$

where g = acceleration of gravity.

The function $G(r)$ can be expanded in a power series in r as follows:

$$G(r) = \frac{1}{\pi}\left(8r + \frac{4r^3}{3} + \frac{r^5}{15} + \frac{r^7}{70} + \frac{5r^9}{1008} + \ldots\right) \tag{11}$$

This series converges quite rapidly for $0 \le r \le 1$.

At $r = 1$, the linear approximation:

$$G_1(r) = \frac{8r}{\pi} \tag{12}$$

differs from $G(r)$ by 15 %.

The cubic approximation:

$$G_3(r) = \frac{1}{\pi}\left(8r + \frac{4r^3}{3}\right) \tag{13}$$

differs from $G(r)$ at $r=1$ by only 1.1 %. That means, if the drag contribution is approximated by the two first terms in the series, the maximum error is only of the order of 1 %.

Hence, Eq. (5) may be recast as follows with good accuracy for most engineering problems:

$$R_{FF}(z,\tau) = \frac{u_D^2(\sigma_{v_x}^0)^4(z)}{\pi}\left[\frac{8R_{v_x v_x}(z,\tau)}{(\sigma_{v_x}^0)^2(z)} + \frac{4R_{v_x v_x}^3(z,\tau)}{3(\sigma_{v_x}^0)^6(z)}\right] + u_I^2 R_{a_x a_x}(z,\tau) \tag{14}$$

The Fourier transform of $R_{FF}(z,\tau)$ gives the spectral density of $F(z,t)$ as:

$$S_{FF}^0(z,\omega) = \frac{u_D^2(\sigma_{v_x}^0)^4(z)}{\pi}\left\{\frac{8S_{v_x v_x}^0(z,\omega)}{(\sigma_{v_x}^0)^2(z)} + \frac{4[S_{v_x v_x}^0(z,\omega)]^{*3}}{(3\sigma_{v_x}^0)^6(z)}\right\} + u_I^2 S_{a_x a_x}^0(z,\omega) \tag{15}$$

where the three-hold convolution of $S_{v_x v_x}^0(z,\omega)$ with itself is given as:

$$[S_{v_x v_x}^0(z,\omega)]^{*3} = S_{v_x v_x}^0(z,\omega) * S_{v_x v_x}^0(z,\omega) * S_{v_x v_x}^0(z,\omega) =$$
$$= \int_{-\infty}^{\infty}\int_{-\infty}^{\infty}[S_{v_x v_x}^0(z,\omega'')S_{v_x v_x}^0(z,\omega'-\omega'')d\omega'']S_{v_x v_x}^0(z,\omega-\omega')d\omega' \tag{16}$$

and the deep water spectrum of $a_x(z,t)$ is defined as:

$$S_{a_x a_x}^0(z,\omega) = \left(\omega^2 \frac{\cosh[k(z+d)]}{\sinh kd}\right)^2 S_{\eta\eta}^0(\omega) \tag{17}$$

4 Nonlinear model of random wave load on pipelines

As shown by Vuković & Kuspilić[4] and by analogy with Eqs. (1), (2) and (15), it follows that the spectra of the horizontal component, $S_{F_x F_x}(z,\omega)$, and vertical component, $S_{F_z F_z}(z,\omega)$, of random wave force per unit length of submarine pipelines, are defined by the equations:

$$S_{F_xF_x}(z,\omega) = \frac{u_D^2 \sigma_{v_x}^4(z)}{\pi}\left\{\frac{8 S'_{v_xv_x}(z,\omega)}{\sigma_{v_x}^2(z)} + \frac{4[S'_{v_xv_x}(z,\omega)]^{*3}}{3\sigma_{v_x}^6(z)}\right\} + u_I^2 S_{a_xa_x}(z,\omega) \quad (18)$$

$$S_{F_zF_z}(z,\omega) = \frac{u_D^2 \sigma_{v_z}^4(z)}{\pi}\left\{\frac{8 S'_{v_zv_z}(z,\omega)}{\sigma_{v_z}^2(z)} + \frac{4[S'_{v_zv_z}(z,\omega)]^{*3}}{3\sigma_{v_z}^6(z)}\right\} + u_I^2 S_{a_za_z}(z,\omega) \quad (19)$$

where:

$$\sigma_{v_x}(z) = \left[\int_0^\infty S'_{v_xv_x}(z,\omega)d\omega\right]^{1/2} \quad (20)$$

$$\sigma_{v_z}(z) = \left[\int_0^\infty S'_{v_zv_z}(z,\omega)d\omega\right]^{1/2} \quad (21)$$

$$S'_{v_xv_x}(z,\omega) = \left(\omega\frac{\cosh[k(z+d)]}{\sinh kd}\right)^2 S'_{\eta\eta}(\omega) \quad (22)$$

$$S'_{v_zv_z}(z,\omega) = \left(\omega\frac{\sinh[k(z+d)]}{\sinh kd}\right)^2 S'_{\eta\eta}(\omega) \quad (23)$$

$$S_{a_xa_x}(z,\omega) = \left(\omega^2\frac{\cosh[k(z+d)]}{\sinh kd}\right)^2 S'_{\eta\eta}(\omega) \quad (24)$$

$$S_{a_za_z}(z,\omega) = \left(\omega^2\frac{\sinh[k(z+d)]}{\sinh kd}\right)^2 S'_{\eta\eta}(\omega) \quad (25)$$

$$[S_{v_xv_x}(z,\omega)]^{*3} = \int_{-\infty}^{\infty}\int_{-\infty}^{\infty}[S_{v_xv_x}(z,\omega'')S_{v_xv_x}(z,\omega'-\omega'')d\omega'']S_{v_xv_x}(z,\omega-\omega')d\omega' \quad (26)$$

$$[S_{v_zv_z}(z,\omega)]^{*3} = \int_{-\infty}^{\infty}\int_{-\infty}^{\infty}[S_{v_zv_z}(z,\omega'')S_{v_zv_z}(z,\omega'-\omega'')d\omega'']S_{v_zv_z}(z,\omega-\omega')d\omega' \quad (27)$$

In the above equations $\sigma_{v_x}(z)$ and $\sigma_{v_z}(z)$ = standard deviations of $v_x(z,t)$ and

Environmental Coastal Regions   307

$v_z(z,t)$, respectively; $S_{v_x v_x}(z,\omega)$ and $S_{v_z v_z}(z,\omega)$ = spectra of $v_x(z,t)$ and $v_z(z,t)$, respectively; $S_{a_x a_x}(z,\omega)$ and $S_{a_z a_z}(z,\omega)$ = spectra of $a_x(z,t)$ and $a_z(z,t)$, respectively; $[S_{v_x v_x}(z,\omega)]^{*3}$ and $[S_{v_z v_z}(z,\omega)]^{*3}$ = three-hold convolutions of $S_{v_x v_x}(z,\omega)$ and $S_{v_z v_z}(z,\omega)$ with itself, respectively; and $S_{\eta\eta}(\omega)$ = deformed, directional wave spectrum.

Substituting the velocity and acceleration spectrum functions from Eqs. (22), (23), (24) and (25) into Eqs. (18) and (19), it follows that:

$$S_{F_x F_x}(z,\omega) = \frac{8}{\pi}\left\{\omega u_D \sigma_{v_x}(z)\frac{\cosh[k(z+d)]}{\sinh kd}\right\}^2 S_{\eta\eta}(\omega) + \frac{4u_D^2[S_{v_x v_x}(z,\omega)]^{*3}}{3\pi\sigma_{v_x}^2(z)} + \left\{\omega^2 u_I \frac{\cosh[k(z+d)]}{\sinh kd}\right\}^2 S_{\eta\eta}(\omega) \quad (28)$$

$$S_{F_z F_z}(z,\omega) = \frac{8}{\pi}\left\{\omega u_D \sigma_{v_z}(z)\frac{\sinh[k(z+d)]}{\sinh kd}\right\}^2 S_{\eta\eta}(\omega) + \frac{4u_D^2[S_{v_z v_z}(z,\omega)]^{*3}}{3\pi\sigma_{v_z}^2(z)} + \left\{\omega^2 u_I \frac{\sinh[k(z+d)]}{\sinh kd}\right\}^2 S_{\eta\eta}(\omega) \quad (29)$$

Thus, in order to calculate wave load spectra according to equations (28) and (29), it is necessary to know the deformed, directional wave spectrum for a partical location for which the load is to be calculated, because each model of the wave load of submarine pipelines laid on the seabed, or close to it, gives load values only in intermediate depth and shallow water, that is for the case of the interaction of waves with the seabed when the deformation of wave parameters occurs.

In the case for two dimensional seabed, a directional wave spectrum, deformed by shoaling and refraction, is defined[4,5] as:

$$S_{\eta\eta}(\omega) = \int_{\theta_{min}}^{\theta_{max}} S_{\eta\eta}(\omega,\theta)d\theta = \int_{\theta_{min}}^{\theta_{max}} K_s^2(\omega,d)K_r^2(\omega,d,\theta_o)\frac{\partial\theta_o}{\partial\theta}S_{\eta\eta}(\omega)G(\theta_o)d\theta \quad (30)$$

Here $K_s(\omega,d)$ = shoaling coefficient of particular spectral wave component of frequency ω at depth d; $K_r(\omega,d,\theta_o)$ = refraction coefficient of particular spectral wave component of frequency ω at depth d and propagation direction θ_o; $G(\theta_o)$ = function of propagation of wave energy or spectral wave components; θ_o = angle of propagation direction of particular spectral wave component in deep water with respect to the normal to isobaths; and θ = angle of propagation direction of particular spectral wave component with respect to the normal to isobaths.

Also, on of the main difficulties in practical application of Eqs. (28) and (29) lies in the evaluation of the convolution integrals, Eqs. (26) and (27), which is a rather time consuming numerical operation.

5 Numerical example

In order to illustrate the previous analysis, a numerical example of random wave load on a hypothetical submarine pipeline laid in the intermediate depth is presented in the text below.

It is assumed that the water depth at the pipeline location, d = 15.0 m; water density, ρ = 1025.0 kg/m³; acceleration of gravity g = 9.81 m/s²; vertical co-ordinates of pipeline horizontal axis, z = - 14.0 m (Fig. 1); angle of random wave propagation with respect to the normal of isobaths θ_o = 45°; pipeline outside diameter, D = 0.8 m; and the values of drag and inertia coefficients, C_D = 1.0 and C_I = 2.0, respectively.

For the wave spectrum we will use Tabain's[6] wave spectrum applicable for the Adriatic Sea conditions, for the deep water significant wave hight, H_s^0 = 6.0 m.

The so selected random wave climate parameters and the diameter of a hypothetical pipeline grants the drag loading regime.

For the purpose of evaluation of the convolution integrals given by Eqs. (26) and (27), an effective computation algorithm is developed by means of the fast Fourier transformation technique.

The calculation results are shown in Fig. 2 (a) to 2 (c). Fig. 2 (a) represents Tabain's wave spectrum. The horizontal force spectra, $S_{F_x F_x}(z,\omega)$, from Fig. 2 (b), and vertical force spectra, $S_{F_z F_z}(z,\omega)$, from Fig. 2 (c), are calculated according to Eqs. (28) and (29), respectively.

(a) Tabain's wave spectrum, $H_s^0 = 6.0$ [m]

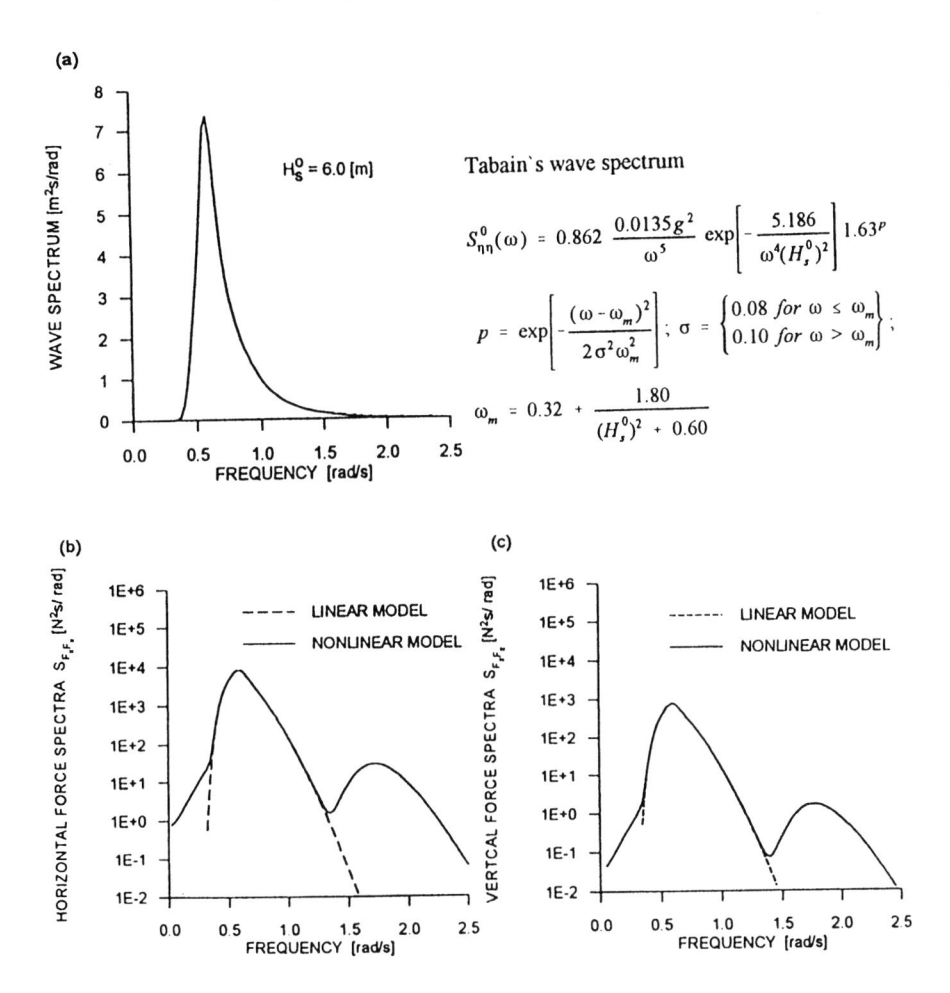

$$S_{\eta\eta}^0(\omega) = 0.862 \frac{0.0135 g^2}{\omega^5} \exp\left[-\frac{5.186}{\omega^4 (H_s^0)^2}\right] 1.63^p$$

$$p = \exp\left[-\frac{(\omega - \omega_m)^2}{2\sigma^2 \omega_m^2}\right]; \quad \sigma = \begin{cases} 0.08 \text{ for } \omega \leq \omega_m \\ 0.10 \text{ for } \omega > \omega_m \end{cases};$$

$$\omega_m = 0.32 + \frac{1.80}{(H_s^0)^2 + 0.60}$$

Figure 2: Wave spectrum and random wave load spectra for a hypothetical pipeline

It can be clearly seen from Fig. 2 (b) and 2 (c) that the influence of the drag force nonlinear term is considerable, both for the increase of the spectra peak values and for the increase of the total area below the spectra. Besides, the occurrence of spectra secondary peaks is evident in the region of higher frequencies, which can in the case of a certain type of structures cause a significantly higher displacement response that cannot be defined by the application of linear model.

This points to the need of taking into account the drag force nonlinear term while calculating the load and response of pipelines located in the drag loading regime.

6 Conclusion

A nonlinear random wave load effects on submarine pipelines are presented by applying the spectral analysis method. The performed analyses and the results of the numerical example point to the considerable influence of the drag force nonlinear term in the calculation of the wave load of pipelines located in the drag loading regime. It is therefore justified to use in engineering problems for such cases a nonlinear model of random wave load.

References

1. Borgman, L. E. Spectral analysis of ocean wave force on piling, *Journal of the Waterways and Harbors Division*, 1967, Vol. 93, WW2, pp. 129-156.
2. Borgman, L. E. Statistical models for ocean waves and wave forces, *Advances in Hydroscience*, Vol.8, Academic Press, London, 1972.
3. Morison, J. R., O'Brien, M. P., Johnson, J. W. & Schaaf, S. A. The force exerted by surface waves on piles, *Petroleum Transactions*, 1950, Vol. 189, pp. 149-157.
4. Vuković, Ž. & Kuspilić, N. Spectral analysis of submarine pipelines wave loading (in Croatian), *Građevinar*, 1990, 42 (11). pp. 463 - 469
5. Karlsson, T. Wave spectrum changes on sloped beach, *Journal of the Waterway, Port, Coastal and Ocean Engineering*, 1982, Vol. 108, No. WW1, pp. 33 - 47.
6. Tabain, T. Forecasting of rolling motions of small ships under simultaneous action of natural waves and wind (in Croatian), Ph.D.Thesis, Faculty of Mechanical Engineering, University of Zagreb, Zagreb, 1985.

Optimal estimation of concentrations in particle-type transport models

J.W. Stijnen, A.W. Heemink & H.X. Lin[1]
H.F.P. van den Boogaard[2]
[1] Faculty of Information Technology and Systems (ITS)
Department of Applied Mathematics
Delft University of Technology, Mekelweg 4
2600 GA Delft, The Netherlands
EMail: j.stijnen@math.tudelft.nl
[2] Delft Hydraulics, Rotterdamse weg 185
2600 MH Delft, P.O. Box 177, The Netherlands

Abstract

When dealing with a pollutant in coastal waters, it is important to know where and how it is transported. Since the initial concentration gradient is very steep, often a particle model is used to simulate the transport of the pollutant. In a particle model the spread of a solute is simulated by means of a large set of tracks (of individual particles) and from this ensemble concentrations must be derived. This last step is far from trivial. Particles represent mass rather than concentrations, and while the particle positions are a finite and discrete set, the observed concentrations generally have a continuous and smooth spatial form.

This paper discusses some ideas, also known in density estimation theory, to solve this problem. Instead of using histogram-like functions (where the number of particles in a gridcell is counted), the concept of the point spread function (psf) is introduced. A psf spreads the mass of a particle across a small interval surrounding its position, thus providing a smoother estimate for the concentration profile of the pollutant. The asymptotic behaviour of the estimation error is analyzed for a large number of particles in both 1D and 2D problems. In the 1D case, optimal point spread functions with smallest estimation errors are derived. The various estimation procedures are then applied to realistic real-life water pollution problems.

1 Introduction

To predict the dispersion of a pollutant in rivers or coastal waters, often the well-known advection diffusion equation is employed (Fisher et al[2]). A possible solution can be found by using the Eulerian approach and solving the equation numerically. Alternatively one can use a Lagrangian approach and develop a random walk model (Heemink[3]). By interpreting the advection diffusion equation as a Fokker-Planck equation, a sochastic differential equation can be derived that is consistent with the advection diffusion equation. This concept lies at the basis of particle models. An advantage of the Lagrangian approach is that it can deal with arbitrarily steep concentration gradients (which is often the case at the point of release). Particle models are also inherently suited for a parallel implementation, which is another big advantage, especially when the model is quite large. Large numbers of particles are released in a certain point to simulate the transport of a pollutant (see figure 1) so that the ensemble of particles may give a good picture of the situation.

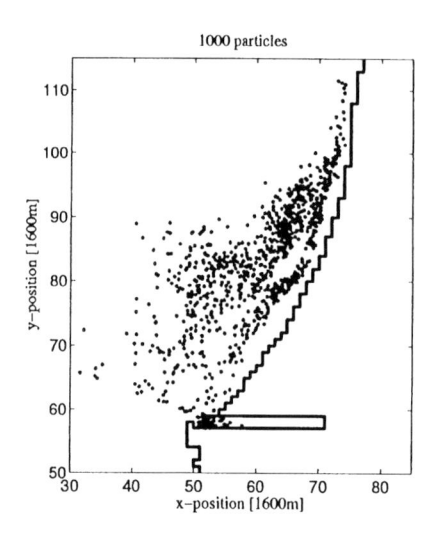

Figure 1: A typical example of a particle model simulation. In this case the model area is part of the Dutch coastal zone. A particle dump in the New Waterway was simulated with 1000 particles, and the result after 6 tidal periods is shown here.

The resulting mass distribution then needs to be converted to a concentration profile. Traditionally this is done by means of histograms. This paper discusses some ideas from density estimation theory (Silverman[4]) to develop accurate methods to estimate the concentration with the aid of *point spread functions*. That way it is possible to reduce the number of particles that has to be used in simulations. See also Van den Boogaard[1].

2 Point spread functions

2.1 The concentration estimator

Suppose that at a certain time the positions $\underline{X}_k = \left(X_k^{(1)}, \cdots, X_k^{(N)}\right)^T$ of K particles in N-dimensional Euclidean space are available. For simplicity we assume the total mass of all particles equals one, and that the mass of every particle is the same and equal to $1/K$. Instead of using a histogram-like approximation of the concentration (counting the number of particles within a certain gridcell) we use a psf that spreads the mass of each particle across neighbouring gridcells. By using psfs a much smoother and more accurate concentration profile can be obtained. When using a psf the two most important factors that determine how well the approximated concentration resembles the real one are *shape* and *width*.

In 1D the psf can have a triangular shape, a block shape, a parabolic shape or a Gaussian shape for example. Logical assumptions for the shape of the psf would be that they need to be positive (concentrations cannot be negative) and direction independent.

The width of the point spread function can be scaled according to the total number of particles. If more particles are used, more information is available. This means that more detail can be obtained in the approximated concentration profile by reducing the width of the psf. Of course when the width of the psf is changed, the height also has to change so that the amount of mass for the particle stays the same. The contribution $\mu_k(\cdot)$ of particle k located at $(X_k^{(1)}, \cdots, X_k^{(N)})$ to the total mass distribution is

$$\mu_k(\underline{x}) = \frac{1}{K}\phi\left(\underline{x} - \underline{X}_k\right) \qquad (1)$$

where $\underline{x}$ is the location where we want to have a concentration estimate. ϕ is an arbitrary psf in N dimensions with

$$\int_{x^{(1)}=-\infty}^{\infty} \cdots \int_{x^{(N)}=-\infty}^{\infty} \phi(\underline{x})dX = 1. \qquad (2)$$

and

$$dX = dx^{(1)} dx^{(2)} \cdots dx^{(N)} \qquad (3)$$

We can also use a scaling parameter $f(K)$ that depends on the total number of particles to obtain a scaled psf:

$$\tilde{\phi}(\underline{\xi}) = \frac{\left(f(K)\right)^N}{K}\phi\left(f(K)\underline{\xi}\right), \qquad \underline{\xi} = \underline{x} - \underline{X}_k \qquad (4)$$

When more particles are used in the simulation, more information about the sought after density is available. This means that the width of the psf can be reduced. The exact concentration $C(\underline{x})$ can be approximated by the following estimator:

$$C_K(\underline{x}) = \sum_{k=1}^{K} \tilde{\phi}\left(\underline{x} - \underline{X}_k\right) \qquad (5)$$

The total amount of mass is of course independent of the type or scaling of the psf.

2.2 The BIAS of the estimator

Since $C_K(\underline{x})$ is an estimator of $C(\underline{x})$, it is possible to evaluate the bias in $C_K(\underline{x})$ under the assumptations that the number of particles is large and that the probability density function $C(\underline{x})$ is sufficiently smooth. We have

$$\text{BIAS}\left\{C_K(\underline{x})\right\} = E\left\{C_K(\underline{x})\right\} - C(\underline{x}) \qquad (6)$$

After some calculations we find the following expression for our bias:

$$\text{BIAS}\left\{C_K(\underline{x})\right\} = \frac{\nabla_x^T \mathbf{SS}^T \nabla_x C(\underline{x})}{2\left(f(K)\right)^2} + \mathbf{O}\left(\frac{1}{\left(f(K)\right)^3}\right) \qquad (7)$$

Note that $f(K) \to \infty$ when $K \to \infty$, and

$$\mathbf{SS}^T = \mathbf{P} = \int_{x^{(1)}=-\infty}^{\infty} \cdots \int_{x^{(N)}=-\infty}^{\infty} \underline{x}\,\underline{x}^T\,\phi(\underline{x})\,dX \qquad (8)$$

The matrix $\mathbf{P}$ is a covariance matrix of the variable $\underline{X}_k$ with psf $\phi(\underline{x})$. In the above derivation we have used the fact that the covariance matrix satisfies $0 < \mathbf{P} < \infty$ and that $\phi(\underline{x})$ is symmetrical in the direction of every axis.

2.3 Calculating the MSE

We can do something similar for the mean square error of $C_K(\underline{x})$:

$$\text{MSE}(\underline{x}) = E\left\{\left(C_K(\underline{x}) - C(\underline{x})\right)^2\right\} \qquad (9)$$

$$= C(\underline{x})\frac{\left(f(K)\right)^N}{K} \cdot \qquad (10)$$

$$\int_{x_k^{(1)}=-\infty}^{\infty} \cdots \int_{x_k^{(N)}=-\infty}^{\infty} \left(\widetilde{\phi}\left(\underline{x}-\underline{x}_k\right)\right)^2 dX_k +$$

$$+ \left(\frac{\nabla_x^T \mathbf{S}\mathbf{S}^T \nabla_x C(\underline{x})}{2\left(f(K)\right)^2}\right)^2 + \text{H.O.T.} \qquad (11)$$

Here we have used the fact that X_k and X_l are mutually independent for $k \neq l$. In order for the asymptotic behaviour of the local mean square error of $C_K(\underline{x})$ to be optimal, the bias and the (co)variance must be in balance. When choosing $f(K) = K^\alpha$ we find that α depends on the dimension of the psf as follows

$$\alpha = \frac{1}{N+4} \qquad (12)$$

The optimal asymptotic behaviour of the local mean square error of $C_K(\underline{x})$ for $K \to \infty$ can now be obtained:

$$E\left\{\left(C_K(\underline{x}) - C(\underline{x})\right)^2\right\} = \mathbf{O}\left(K^{\frac{-4}{N+4}}\right). \qquad (13)$$

The parameter that can be used to scale the psfs in one, two or three dimensions is respectively the number of particles K raised to the power $\frac{1}{5}$, $\frac{1}{6}$ and $\frac{1}{7}$. The order of the mean square error will in those cases be respectively $K^{-\frac{4}{5}}$, $K^{-\frac{2}{3}}$ and $K^{-\frac{4}{7}}$. This means that convergence of the error in 1D is slow in comparison with numerical models, but gets better in 2D and 3D situations. In a 1D numerical model with a first order scheme the amount of gridpoints needs to be doubled in order to half the error. In a one-dimensional particle model the number of particles needs to be multiplied by a factor of approximately 2.38 in order to half the error. In 2D and 3D this factor is respectively equal to 2.83 and 3.36, while in numerical models this factor is 4 or 8 with a first order scheme. These results correspond with the results of Vos[5], while similar results have been found in density estimation theory (see Silverman[4]).

3 Optimal psfs

The two important characteristics of a psf are its width and shape. The optimal scaling parameter determining the width has been calculated and is assumed to be $f(K) = K^\alpha = K^{\frac{1}{N+4}}$. We now determine the optimal shape of the psf by minimizing the mean square error with respect to $\phi(\underline{x})$. It is found that the optimal psf should satisfy

$$\phi(\underline{x}) = -\frac{1}{2}\underline{x}^T \mathbf{A}\underline{x} - \frac{1}{2}\lambda \qquad (14)$$

$$\int\cdots\int_{\mathcal{G}} \phi(\underline{x})\,dX = 1 \qquad (15)$$

$$\int\cdots\int_{\mathcal{G}} \underline{x}\,\underline{x}^T \phi(\underline{x})\,dX = \mathbf{P} \qquad (16)$$

This means that the optimal form of the psf is *quadratic*. The unknown values of the symmetric matrix $\mathbf{A}$ are found with eqn. 16 for the covariance $\mathbf{P}$. The parameter λ is also unknown, but follows from eqn. 15 for the volume.

4 1D example

To obtain the optimal asymptotic behaviour of the local mean square error in 1D the pfs should be scaled with the total number of particles raised to the one fifth power (see eqn 12). The order of the mean square error would then be $\mathbf{O}(K^{-\frac{4}{5}})$ versus a histogram mean square error of $\mathbf{O}(K^{-\frac{2}{3}})$. This shows that psfs are better estimators than histogram functions. An example of quadratic psfs and their asymptotic behaviour in one dimension as well as the histogram approximation are given below (see figure 2). The dotted line represent the standard Gaussian distribution and the solid lines are the approximations.

Environmental Coastal Regions   317

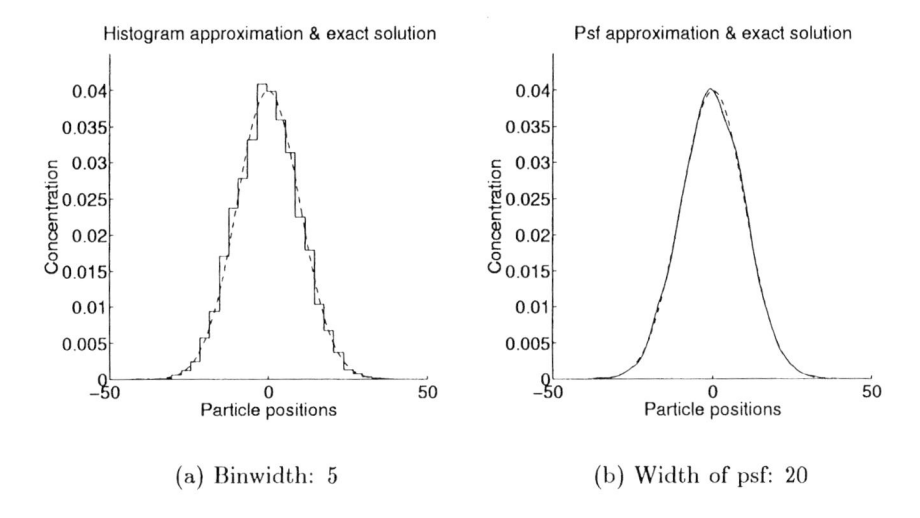

(a) Binwidth: 5

(b) Width of psf: 20

Figure 2: (a) Histogram approximation, and (b) quadratic point spread function approximation using 10000 particles.

5 2D example

For an example in two dimensions we take a psf that is rotationally symmetric with respect to the center, and its volume equals one. Some examples of such psfs in two dimensions are shown in figure 3. The concentration profile is mapped on the vertical axis.

If we choose the maximum radius to be $\alpha = \sqrt{6}$ the expression for a paraboloid becomes

$$\phi(R) = \begin{cases} \dfrac{1}{3\pi}\left(1 - \dfrac{R^2}{6}\right) & \text{if } 0 \le R \le \sqrt{6}, \\ 0 & \text{elsewhere} \end{cases} \quad (17)$$

We find that for $\alpha = \sqrt{6}$ the (co)variance matrix simplifies to (see eqn 16):

$$\mathbf{P} = \begin{pmatrix} 1 & 0 \\ 0 & 1 \end{pmatrix} \quad (18)$$

The approximation term for the bias in this case is

$$\text{BIAS}\{C_K(\underline{x})\} \approx \frac{1}{2(f(K))^2}\left(\frac{\partial^2 C(x,y)}{\partial x^2} + \frac{\partial^2 C(x,y)}{\partial y^2}\right) \quad (19)$$

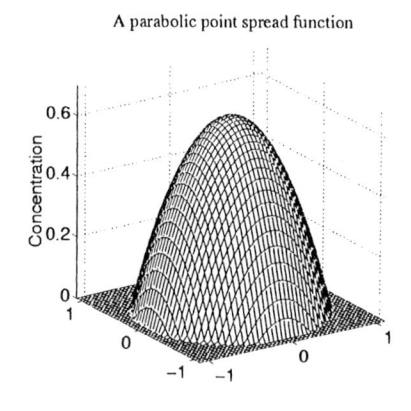

A parabolic point spread function

A conic point spread function

(a) Concentration resulting from a parabolic point spread function in 2D (optimal shape).

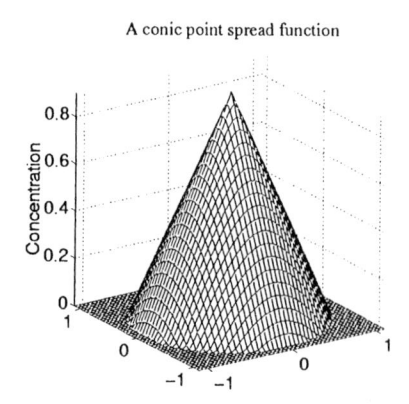

(b) Concentration resulting from a conic point spread function in 2D.

Figure 3: Examples of psfs in 2D

Furthermore we have that

$$\int_{y=-\infty}^{\infty} \int_{x=-\infty}^{\infty} \left(\phi(x,y)\right)^2 dx dy = \frac{2}{9\pi} \approx 0.0707 \qquad (20)$$

The expression for the mean square error now becomes

$$\text{MSE}(x,y) \approx K^{-\frac{2}{3}} \left(0.0707 + \frac{1}{4} \left(\frac{\partial^2 C(x,y)}{\partial x^2} + \frac{\partial^2 C(x,y)}{\partial y^2} \right)^2 \right)$$

6 Real-life applications

The method of using psfs to convert particle masses to concentrations instead of histograms has been applied to different real-life particle models. Results are shown from the two dimensional model SIMPAR, which is derived from the depth-averaged three dimensional advection diffusion equation. It is a program that calculates the transport of pollutants around the Dutch coastal zone. The underlying grid for the interpolation of velocities and diffusion coefficients is curvilinear. The model is developed at the National Institute for Coastal and Marine Management (RIKZ). Some particle dumps were simulated in the Nieuwe Waterweg after which the particles were transported for 7 tidal periods (figure 4). Clearly, the picture on the righthand-

(a) Particle simulation in the Dutch coastal area

(b) Psf concentration estimate for the same simulation

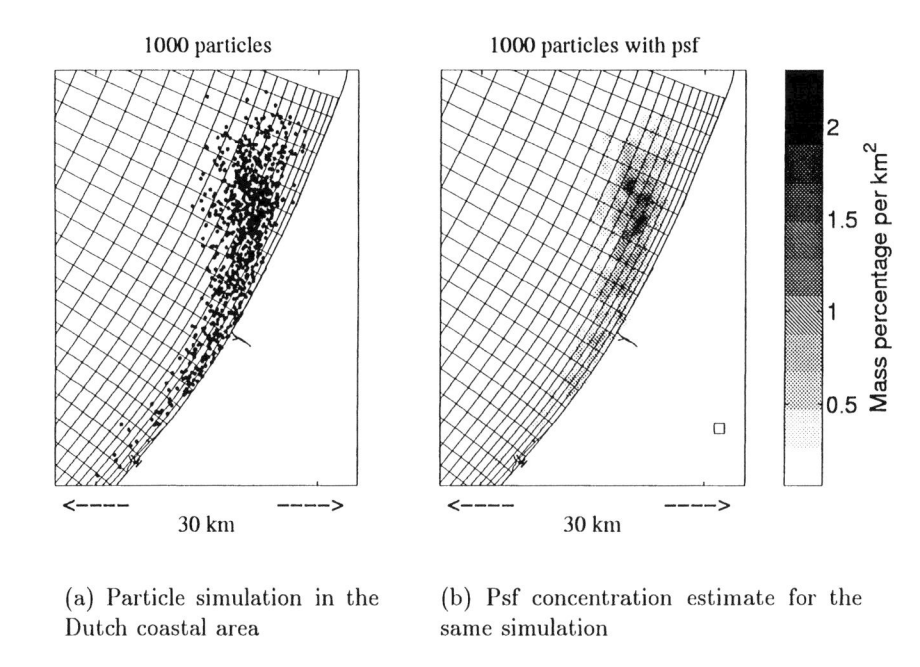

Figure 4: (a) Particle simulation and (b) psf concentration result.

side of figure 4 shows the areas with different concentration levels much better than the pictures on the lefthandside. For other real-life applications with Gaussian psfs see for example Vos[5].

7 Conclusions

Particle models are used to calculate the spreading of pollutants by means of particle tracks. From the ensemble of particles a concentration profile can be derived, which is traditionally done with the use of histograms. To obtain a smoother, more accurate profile, the concept of the point spread function was introduced, which spreads the mass of a particle across the neighbouring cells.

The asymptotic behaviour of the point spread functions was analysed in arbitrary dimension. In dimension one, two and three the order of the mean square error is respectively the number of particles raised to the power $\frac{4}{5}, \frac{2}{3}$ and $\frac{4}{7}$.

Point spread functiona with a quadratic form were found to have the optimal asymptotic behaviour, i.e. minimizing the mean square error between the estimated concentration and the exact solution for large numbers of particles. Differences between different types of psfs are however minimal.

Another important aspect of the point spread function is the choice of the width. Results with a constant width for the point spread function have shown good results. Fewer particles are needed providing a better result than with histograms.

Next, the width of the point spread function was scaled with the number of particles. Increasing the number of particles gives more information, so the width of the point spread function may be reduced. Results of this approach have also been shown to give good results, both in test scenarios and real-life applications.

Future work focusses on enhancing the psf-concept even further by optimally adapting the widths of the psf. This means that when the concentration of a pollutant is locally strongly varying in space, the error in the bias will be large. Chosing a smaller variance for the psf (and consequently a smaller width) will reduce the bias error and thus the mean square error. Another possibility to reduce the error would be to locally generate more particles with smaller masses, or to remove particles while increasing the mass of others.

References

[1] Boogaard, H.F.P. van den, Hoogkamer, M.J.J., and Heemink, A.W., *Parameter identification in particle models,* Stochastic Hydrology and Hydraulics 7, pp. 109-130, 1993.

[2] Fisher, H.B., List, E.J., Koh, R.C.Y., Imberger, J., and Brooks, N.H., *Mixing in Inland and Coastal Waters,* Academic Press, New York, 1979.

[3] Heemink, A.W., *Stochastic modelling of dispersion in shallow water* Stochastic Hydrology and Hydraulics 4, pp. 161-174, 1990.

[4] Silverman, B.W. *Density Estimation for Statistics and Data Analysis,* Chapman and Hall, London and New York, pp. 38-48, 1986.

[5] Vos, R.J., Boogaard, H.F.P. van den, *Point spread functions in DELPAR,* Delft Hydraulics report T1195, 1994.

Physical modelling of the hydrodynamics around a group of offshore structures

S. Pan, N.J. MacDonald, B.A. O'Connor & J. Nicholson,
*Department of Civil Engineering, The University of Liverpool
PO Box 147, Liverpool, UK L69 3BX
Email: S.Pan@liv.ac.uk*

Abstract

This paper presents results of physical model experiments carried out to study the hydrodynamics around a group of offshore breakwaters in the UK Coastal Research Facility (UKCRF) at HR Wallingford. The results of tests with oblique monochromatic (MON) and long-crested random (LCR) sea states are presented of a fixed bed, 1:28 scale model of the Elmer site in southern England. The measured wave and flow results show the three-dimensional complexity of the system due to wave reflection from the breakwaters and, especially, wave transmission and flow through the breakwaters. The results highlight the necessity to consider the permeability of coastal structures in the numerical modelling of nearshore circulation.

1 Introduction

Numerical modelling has played an increasingly important role in solving coastal engineering problems. However, the success of the application of numerical models strongly depends on the availability of reliable and detailed experimental data sets to permit model validation. In order to validate and improve the University of Liverpool's wave-period-averaged (WPA) nearshore circulation models, a series of physical model experiments has been carried out in the UKCRF. Following on from two previous projects, LUCI (which studied wave-current interaction over complex bathymetry (O'Connor et al, 1995)) and LUCIO (which studied wave-current interaction in the presence of river outflows and tides

322   Environmental Coastal Regions

(O'Connor et al, 1997)), the aim of the present LUPY project is to investigate the hydrodynamics and morphodynamics associated with offshore structures. The first phase of the project will study the wave and currents fields using a fixed bed model. The second phase will employ a mobile bed model to examine the morphological development of the same layout.

Due to the availability of field data collected by the University of Plymouth (LUPY project co-investigators), the breakwaters at Elmer in southern England were selected to be modelled in the UKCRF basin. The layout of the breakwaters and the adjacent bathymetry, reproduced at a scale of 1:28 in the model, are shown in Figure 1.

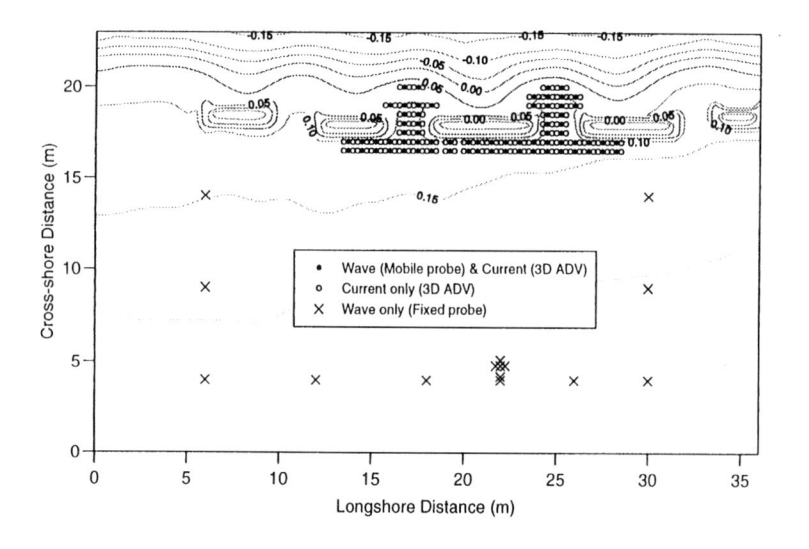

Figure 1 Bathymetry and measurement locations for Elmer model

2 Experiment set-up

The UKCRF is a hydraulic model basin measuring 36 m in the longshore direction and 23 m in the cross-shore direction. Waves are generated by 72 piston-type wave paddles. A 3D programmable positioner (known as the *bridge*) is used to deploy instruments in a large area in the centre of the basin. The facility also provides advanced data logging and analysis equipment, see Simons et al. (1995) for details.

The maximum water depth in the Elmer model was set at 0.32 m for all tests. The breakwaters were constructed of armour stone with nominal size of 5 cm.

2.1 Test conditions

In order to generate reasonably large longshore currents in front of the breakwaters and to avoid wave over-topping of the breakwaters, three sea states were used in the experiments (see Table 1). However, only results obtained from Tests 1 and 3 are presented and discussed in this paper.

Table 1 Wave conditions used in the experiments

	Wave Type	Wave Height (m)	Wave Period (sec)	Wave Direction[*] (deg)
Test1	MON	0.07	1.2	20
Test2	MON	0.06	1.6	20
Test3	LCR	0.05	1.2	20

[*] Wave direction measured clockwise from shore-normal

2.2 Instrumentation

Three Nortek acoustic Doppler velocimeter (ADV) probes were used to measure the three components of the current velocity. Due to limited water depth in the nearshore areas and the size of the ADV probes, measurements were made at one level only in the water column (approximately 0.4 water depth above the bed) in an area covering two bays in the centre of the basin, see Figure 1. At a later stage in the experiments, a 2D ADV probe was installed which, because of its small size, allowed the vertical variation of the horizontal current profile to be measured. Measurements at five points with a vertical interval of 1 cm were achieved with the lowest measuring point being 1.5 cm above the bed.

Fifteen fixed wave probes were used to measure the offshore wave heights and an array of six wave probes was placed in the centre of the basin to determine the wave direction. The wave generators were also equipped with a newly-developed wave absorption system to minimise the effects of wave reflection from the breakwaters. Wave heights in the nearshore region were measured using a mobile wave probe at the same locations as the velocity measurements.

Three ADV probes and the mobile wave probe were mounted on the bridge. Several bridge routes were set up to deploy the instruments to the desired positions. The ADV probes were arranged with their measuring volumes 25 cm apart from each other in the longshore direction. In total, wave measurements were taken at 68 positions in the nearshore area and the current velocities at 204 points.

2.3 Data acquisition

Offshore wave data were logged using the DPlus V2.0 Data Acquisition System at 20 Hz frequency. Nearshore wave and current data were collected at 50 Hz frequency. The logging time was approximately 150 waves for the monochromatic waves and 500 waves for the random waves. The experimental data were archived in a CD-ROM for future analysis. In addition, float tracking and dye visualisation tests were also carried out during the experiments. Overhead cameras were used to record the images in both Beta and super-VHS format for future analysis. No results from these measurement are presented here.

2.4 Data analysis

The root-mean-square wave heights were calculated from the water level elevation time series using the zero up-crossing method for both monochromatic and random waves. The mean velocity was obtained by simple mathematical averaging. Signals were checked before the analysis were carried out.

3 Experimental results

The results present here are restricted to the nearshore area around the breakwaters, particularly the two bays in the centre of the basin where most measurements were taken. As mentioned previously, only the results for Test 1 and Test 3 will be presented and discussed in this paper, where emphasis will be given to the wave-induced currents.

Environmental Coastal Regions   325

3.1 Wave height

The measured wave heights for Test 1 are shown in Figure 2, where the length of each bar represents the wave height and the dot in the centre of the bar shows the measuring position. It was found that in the left-hand bay, waves propagated through the gap with little change in wave height until a plunging-type breaking occurred. In the right-hand bay, waves also propagated through the gap. However, due to greater water depths in the gap, the waves broke nearer the shoreline. The shadow area behind the breakwater caused by wave diffraction can also be seen clearly. It was also found that the wave heights along a line at 16.5 m (the most offshore wave height measurements) were higher than those at 17 m in front of breakwaters. This is believed to be due to partial wave reflection from the breakwaters.

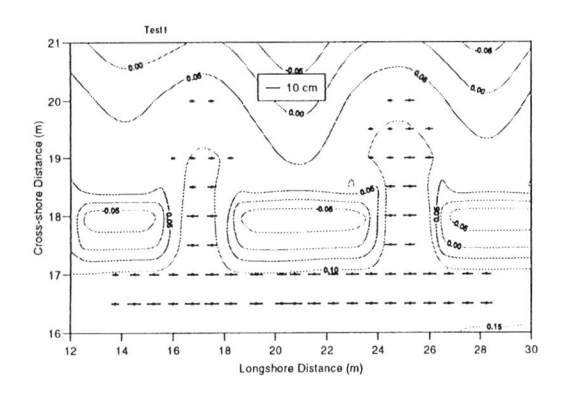

Figure 2 Measured wave height (Test 1)

Figure 3 shows the wave height distributions for Test 3 (LCR waves). In both bays, the wave heights had similar patterns. Waves propagated through the gap and maintained their heights in the bay and finally broke near the shoreline.

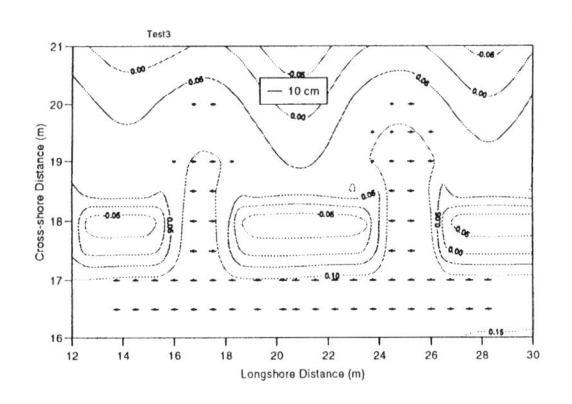

Figure 3 Measured wave height (Test3)

326 Environmental Coastal Regions

Wave reflection in front of the breakwaters was found less significant than that in the monochromatic wave test.

3.1 Wave-induced currents

Figures 4 and 5 show the measured velocities for Tests 1 and 3, respectively. Since the velocity measurements were taken at 0.4 of the water depth above the bed, the results represent the flow in the lower layer. These figures clearly show that the wave-induced currents formed a gyre in both bays. Due to the obliquely incident waves, only one gyre can be seen in each bay in contrast to the two gyres found behind each breakwater in normally incident wave cases. The monochromatic waves were found to induce stronger currents than random waves, both in the gaps and in the bays.

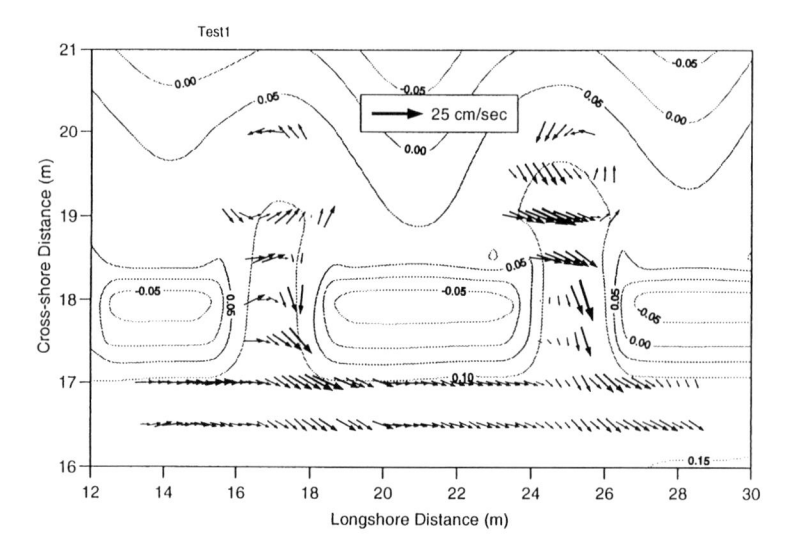

Figure 4 Current vectors (Test 1)

Figure 4 also clearly shows the different location of the centre of the gyre in the both bays. The centre of the gyre in the right-hand bay was found to be further downstream than that in the left-hand bay. This was due to the greater flow transmitted through the central breakwater. Since the water depth in the right-hand bay is deeper than the left-hand bay, there was a significant transmission of wave energy through the breakwater. Consequently, these transmitted waves induced a longshore flow behind the breakwater. This flow then joined the current driven by the incident waves in the bay which resulted in a strong flow re-circulation in the bay and outflow in the gap.

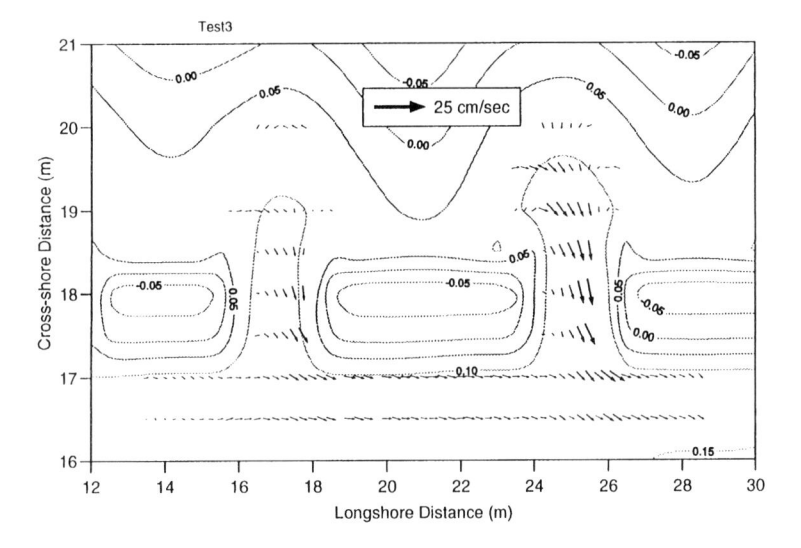

Figure 5 Current vectors (Test 3)

Figures 4 and 5 also show a strong outflow on right-hand side of the gaps between the breakwaters for both tests. These outflows are believed to be the seaward return of the wave-induced shoreward flow through the permeable stone breakwaters. This can be illustrated through a simple mass balance calculation for Test 1. Through dye tracing tests performed during the experiments, the shoreward flow through the breakwaters was estimated to be approximately 0.05 m/s. Over a 5 m length of breakwater in 0.10 m depth, this represents a shoreward discharge of 0.025 m^3/s. In order to estimate the seaward discharge through the gap, the wave-induced cross-shore flow component must be subtracted from the measured flow velocities shown in Figure 6. (This is required because the velocity measurement locations do not extend above wave trough level, and so will only show the seaward-directed undertow and not the shoreward-directed mass transport between the wave trough and crest). Based on the results of a simple cross-shore flow model (O'Connor et al., 1998), it was estimated that the measured velocities should be reduced by 0.05 m/s. Therefore, the seaward discharge through the 0.15 m deep gap can be estimated, assuming a depth uniform but horizontally triangular

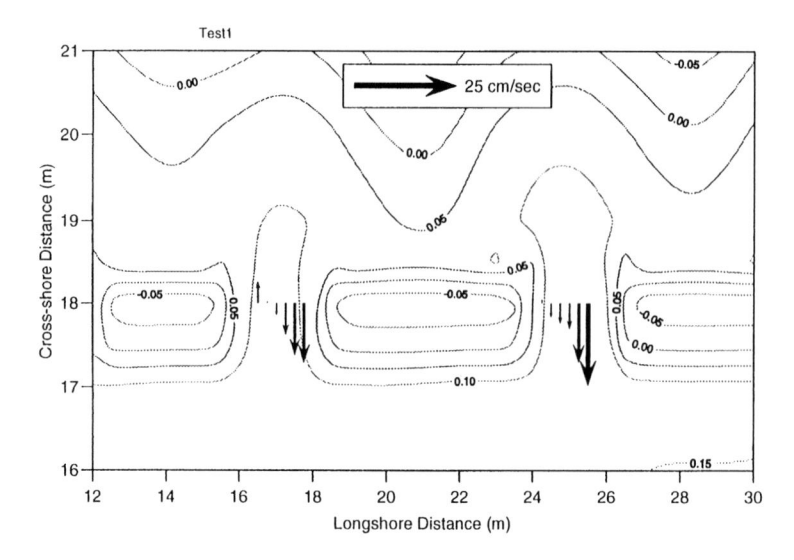

Figure 6 Cross-shore flow components across breakwater gaps (Test 1)

distribution of flow (see Figure 6) with a maximum flow velocity of 0.15 m/s, as approximately 0.035 m³/s, which is of the same order of magnitude as the flow through the breakwater.

Failure to take proper account of the permeability of coastal structures in nearshore circulation models will result in the failure to reproduce the correct circulation pattern and will underestimate the wave-induced flushing of the nearshore system.

In addition, Figures 4 and 5 show an offshore component of flow on the seaward side of the breakwaters. This is believed to be due to the plunging-type of wave breaking occurring on the front of breakwaters, which generates an offshore flow in the bottom layer.

3.2 Vertical distribution of wave-induced current

As mentioned previously, the flow patterns around the breakwaters exhibited significant variation in the vertical. To examine these profiles, measurements were made using the smaller 2D ADV probe at five points through the water column at five locations in the left-hand bay.

Figure 7 shows the velocity vectors at different levels in the water column

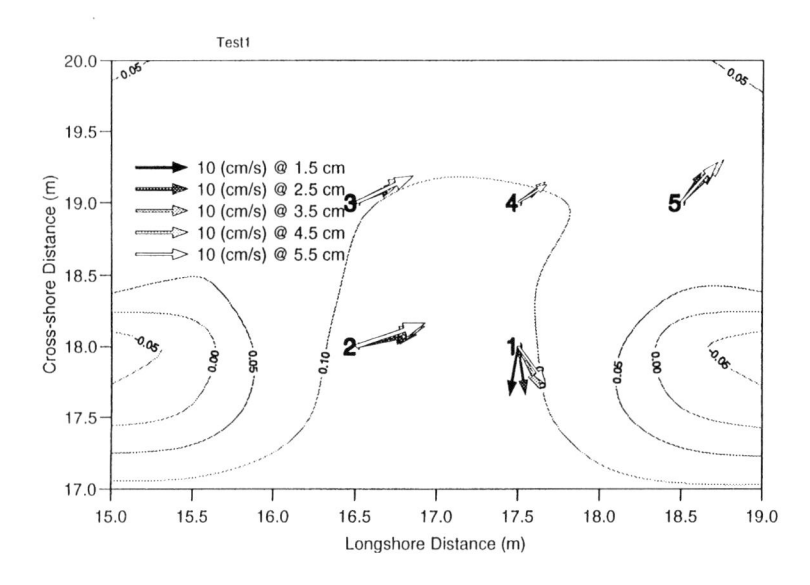

Figure 7 Current vectors at different water depths (Test 1)

for Test 1. The lowest point was 1.0 cm above the bed with the others at 1 cm intervals above that. The darker the velocity vector the lower the measurement level.

Of these five points, Point 1 shows the greatest change in flow direction from the bed to the surface. The surface flow is skewed further to the right due to the wave motion, while the velocity near the bed appeared to be more influenced by the bed topography.

4 Conclusions

This paper describes the set up and results of a hydraulic model experiment carried out on a group of offshore breakwaters. The experimental results clearly show the complex three-dimensionality of the flow circulation in the bays and the wave-induced currents around the offshore structures. It was found that wave transmission and flow through the breakwaters has a significant influence on the flow circulation in the bays. These results suggest that the failure to take proper account of the permeability of coastal structures in nearshore circulation modelling will result in inaccurate circulation patterns and will underestimate the wave-induced flushing of the nearshore system. The experimental results also provide valuable and detailed data sets which can be used to achieve better understanding of the underlying processes and to validate numerical models.

Acknowledgements

The authors would like to acknowledge the financial support by the UK EPSRC for this project and HR Wallingford for the co-sponsorship of the UKCRF under contract of GR/K48600 and the Commission of the European Communities Directorate General for Science and Education, Research and Development under contract numbers MAS3-CT95-0002, MAS3-CT97-0086, MAS3-CT97-0097 and MAS3-CT97-0106. The authors would also like to thank Dr A. Chadwick, Ms S. Ilic and Mr B. Chapman of the University of Plymouth for their help during the experiments. Thanks are also due to Dr R. Whitehouse, UKCRF Manager and Mr A. Channel, UKCRF Operator for their assistance.

References

O'Connor, B.A., Sayer, P.B. and MacDonald, N. J. (1995): Combined refraction-diffraction wave-current interaction over a complex nearshore bathymetry, *Proc. Coastal Dynamics '95*, Gdansk Poland, ASCE, (Ed. Dally, W.R. and Zeidler, R.B.), pp 173-184.

O'Connor, B.A., Pan, S. and MacDonald, N.J. (1997): Modelling of wave-current interaction at a river mouth, *Coastal Dynamics '97*, (in press).

O'Connor, B.A., Pan, S., Nicholson, J., MacDonald, N.J. and Huntley, D.A. (1998): A 2D Model of Waves and Undertow in the Surf Zone. *Abstracts of 26th Int. Conf. Coastal Engrg.*, Copenhagen, Denmark, ASCE. (in press).

Simons, R., Whitehouse, R., MacIver, R., Pearson, J., Sayers. P., Zhao, Y. and Channel, A. (1995): Evaluation of the UK Coastal Research Facility. *Proc. Coastal Dynamics '95*, Gdansk, Poland, ASCE, (Ed. Dally, W.R. and Zeidler, R.B.), pp 161-172.

A Lagrangian 3D Numerical Model of Pollutant Dispersion in Coastal Waters

J.P. Sierra[1], M. Mestres[1], A. Rodríguez[2] and A. S. Arcilla[1]

[1] *Laboratori d'Enginyeria Marítima, Universitat Politècnica de Catalunya, C/ Gran Capità s/n, 08034 Barcelona, Spain
E-mail: sierra@etseccpb.upc.es*

[2] *Laboratorio de Hidráulica, Universidad Nacional de Córdoba, Av/ Velez Sarsfield 299, 5000 Córdoba, Argentina*

Abstract

Two-dimensional numerical models have been extensively used in the far field to solve hydrodynamics and the associated pollutant transport. However the results obtained from these models may not be suitable when used in the near field or in areas where the hydrodynamics has three-dimensional features, such as wind-induced currents in the nearshore region or wave-induced currents in the surf zone. A 3D Lagrangian numerical model has been developed to solve the three-dimensional convection-diffusion equation in order to simulate pollutant transport in non-uniform and stratified hydrodynamic flows. In this paper the model characteristics and its calibration process are described and posteriorly it is applied to the case of the Besòs marine outfall, near Barcelona (Spanish Mediterranean coast) and to simulate tracer clouds dispersion in the surf-zone.

1 Introduction

The simulation of marine outfall discharges must take into account the existence of two differentiated regions in which the pollutant evolution is governed by different physical mechanisms. Nearest the source, in the near-field, transport is driven by the discharge characteristics, such as initial momentum or density differences, as well as the nozzle geometry and orientation. On the other hand, in the far-field, dispersion is due exclusively to the hydrodynamic velocity field and environmental turbulence.

In this latter region, the net substance transport in one spatial direction can often be neglected in comparison to the existing transport in the other two. The problem can then be solved by averaging the transport equation along that direction. This is the basis for 2DH models which generally solve the vertically-averaged form of the transport equation.

Nevertheless, this averaging procedure can not be carried out in the near-field, since pollutant transport is equally important in each direction. The application of two-dimensional models in this region may lead to unreliable results, and as a consequence the use of fully 3D algorithms becomes necessary. Moreover, when the hydrodynamics (circulation and turbulence) changes significantly in the vertical direction, as in the case of nearshore wind-induced currents or wave-induced currents inside the surf-zone, it is suitable to use 3D dispersion models even in the far-field.

This paper presents a recently developed 3D Lagrangian numerical model for the transport of pollutant substances (Sánchez-Arcilla et al. [6]). Several reasons exist to justify the use of a Lagrangian approach, such as the concentration of computational effort in the region where pollutant accumulates, the capability to simulate dispersion of different constituents and the simplicity arising in the formulation and implementation of the various physical processes.

2 Model formulation

The model solves the three-dimensional equation for mass transport due to diffusion and advection, averaged over the time-scale of turbulence (Reynolds average):

$$\frac{\partial C}{\partial t} + u\frac{\partial C}{\partial x} + v\frac{\partial C}{\partial y} + w\frac{\partial C}{\partial z} =$$

$$\frac{\partial}{\partial x}\left(D_x\frac{\partial C}{\partial x}\right) + \left(D_y\frac{\partial C}{\partial y}\right) + \left(D_z\frac{\partial C}{\partial z}\right) + DS \quad (1)$$

where C is the concentration, u, v and w the components of the velocity field, D_x, D_y, D_z the diffusion coefficients in the x, y, z directions respectively and DS is a source/decay term.

2.1 Transport mechanisms

The Lagrangian approach herein followed assumes that the pollutant mass is divided into small particles (Lagrangian elements) which are moved inside a water body according to the different physical mechanisms involved in the transport. Eventually, these particles may disappear if microbiological decay exists, or if they become trapped at physical boundaries (bottom or dry edges).

Under this hypothesis, each particle is displaced a certain distance (Δx, Δy, Δz) during a timestep Δt, with a velocity which depends on the different processes involved in the overall dispersion. The individual contribution to the total velocity arising from each mechanism is calculated in the following way:

$$x^{n+1} = x^n + (u_C + u_w + u_o + u_{D_L}\cos\theta - u_{D_T}\sin\theta + u_{D_M})\Delta t$$

$$y^{n+1} = y^n + (v_C + v_w + v_o + u_{D_L}\sin\theta - u_{D_T}\cos\theta + u_{D_M})\Delta t$$

$$z^{n+1} = z^n + (w_C + w_w + w_o + w_{D_V} + w_{D_M} + w_S + \frac{1}{2}w_B)\Delta t \quad (2)$$

where u_C, v_C, w_C are the velocity components from the hydrodynamic model; u_w, v_w, w_w the wind induced velocity components; u_o, v_o, w_o the wave-induced transport velocity components; u_{D_L}, u_{D_T}, w_{D_V} the turbulent diffusion velocity components in the longitudinal, transversal and vertical direction respectively; u_{D_M} the molecular diffusion velocity; w_B, w_S the buoyant and settling velocity components and θ the angle between the x-axis and the current lines.

The components of the 3D velocity field are computed with the help of a 2DH hydrodynamic model (Sánchez-Arcilla et al. [5]) which solves the vertically integrated mass and momentum conservation equations, including wave and wind effects, but without considering the circulation due to density gradients. Once the 2DH mean velocities are determinated, they are vertically modified assuming a logarithmic profile in the bottom boundary layer and a parabolic profile elsewhere. Then the vertical velocity component can be calculated by solving the mass conservation equation.

The oscillatory flow effects must be taken into account because the surface waves affect the pollutants transport in two different ways: generating a net advective transport and enhancing the local turbulence level which stimulate the vertical turbulence mixing. The transport by oscillatory flow is included in the model distinguishing between outside and inside the surf zone.

Outside the surf zone the effects are modelled using a formulation proposed by Davydov [2] in which the wave field is described by a linear theory. The pollutant flow in the horizontal directions is

$$q_j = \delta_j \frac{\partial C}{\partial z} \tag{3}$$

and in the vertical direction is

$$q_z = \delta_j \frac{\partial C}{\partial x_j} \tag{4}$$

where the subscript j indicates the components in the x and y directions and δ_j is

$$\delta_j = \frac{H^2 \pi \sinh(2kz) k_j}{8T \sinh(kh)|\vec{k}|} \tag{5}$$

where $\vec{k}$ is the wave number vector, with components k_j, T the characteristic wave period and H the wave height. If it is assumed that q_i (with $i = 1, 2, 3$) represents the pollutant flux per unit area and unit time in the x_i direction, then

$$q_i(l, m, n) = u_{oi}(l, m, n) C(l, m, n) \tag{6}$$

and a "psuudo-velocity" generated by the oscillatory flow can be deduced for each particle contained inside a certain cell (l, m, n) with concentration $C(l, m, n)$.

In order to model diffusive processes, a diffusive velocity can be defined as

$$u_D = (2R_{01} - 1)\sqrt{\frac{6D}{\Delta t}} \tag{7}$$

where D is the diffusion coefficient and R_{01} is a random number $(0 < R_{01} < 1)$.

Outside the surf zone, the turbulent dispersion is splitted into vertical and horizontal components. The latter component is further decomposed into a diffusion paralel to the horizontal current lines and a lateral diffusion. Following Holly [3], the corresponding diffusivities are given by

$$D_L = c_L u_* h$$

$$D_T = c_T u_* h \tag{8}$$

where u_* is the shear velocity ($u_* = (\tau/\rho)^{1/2}$, τ is the bottom shear stress), h is the local depth and c_L, c_T are empirical coefficients with typical values of $c_L = 5.93$ and c_T between 0.07 and 0.23.

On the other hand, an accepted estimation for the vertical diffusivity outside the surf zone is

$$D_z = c_V u_* h \qquad (9)$$

with c_V ranging from 0.16 to 0.23.

Inside the surf zone, additional mixing due to wave breaking has to be considered. The horizontal diffusivity, assumed isotropic, can be estimated with a number of expressions, like the one proposed by Battjes [1],

$$D_H = c_H \left(\frac{D_b}{\rho}\right)^{1/3} h \qquad (10)$$

where c_H is an experimental coefficient, usually taken as 4.5 and D_b is the wave breaking dissipation, obtained from the wave propagation module in the hydrodynamic model.

The vertical diffusion coefficient is computed following Nadaoka and Hirose [4]:

$$D_z = 0.18 \left(\frac{H}{h}\right)^{7/3} L^{1/3} h^{2/3} (gh)^{1/2} i^{1/3} \qquad (11)$$

where i is the bottom slope and L the wave length.

Even though its effects are small compared to turbulent diffusion, molecular diffusion is also included in the computations for model completitude. In this case, the diffusivity is assumed isotropic and constant and depends on the discharged substance. Other effect taken into account into the model is the particle settling when dealing with suspended sediment.

For the treatment of discharges whose density ρ is different from that of the receiving water body ρ_0 (due either to salinity or temperature), or in situations where environment stratification exists, buoyancy effects are included by means of a buoyancy velocity

$$w_B = \frac{\rho_0 - \rho}{\rho_0} g \Delta t \qquad (12)$$

Endly, the effect of microbiological decay on dispersion is also included in the transport calculations when urban sewage is discharged directly, or by marine outfalls, into the waterbody. The concentration $C(t)$ at time t is

$$C(t) = C(t_0) e^{-K_d(t-t_0)} \qquad (13)$$

where $C(t_0)$ is the concentration at time t_0 and K_d is a decay coefficient which depends on a number of factors (organism type, temperature, solar radiation, etc.). This coefficient can be evaluated from several expressions with different accuracy.

2.2 Discrete-continuous transformation

A transformation scheme is required to obtain the concentration field from the modelled cloud of Lagrangian particles. The model allows to perform this transformation with three different methods:

- The classical particle-in-cell (PIC) method which consists in dividing the volume affected by the pollutant in "concentration cells" and counting the number of particles in each one. The concentration in an individual cell is obtained dividing the total mass in the cell by its volume.
- The smoothed-particle hydrodynamics (SPH) method, based in the hypothesis that each particle can be replaced by a smoothed-out density distribution W. The mass concentration at a point is found by adding all the contributions from individual particles.
- The hybrid (kPIC) method is conceptually similar to the first one, but in this case, the mass of a particle is spread out amongst the cell which contains the particle and its 26 neighbours, according to a predefined mass distribution.

3 Model validation

The model has been validated with a number of simplified analytical cases, two of which are shown in the present paper.

The first validation test is an instantaneous release of conservative pollutant in a stagnant environment, in which mass transport is driven exclusively by diffusion. The analytical solution in a three-dimensional space is

$$C(x,y,z,t) = \frac{C_0}{(4\pi D_x D_y D_z t)^{3/2}} e^{-\frac{(x-x_0)^2}{4D_x t} - \frac{(y-y_0)^2}{4D_y t} - \frac{(z-z_0)^2}{4D_z t}} \qquad (14)$$

The concentration distribution resulting from a 15 s diffusion process has been computed using the PIC and kPIC methods, with a cloud of 100,000 particles. The obtained distributions are plotted in figure 1, showing a relative error of about 2%.

The second validation test shown here corresponds to a continuous release of infinite duration, whose 1D analytical solution for a conservative substance is

$$C(x,t) = \frac{C_0}{2}\left[erfc\left(\frac{x-ut}{2\sqrt{Dt}}\right) + erfc\left(\frac{x+ut}{2\sqrt{Dt}}\right)\exp\left(\frac{ux}{D}\right)\right] \quad (15)$$

where $erfc$ is the complementary error function. For this test, 17 particles were discharged every 0.05 seconds during a period of 30 seconds. The results obtained with the PIC method are shown in figure 2, where it can be seen a good agreement between the analytical function and the numerical results.

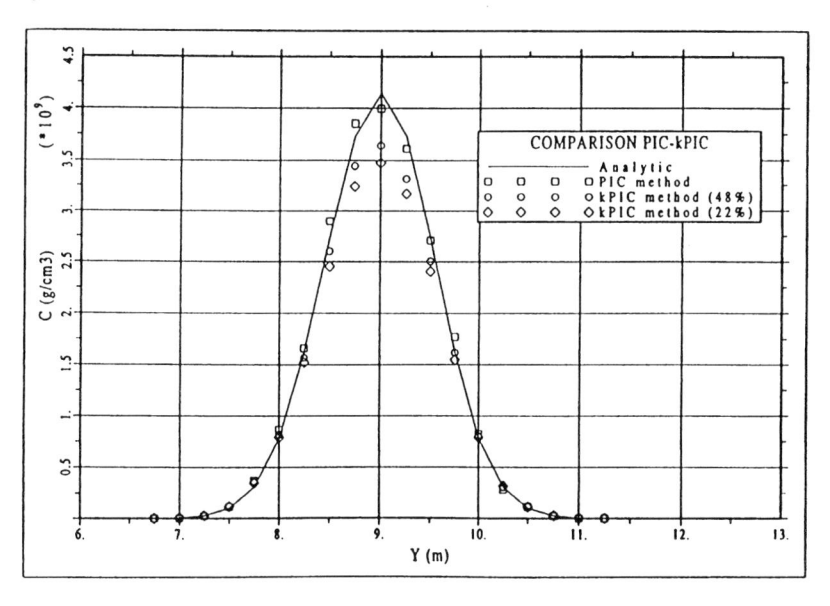

Figure 1. Model validation. Instantaneous release.

4 Model application

The model was firstly applied to simulate the sewage discharges from the Besòs marine outfall (Barcelona, Spanish Mediterranean Coast). In figure 3, the results of the discharge from 10 diffusers are shown. In this case, a stratified environment was assumed, which is a typical summer condition in this area. The plume has been computed by releasing 36,060 particles for a period of 30 minutes, with a 3 seconds computational timestep and an effluent density of $\rho = 1.026$ g/cm^3.

338 Environmental Coastal Regions

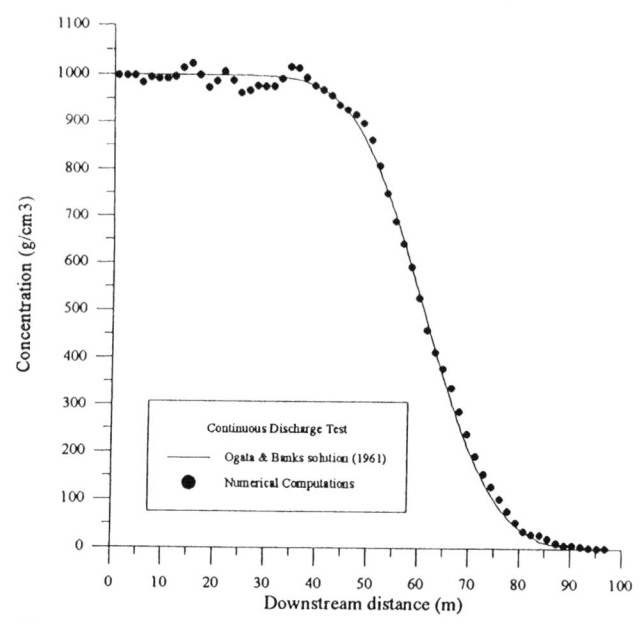

Figure 2. Model validation. Continuous release

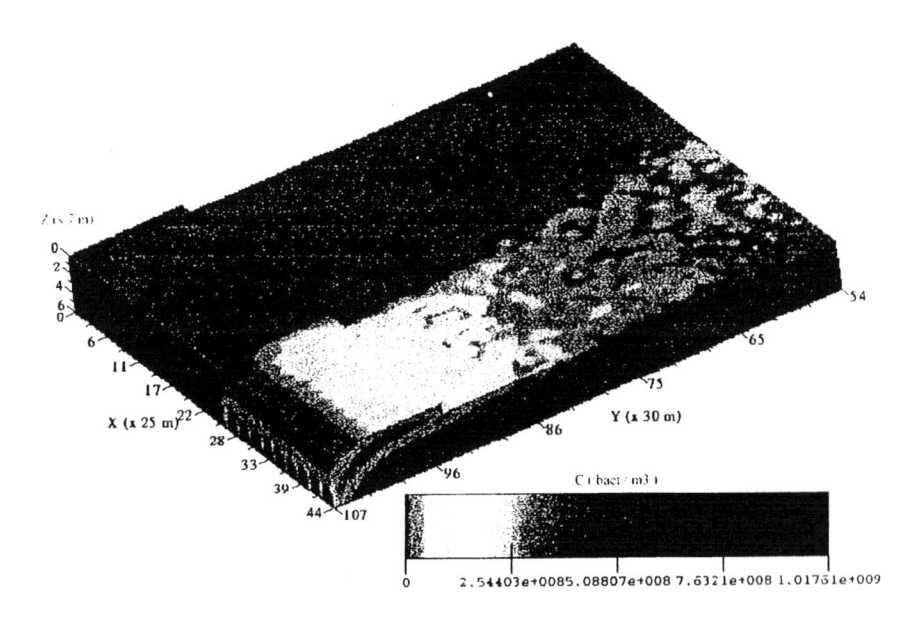

Figure 3. Simulation of Besòs marine outfall discharge

In the second case of model application, the horizontal length scales of a tracer cloud released in the surf-zone (in the Ebro Delta area) have been modelled as a first step towards reproducing the tracer overall behaviour. Approximate longitudinal and transverse dispersion coefficients have been calculated using the random-walk step equation

$$r_{max} = \sqrt{6K\Delta t} \qquad (16)$$

where r_{max} is the maximum step a particle can jump in a timestep. The patch diameters were obtained from a series of digitised video images. These coefficients have then been used to model the time evolution of the released trazer and to estimate the size of the resulting patch. Figure 4 compares the modelled longitudinal patch diameters for different dispersion coefficients with that obtained from video images.

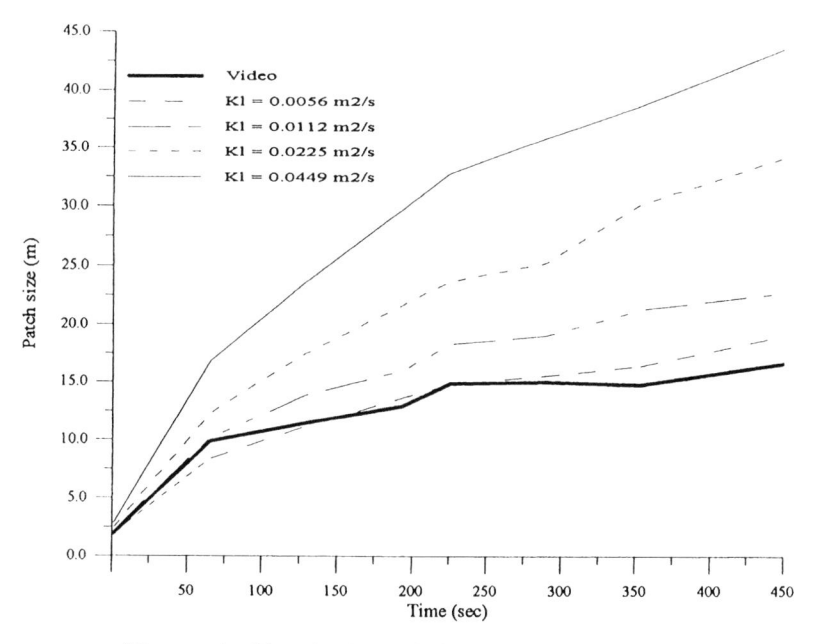

Figure 4. Simulation of the tracer cloud size

5 Summary and conclusions

A general 3D model for the dispersion of pollutants has been developed, based on a Lagrangian formulation. The model allows to describe the evolution of marine discharges both in the near and far-field.

The model has been validated with a series of simple analytical cases showing a good agreement, and has been finally used to simulate two real cases: the Besòs (Barcelona, Spain) marine outfall discharge and the dispersion of a tracer cloud released in the surf-zone. In this latter case, for a $K_L = 0.0056$ m^2/s, a good fit of the modelled results to the experimental ones is observed.

The model requires a thorough calibration process, but the first results illustrate the possibility of considering the model a useful engineering tool to verify existing outfall designs and to predict future or hypothetical situations.

Acknowledgements

This work was jointly sponsored by the FANS project (in the framework of the MAST-III Programme under contract no. MAS3-CT95-0037) and the Spanish CICYT (CYTMAR Programme, project MAR96-1856). The authors are also grateful to F. Collado and EMMSA for their collaboration.

References

[1] Battjes, J., Surf-zone turbulence, *Proc. XX IAHR Congress,* Moscow, pp. 137-140, 1983.
[2] Davydov, L., On the role of surface waves in the transport of pollutants, *Proc. XXIII IAHR Congress,* Ottawa, pp. 355-359, 1989.
[3] Holly, F., Dispersion in rivers and coastal waters, chapter 1 in *Developments in Hydraulic Engineering,* ed. Novak, Elsevier, London, 1985.
[4] Nadaoka, K. & Hirose, F., Modelling of diffusion coefficient in the surf zone based on the physical process of wave breaking, *Proc. 33rd Japan Conference on Coastal Engineering,* pp. 26-30, 1986.
[5] Sánchez-Arcilla, A., Collado, F. & Rodriguez, A., Vertically varying velocity field in Q3D nearshore circulation, *Proc. 23rd Int. conf. on Coastal Eng.,* Venice, ASCE, pp. 2811-2838, 1992.
[6] Sánchez-Arcilla, A., Rodriguez, A. and Mestres, M., A three dimensional simulation of pollutant dispersion for the near and far field in coastal waters, *J. Marine and Environmental Engineering* (in press), 1998.

Analysis of the hydrodynamical and physico-chemical behaviour of a natural wetland in the Lagoon of Venice

Zitelli A., Bergamasco A.A., Zampato L., (*) Umgiesser G., (*) Bergamasco A.
Ministry of the University and Scientific and Technological Research (MURST); Venice Lagoon System Project- DAEST-IUAV; (*) CNR-ISDGM Venice.
EMail: andreina@iuav.unive.it

Abstract

The paper presents the results of the monitoring phase and the main findings of the modelling activities carried out on a sub-basin of the inner Lagoon of Venice, called Palude della Rosa. Due to the beginning of a national project aiming at the environmental restoration of the Lagoon, a detailed monitoring plan has been carried out during the years 1993-1994-1995. Furthermore a very high resolution shallow water equations finite element model has been set-up for the same sub-basin in order to analyse the circulation patterns and the renewal rate of the masses of water.

1 Introduction

The loss of coastal wetlands can be attributed mainly to changes in land use as a result of anthropic development, to hydrological and water quality changes and to the economic activities associated with shipping and their impact on natural systems.

Coastal wetlands have gradually disappeared as a consequence of dredging and re-filling operations, to be replaced by urban, industrial or port activities. The deepening of shipping channels increases the hydrodynamic exchange with the sea and triggers erosive processes which lead inevitably to changes in the extent and morphology of the wetlands concerned.

As a direct consequence of these rigid land transformations, the fundamental natural functions of transition habitats are continually being lost; increased awareness of their irreplaceable role has now led to a conviction of the need for corrective actions. They pass through the comprehension of the physical and ecological mechanisms that govern the functioning of the natural system and arrive then to the management of it. Essential components in this process are monitoring, modelling and GIS [2].

With these premises in the previous years a very complex pilot project has been carried out in a sub-basin of the Venice Lagoon in order to lay the bases for understanding the behaviour of these natural systems and triggering the reversal of the damaging processes [3].

2 The natural wetland

The pilot project has focused on a sub-basin of the northern Lagoon of Venice called Palude della Rosa. It is a shallow brackish 3.5 km²-wide waterbody with an average depth of 0.5m. This basin has the morphological characteristics of a transition wetland surrounded by marshes [6]. It communicates with the external Lagoon through several entrances and is subject to the influence of riverine inflows coming from the mainland (Fig. 1a).

The hydraulic and riverine system that forces freshwater inputs in the Palude includes natural rivers and artificial drainings, fed by a complex water-scooping machines system serving a wide reclaimed area (Fig. 1b). In particular there are:
- the Canale Silone, that allows the passing of boats from the River Sile to the Lagoon through a lock and receives the waters coming from the river Vallio-Vela. The average outflow of this mouth is about 7 m³/s;
- the Canale Siloncello, that discharges in the Lagoon the waters of the River Sile through an old and unused lock; after its meeting with the Silone it runs along the borders of the Palude. The average outflow of this mouth is about 1 m³/s;
- the river Dese that flows in an area devoted to agriculture; after its meeting with the river Zero it receives the urban and treated waters of the Canale Osellino coming from the town of Mestre; then it enters in the Lagoon in the neighbourhood of the Palude. The average outflow of this mouth is about 4.5 m³/s.

Environmental Coastal Regions 343

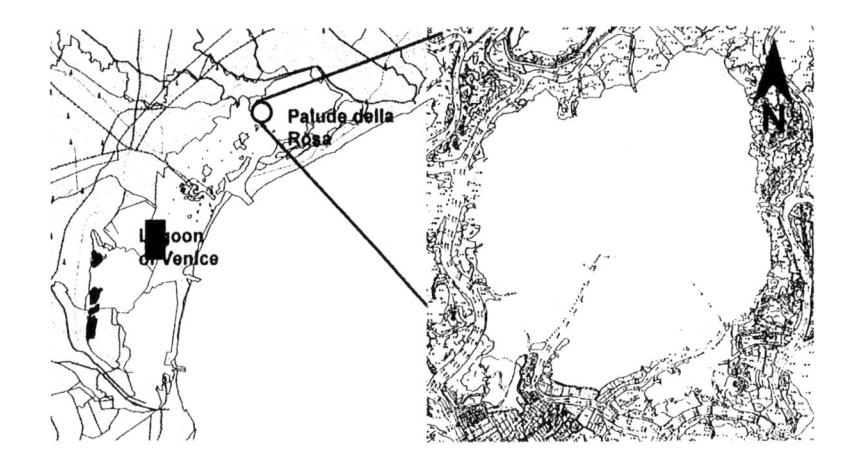

Fig. 1a: The localization of the Palude della Rosa inside the Lagoon of Venice

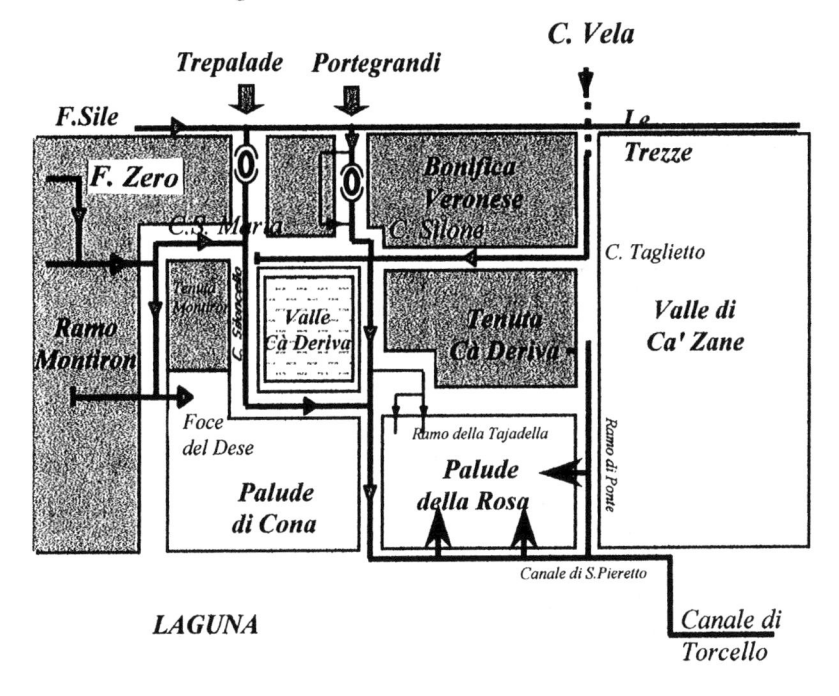

Fig. 1b: The hydraulic scheme of the Palude and its surrounding area

The overall freshwater inflow in this area of the Lagoon of Venice is therefore about 10-12 m³/s, average conditions.

The physical monitoring of the hydrodynamic features of the Palude has highlighted
- an important influence of the tidal signal coming from the Adriatic sea and reaching the Palude with a delay of about 100 minutes with respect to the Lido inlet and with an attenuation of about 20-25%;
- gradients of the free surface elevation up to 15 cm have been recorded inside the Palude due to intense Bora winds (north-east)
- a different hydrodynamic functioning with respect to different mean sea levels due to the strong interaction between the morphology of the intertidal and the hydrodynamical circulation.

The Palude is a natural wetland with peculiar environmental features resulting from the interaction between marine and riverine waters, whose quality together with the hydrodynamics determines the tropic characteristics of the Palled itself. During the monitoring the following phenomena have been observed:
- a clear stratification of the freshwater over the saltwater and the intrusion of the salt wedge in the channels
- a modulation of the relative roles depending on the tidal cycles, the season and the rainfall;
- a variation of the nutrient loading depending on the above-listed physical forcings and on the agricultural season practices.

3 Monitoring results

Due to the beginning of a national project aiming at the environmental restoration of the Lagoon, a detailed monitoring plan has been carried out under the Supervision of IUAV-DAEST of Venice in order to improve the understanding of the behaviour of the Palude della Rosa and in general of a natural wetland [4].

The research focus was the study of the mechanisms of functioning of the ecosystem both from the physico-chemical and the biological point of view. Hydrology and sediment properties have been put in relationship with the hydrobiological features of the area. The natural trends have been followed during the years 1993-1994-1995 through a series of short (48 hours) seasonal campaigns carried out on several points both inside the Palude and in the neighbourhood.

The interpretation of the collected data has pointed out that the Palude, together with the inflowing watercourse network that feed it and force the hydrodynamic and trophic climates, is a representative sub-basin of the northern lagoon: it is in fact an estuarine system showing its own typical features, resulting from its morphology and the interaction and origin of the external water inputs.

The key role played by the freshwaters coming from the mainland has been highlighted through the estimation of the mass balances of the Palude for water and nutrients: the relative weight of these loadings with respect to the nutrient loadings brought by the saltwater inflows has been studied and it has been demonstrated that they introduce an unbalance in the trophic state of the Palude and contribute to the high macroalgae production of the north-east area of the Palude itself, featuring in most cases a very low hydrodynamic circulation (Fig.2).

The vivification effect induced by the seawater entering the Palude at each tide has been highlighted through the study of plancton. Different species of zooplankton have been selected in order to be used as indicators of the origin of the water masses in the districts of the Palude and some form of competition between phytoplancton and macroalgae for the utilization of the available nutrients has been revealed. This competition seems generally to benefit the macroalgae along the monitoring period, in fact the macroalgae spread out their presence so much that they interacted with other ecological niches (e.g. seagrass beds of *Zostera noltii*).

4   Hydrodynamical model findings

A very high resolution shallow water equation finite element model was set up for the whole Lagoon of Venice by the CNR-ISDGM in the past [1].

The standard mesh for the whole Lagoon ensures a spatial resolution varying from 1 km along the mudflats to 40 m in the channels. In order to better reproduce the circulation patterns of the Palude della Rosa, the local resolution of the mesh has been increased, so reaching an average element size of 100 m inside the Palude and 20 m in the channels connecting the Palude with the external Lagoon. Along the boundary and in the neighbouring region the original mesh has been then tailored and adjusted to link it with the new one.

To analyse the hydrodynamic behaviour of the Palude, a series of 18 simulations have been set-up, including short-term (48 hours) runs with variously combined external physical forcing: tidal regime (neap and

346   Environmental Coastal Regions

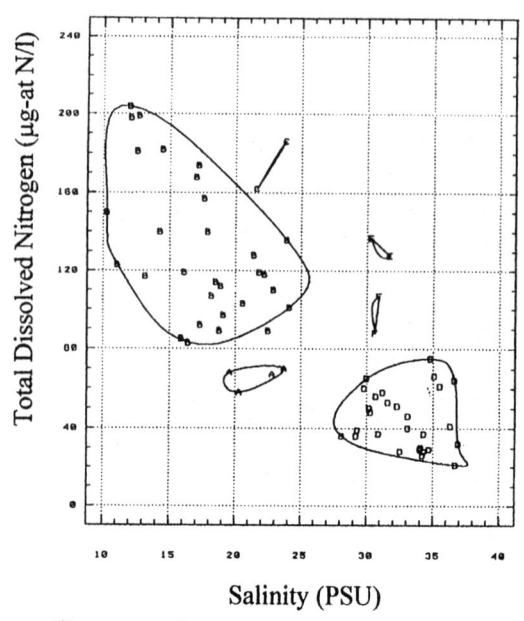

Fig. 2   Cluster analysis: Canale Silone - TDN vs Salinity (68 measurements).

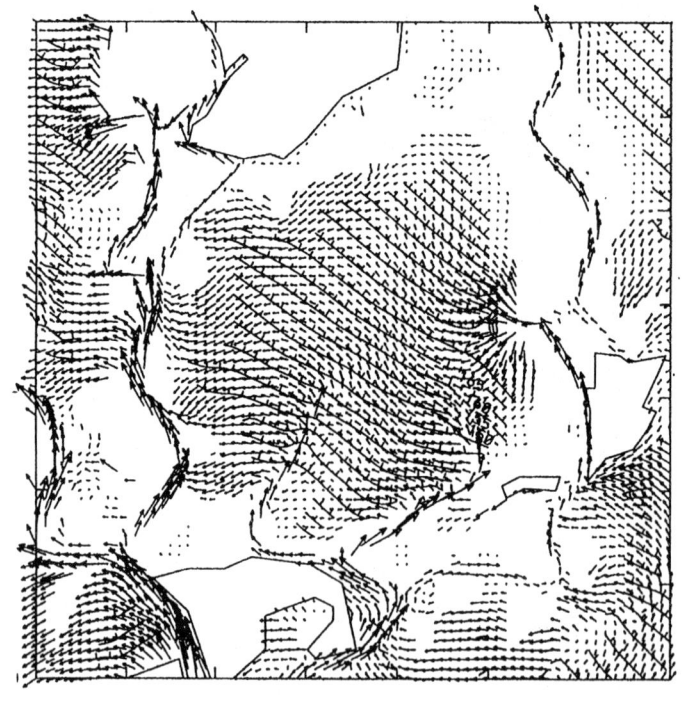

Fig. 3 Circulation pattern during floodings. Forcings: Spring tide, Bora, mean sea-level: high. Wide surrounding tidal marshes are exposed, the wind stress induces a 10cm gradient of the free surface elevation.

spring tide), winds (Bora east north-east wind, Scirocco and calm) and mean sea level (high, medium, low) [5].

The circulation patterns are very different depending on the wind conditions. In particular without wind the water flows along the water level gradient. The free surface presents a weak slope with a maximum sea level difference of about 3-4 cm over the Palled. Depending on the mean sea level, more or less intertidal elements are submerged so introducing a modulation in the dominance of the effect of morphology on the hydrodynamics. As expected, the most intense exchanges of water between the Palude and the surrounding Lagoon take place during spring tides, whereas during neap tides the circulation is weaker.

The hydrodynamic behaviour of the Palude is quite different when blowing the wind. The water flows parallel to the sea-level isolines and sometimes it flows up the gradient. The free surface presents a steeper slope with maximum sea-level differences up to 16 cm.

Moreover the Bora winds sweep the water out of the Palude so inducing a lowering of the inner mean sea-level (up to 15 cm) with respect to the cases without wind. On the other hand the Scirocco pushes the water inside the Palude, so that the inner mean sea-level increases of up to 10 cm. Furthermore even in case of medium and low mean sea-level, the Scirocco wind induces a counterclockwise circulation in the north-eastern area of the Palude, almost stagnant in the other conditions.

The numerical simulations describe the hydrodynamical response of the considered natural wetland to different external forcings (tidal regime, mean sea-level, winds).This induces a variability of the renewal rate of the Palude, defined as the rate between the total entering water flux and the average water volume stored in the Palude. The inverse of this quantity, i.e. the residence time Tr of the water masses, has been therefore chosen as indicator of the variability of the exchanges between the Palude and the external Lagoon.

The statistical analysis has highlighted two clusters, related to the tidal phases, spring and neap. The following table shows that Tr in neap is more than twice with respect to spring tide.

Tr (h)	min	ave	max	st. dev
spring	4.18	8.60	12.76	2.50
neap	14.02	18.28	22.92	2.84

Furthermore Tr is higher and less varying during ebb tides with respect to floods, when the exchanges along the borders are greater: this would indicate a shorter permanence of the marine waters with respect to freshwaters, called back in the Palude during ebb.

Tr seems to be almost unaffected by the sea-level conditions while winds increase the system variability: in particular Bora reduces Tr and Scirocco induces higher values of it.

5 Conclusions

The paper presented the monitoring results and the main findings of the hydrodynamical modeling activities carried out on a sub-basin of the inner Lagoon of Venice, where the tidal signal arrives delayed and reduced in amplitude. The main objective was to better understand the variability of the physical processes such as the circulation patterns and the renewal rate of a natural wetland, even in relation to the diffusion of nutrients coming from the mainland. These issues become more and more important in sight of the environmental impact assessment of a mobile closing system project to be built at the three Lagoon inlets in order to separate the city of Venice ad the Lagoon from the sea against flooding due to high tides.

References

[1] Umgiesser G., Bergamasco A., *A staggered grid finite element model of the Venice Lagoon*, in *Finite Elements in Fluids*, K. Morgan, E. Ofiate, J. Periaux and O. C. Zienkiewicz Eds., Pineridge Press, 1993.

[2] Zitelli A., Cardinaletti M., *Marine coastal habitats: mitigation, restoration, creation*, Thetis News, n.6, Venice (Italy),1993.

[3] Zitelli A., *Intervento pilota sperimentale per il recupero biologico della Laguna Veneta: la Palude della Rosa*, on Cà Tron Cronache IUAV, vol.4, pgg. 45-63, Venice, 1994.

[4] Consorzio Thetis – DAEST-IUAV, care of A.Zitelli, Fossato V. and Bergamasco AA. *Verifica di Tecniche di Arresto e Inversione del Degrado della Laguna - Intervento in Palude della Rosa, Rapporto di Sintesi*. Internal Report n.94.T405-REL-G083; Rev.2, 30.9.1995, Venice, 1995.

[5]. Zampato L., Bergamasco A.A., Bergamasco A., Umgiesser G., *Hydrodynamic variability in the 'Palude della Rosa,* (Venice Lagoon, Italy), National Council of Research, T.R. 212, Venice (Italy), 1997.

[6]. Zitelli A., Ciceri G. and Ceradini S., *Nutrient dynamic and regeneration in a shallow brackish marsh (palude della Rosa) of the northern lagoon of Venice: part1 – description of the site studied and pore water chemistry.* Journal of Analytical and Environmental Chemistry (submitted), 1997

Activity distribution of fission products in the surface water of the Adriatic Sea

Z. Franić and G. Marović
*Institute for Medical Research and Occupational Health,
Radiation Protection Unit
HR-10000 Zagreb, Ksaverska cesta 2, PO Box 291,
CROATIA
E-mail: franic@imi.hr*

Abstract

The activity concentrations of ^{90}Sr and ^{137}Cs in the surface waters of the Adriatic Sea were investigated twice a year (in April and October, if possible) over the 1963 - 1997 period at four locations (towns of Rovinj, Rijeka, Split and Dubrovnik). The fallout samples were collected monthly in the town of Zadar. An exponential decline of radioactivity in sea water as well as in fallout samples followed the nuclear moratorium. After the nuclear accident at Chornobyl, higher levels of ^{137}Cs were detected again, while the activity concentrations of ^{90}Sr, that is less volatile than radiocaesium, have not been significantly increased. The coefficient of correlation between ^{90}Sr activity concentrations in fallout and sea water was found to be fairly good, representing the fact that approximately 85% of man-made radioactive contamination in the Mediterranean Sea comes from fallout. The long term measurements were used to model the mean residence time of ^{90}Sr in sea water. No significant variations of mean residence time on different locations were found, implying the uniform distribution of ^{90}Sr through the mixed layer of the Adriatic Sea.

352 Environmental Coastal Regions

1 Introduction

The Adriatic Sea is the northernmost extension of the Mediterranean Sea. (Figure 1). The elongated shape (approximately 800 by 200 km) and almost landlocked position of the Adriatic Sea play an important role in dynamics of its water. The Strait of Otranto joins the Adriatic to Ionian Sea, another arm of the Mediterranean. The area of the Adriatic Sea is 140,000 km^2, average depth 160 km and the volume 35,000 km^3. The Adriatic Sea is a rather shallow, especially in the northern part where the depth does not exceed 80 m. It is under the strong impact of the Po River as the major source of fresh water. The run-off (precipitation that the ground does not absorb and that ultimately reaches rivers, lakes or oceans) from the large area of Northern Italy introduced in the Adriatic Sea by the Po River is dispersed

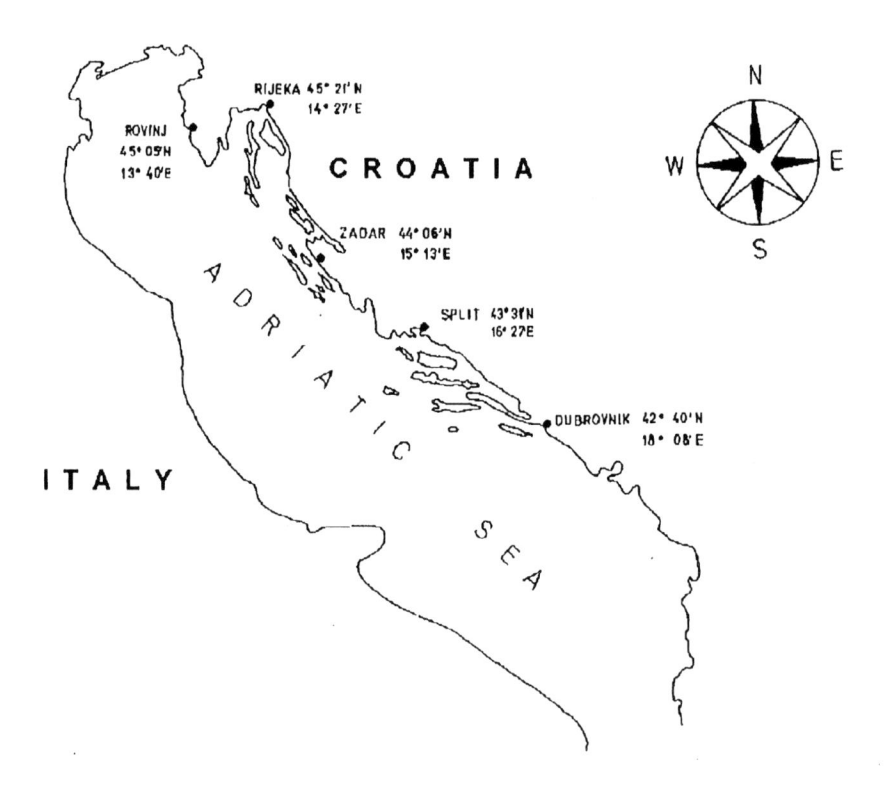

Figure 1: Map of the Adriatic Sea with sampling locations

by the wind-driven currents through the whole Northern and Middle Adriatic. Generally, the Adriatic Sea is characterized by low precipitation, high evaporation, low tidal action, low nutrient content, low suspended load, and low biological productivity. While these features result in hydrographical conditions quite different from those in other seas, the fundamental biogeochemical processes taking place in water columns are not to be considered different from other seas [1]. Investigations of the distribution and fate of natural, weapon-produced and reactor-released radionuclides in the Adriatic Sea have been conducted in the Unit of Radiation Protection of the Institute for Medical Research and Occupational Health, as a part of an extended monitoring program. Strontium investigations have been going on since 1963 and those of caesium and some other radionuclides from 1978. The results are published yearly [2,3,4]. Investigation of the fate of radionuclides that enter the Adriatic Sea either as a run-off or through the air-sea interface can provide valuable radioecological data as well as additional data needed for studies on water mass circulation and some other properties.

2 Experimental

Sea-water samples, 150 l each, were collected twice a year (in May and October, if feasible) 3 km from the shore, at a depth of 0.5 m, at four sampling sites (towns of Rovinj, Rijeka, Split and Dubrovnik). Fallout samples were collected at the city of Zadar on the Adriatic coast, monthly for the 1963-1982 period and from 1982 to 1996 quarterly. For the determination of strontium and caesium, were used radiochemical methods. The radioactivity of ^{90}Sr was determined by beta-counting its decay product (^{90}Y), in a low-background anti-coincidence, shielded Geiger-Müller counter. A gamma-ray spectrometry system based on a Ge(Li) detector (FWHM 1.82 keV at 1.33 MeV) coupled to a computerized data acquisition system (a 4096-channel pulse height analyser and a personal computer) was used to determine radiocaesium levels in the samples from their gamma-ray spectra. The detector is shielded with 10 cm thick lead that is lined with 2 mm of cadmium and 2 mm of copper. Samples were measured in cylindrical plastic containers of appropriate volume placed directly on the detector. Counting time depended on sample activity concentration, but was never less than 60,000 s. Efficiency calibrations were carried out using sources provided by the International Atomic Energy Agency (IAEA) and World Health Organization (WHO).

3 Results and discussion

Radioactive fallout, which resulted from the large-scale nuclear weapon tests in the atmosphere conducted during the 1960s, followed by similar, but smaller scale tests by the Chinese and French in the 1970s and afterwards, represented the dominant route for the introduction of the artificial radionuclides in the Mediterranean and therefore Adriatic Sea as well, until the 1986 nuclear accident in Chornobyl. The activity concentrations of ^{90}Sr and ^{137}Cs are in good correlation with the activity concentrations of these radionuclides in fallout. That is consistent with estimation that 85% of all man-made radioactive contamination in the Mediterranean comes from fallout[1]. Over the observed time period, ^{90}Sr activity concentrations exponentially decreased from 14.8 ± 2.3 Bqm^{-3} in 1963 to only 2.8 ± 0.2 Bqm^{-3} in 1996 [5,6] (Figure 2).

3.1 ^{90}Sr mean residence time in the Adriatic Sea water

By fitting the experimental data for ^{90}Sr activity concentrations in sea water to the exponential equation:

$$A(t) = A(0)\, e^{kt} \qquad (1)$$

where A(t) and A(0) are ^{90}Sr sea-water activity concentrations in times t and 0 respectively, constants k for respective locations were found. The reciprocal value of the constant k is the observed mean residence time of ^{90}Sr in the sea

Environmental Coastal Regions   355

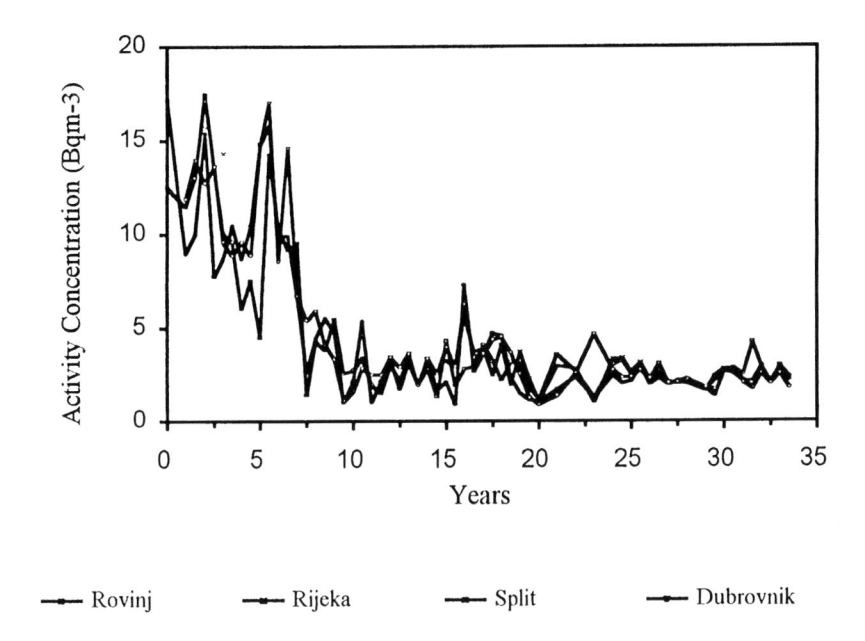

Figure 2: ^{90}Sr activity concentrations in the Adriatic Sea

water, i.e.:

$$T_{ob} = \frac{1}{k} \qquad (2)$$

From (1) and (2) the average of four locations for T$_{ob}$ is 10.9 ± 0.7 y. However, to get the idea about the upper limit of the mean residence time, T$_m$, of non-radioactive pollutants in the sea water, the effect of radioactive decay should be excluded. It can be shown that:

$$T_m = \frac{1}{k_m} = \frac{1}{k - \lambda} \qquad (3)$$

where $\lambda = \ln(2)/T_{1/2}$ is radiological decay constant for ^{90}Sr (y^{-1}). From (3), as T$_{1/2}$ for ^{90}Sr is 29.12 y for T$_m$ is obtained value of 14.6 y. This, relatively high value also represents the turnover time of the Adriatic Sea water as it is exchanged with the water from the Ionian Sea. The areas of short mean residence times, would rapidly respond to potential discharges of radioactive contaminants. As no significant variations of the ^{90}Sr mean residence times for considered locations were found, that implies the similarity of oceanographic factors and uniform distribution of ^{90}Sr in the surface water.

3.2 ^{137}Cs activity concentrations

The base line level of ^{137}Cs in the Adriatic Sea water in 1978 was 4.5 ± 1.7 Bqm^{-3}. The essentially constant levels of ^{137}Cs in the pre-Chornobyl period have increased by two orders of magnitude immediately after the accident, decreasing ever since. Unlike the debris from the atmospheric testing of nuclear weapons, the radionuclides that originated from the damaged reactor at Chornobyl were not released directly into the upper atmosphere. As the result of the release mechanism (continuing releases over 10-day period, steam explosions and fire of the graphite moderator) and the prevailing meteorological conditions at the time, the more volatile components of the Chornobyl debris (e.g. ^{137}Cs) were subjected to the global dispersion processes, contrary to less volatile radionuclides like ^{90}Sr. The latter ones were deposited to the Earth's surface within a short period after the accident (days to weeks). Generally, changing meteorological conditions with wind of different directions at various altitudes and prolonged releases resulted in a very complex dispersion pattern over Europe. Consequently, the ^{137}Cs activity concentrations in the sea water considerably varied from location to location. The same procedure as for ^{90}Sr was used for calculation of observed residence times of ^{137}Cs for the post-Chornobyl period. The values of 1.2, 1.8, 0.8 and 1.1 years were obtained for Rovinj, Rijeka, Split and Dubrovnik, the average being 1.2 ± 0.3 years. Coefficient of correlation, is r>0.95 except for Rijeka, the r value being 0.65, leading to higher standard error. Therefore, the residence time of 1.8 years for the location Rijeka might be overestimated.

3.3 Further modeling

To obtain better assessments of the ^{90}Sr mean residence time in the sea water, the effect of fresh fallout that continuously enters the sea should be excluded. The Adriatic Sea is the major source of the densest water in the Eastern Mediterranean, the *Eastern Mediterranean Deep Water* (EMDW) [7]. The meteorological conditions favorable to dense water formation occurring during Bora events (the dominant winter wind blowing from Northeast), cause rapid mixing of surface waters with the intermediate water layer. Consequently, they can be considered as a single reservoir. Therefore, the activity concentrations could be modeled by the equation

$$\frac{dA_s(t)}{dt} = -(k_s + \lambda)A_s(t) + I(t) \qquad (4)$$

where

$A_s(t)$ is the total, time-dependant ^{90}Sr activity (Bq) in the entire Adriatic consisting of surface and intermediate water, obtained by multiplying the observed activity concentration data by the volume the reservoir (29,000 km³)

λ is the decay constant for ^{90}Sr (y^{-1}),

$1/k_s$ is the mean residence time (y) of ^{90}Sr in a reservoir and

$I(t)$ is the total annual ^{90}Sr input into the Adriatic Sea (Bqy^{-1}).

However, in order to solve the equation (4) function describing I(t) must be found. It consists of the direct input by fallout and the Po River run-off. While the long-term fallout measurements are performed in the city of Zadar on the Croatian coast, long term data for the Po River are unavailable. By solving the equation (4) and fitting the obtained function to the experimental data, more reliable value for k_m could be hopefully found. Presumably, it would be less than 15 years which was calculated as the upper limit. By studying water flows of the Adriatic Sea water through the strait of Otranto the turnover time has been estimated to be approximately 5 years [7].

4 Conclusions

Generally, the marine environment in Croatia is not contaminated by fission products. The baseline levels of ^{90}Sr and ^{137}Cs in the Adriatic Sea surface water are around 3 and 5 Bqm^{-3} respectively, as in the rest of the Mediterranean Sea. The impact of Chornobyl fallout was restricted to the first and second year after the accident, which was especially true for the less volatile radionuclides. The upper limit of the mean residence time of ^{90}Sr in Adriatic Sea water, which also reflects the Adriatic Sea water turnover time, was estimated to be approximately 15 years. For more precise assessment of turnover time, the mathematical model that takes into account the continuous fallout input should be developed.

References

[1]   UNEP - United Nations Environment Programme, Mediterranean action plan: Assessment of the state of pollution in the Mediterranean Sea by radioactive substances. UNEP, Athens 1991.

[2]   Popović V. Environmental radioactivity in Yugoslavia 1963 - 1977. Summary reports 1963 - 1977. Federal Committee for Labour, Health and Social Welfare, Belgrade 1964 - 1978. (In Croatian).

[3]   Bauman A, Cesar D, Franić Z, Kovač J, Lokobauer N, Marović G, Maračić M, Novaković M. Results of environmental radioactivity measurements in the Republic of Croatia 1978 - 1990. Summary reports 1978 - 1991. Institute for Medical Research and Occupational Health, Zagreb 1979 - 1992. (In Croatian).

[4] Kovač J, Cesar D, Franić Z, Lokobauer N, Marović G, Maràčić M. Results of environmental radioactivity measurements in Republic Croatia 1992 - 1996. Summary reports 1992 - 1996. Institute for Medical Research and Occupational Health, Zagreb 1993 - 1997. (In Croatian).

[5] Franić Z, Bauman A. Activity of ^{90}Sr and ^{137}Cs in the Adriatic Sea. Health Phys **64**, pp.162-169, 1993.

[6] Franić Z. The Adriatic Sea-water Radioactivity, Report of the International Atomic Energy Agency (IAEA) Research Contract No. 302-K4-CRO-8844, B5-CRO-24175. 1996.
Available at: <http://mimi.imi.hr/~franic/adriatic.html>
(URL checked 1998 April 28)

[7] Orlić M, Gačić M, La Violette PE. The currents and circulation of the Adriatic Sea. Oceanologica Acta, **15**, pp.109-123, 1992.

Section 8:
Coastal Structures

Investigation of influences of very large floating structures on exchange of sea water

S. Tabeta
Division of Artificial Environment and Systems,
Yokohama National University,
79-5 Tokiwadai Hodogaya-ku Yokohama 240-8501, Japan
EMail: beta@hotei.shp.ynu.ac.jp

Abstract

Recently in Japan, there are some projects of constructing very large floating structures (VLFS) such as ocean airports. The exchange ratio of sea water can be an important index in evaluating influences of VLFS on marine environment. The main purpose of the present study is to investigate the influences of large-scale ocean structures on exchange of sea water by means of numerical experiments. At first, the results of numerical calculations were compared with field observation and hydraulic model experiments in order to validate the numerical model. The exchange ratio obtained by the calculation agreed well with the observation, and the calculation could also reproduce the influences of breakwater appeared in the hydraulic model experiments. Thus it is concluded that the exchange of sea water can be evaluated properly by the numerical model used in the present study. By using the same model, some numerical experiments were carried out in order to investigate the influence of VLFS on tidal exchange of sea water. The results showed that the exchange is promoted by the circulation caused by the structure, while it is suppressed because of the reduction of section area of the bay when a large-scale ocean structure is installed.

1 Introduction

The concept of the exchange of sea water is often used when the water quality problem of a bay is considered. Recently in Japan, there are several projects of constructing very large floating structures (VLFS) such as ocean airports, and some studies have been carried out for investigating the environmental impacts of VLFS [1,2]. The exchange ratio of sea water can be an important index in evaluating influences of VLFS on marine environment. The main purpose of this study is to investigate the influences of large-scale ocean structures on exchange of sea water by means of numerical simulations.

At first, in order to validate the numerical model, the results of the calculations were compared with field observation and hydraulic model experiments. There have been a lot of studies about the numerical simulation of coastal currents, but many of them examined about only the characteristic current pattern under constant boundary conditions. From the viewpoint of the reliability of the numerical simulations, it is necessary to investigate more whether the model can reproduce real phenomena in a certain period. In the present study, a numerical simulation of exchange of sea water in Tokyo Bay was carried out using real changing atmospheric conditions. The tidal exchange ratio obtained by the calculation was compared with that by the field observation. The numerical model was also applied to the calculations of tidal exchange of a bay with or without a breakwater. In these simulations, it was examined if the influences of large-scale ocean structures on tidal exchange can reproduce. The numerical results were compared with those of hydraulic model experiments and validity of the numerical model was discussed.

By using the same numerical model, some numerical experiments were carried out to investigate the influence of VLFS on exchange of sea water. The experiments of tidal exchange were carried out for VLFS with some different drafts. The spread of freshwater from the river was also investigated under the influences of artificial islands.

2 Validation of the Numerical Model

2.1 Comparison with the field observation

Matsumoto et. al.[3] made a field observation in the winter season in 1974 to investigate the actual state of the exchange of sea water in Tokyo Bay. The observation was made along the section shown in Figure 1. Currents were measured continuously at 4-6 positions in the vertical direction at each point of B, D, F during 15 days from February 1st to 16th. Salinity at each point of A-F and at 15 points of outer ocean was also measured during one day from Feb. 10 to 11.

Environmental Coastal Regions 365

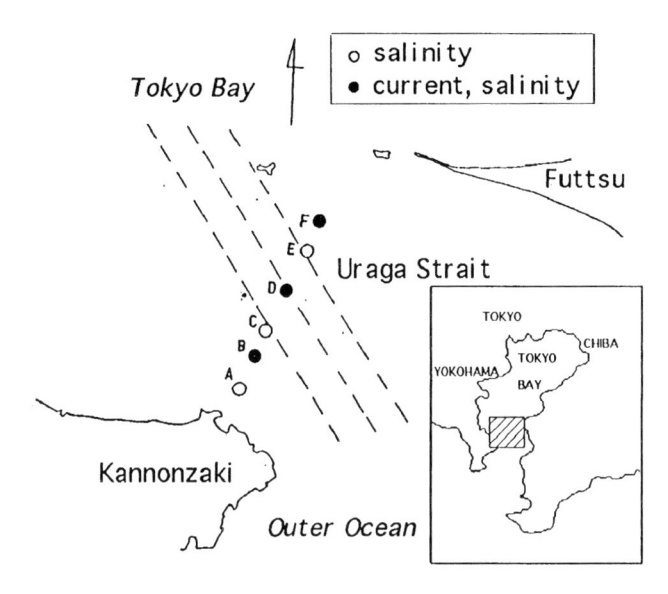

Figure 1: The measurement site in Tokyo Bay

A real time simulation was carried out for the same period as the observation, February 1-16, 1974. The numerical model used in this study is a multilevel model based on three dimensional momentum equations with hydrostatic approximation, the equation of continuity, transport equations of heat and salinity, and the equation of state [4]. For the simulation of Tokyo Bay, the horizontal resolution was 1km and the vertical resolution was 7 layers. The observed time depending atmospheric data such as wind, atmospheric temperature, vapor pressure, cloud amount, and precipitation were used for calculating surface flux.

Time variation of 25-hours' moving average of the current velocity in the direction normal to the observation section is shown in Figure 2. The characteristics of the variation of observed currents are well reproduced by the simulation. In their study about the tidal exchange through the cross section at the Golden Gate in San-Francisco Bay, Parker et al. [5] defined the exchange ratio R by the following formula;

$$R = \frac{Q_0}{Q_F} \sim \frac{\bar{C}_F - \bar{C}_E}{C_0 - \bar{C}_E} \qquad (1)$$

where Q_F is the amount of total flow entering the inner bay during the flood tide, Q_0 is the amount of newly entering water from the outer ocean, and C_0 is the salinity of outer ocean. $\bar{C}_E$ and $\bar{C}_F$ denote the average salinity during ebb and flood tide respectively, and can be expressed as follows;

$$\bar{C}_E = \frac{\int Q_e C_e dt}{\int Q_e dt}, \qquad \bar{C}_F = \frac{\int Q_f C_f dt}{\int Q_f dt} \qquad (2)$$

where Q_e, C_e are the rate of flow and the salinity during the ebb tide, and Q_f, C_f are the rate of flow and the salinity during the flood tide. Therefore, R can be obtained by measuring the rate of flow and salinity during one tidal cycle. The tidal exchange ratio estimated by Eqs.1 and 2 from the observation is 8%, while that from the numerical simulation is 7.3%.

2.2 Comparison with the hydraulic model experiment

Murakami[6] examined the influence of a breakwater on the tidal exchange of sea water by hydraulic model experiments. The experiments were carried out by Suzaki Bay model as the following procedure. At first, the water filled in the model is divided into inner bay water and outer ocean water by the barrier board installed at the cross section of bay entrance. In the cases with density difference, the density of water inside the barrier board is set lower than that of outside water. The tracer (fluorescein sodium dye) is put into the inner bay water and mixed up sufficiently until the water has homogeneous concentration of tracer. Next, the barrier board is removed and the operation of the pneumatic tide generator starts simultaneously. After then, water samples are picked up from the model basin to measure the concentration of the tracer in every several tidal cycles. In order to examine the influences of the density current and the breakwater installed at the mouth of the bay, the experiments were carried out under four different conditions (EXP1-4 in Table 1).

Some numerical calculations were carried out by using the simple rectangular bay with constant depth as shown in Figure 3 to examine if the influence of a breakwater on tidal exchange can be reproduced by the numerical model. At the beginning of the calculations, the concentration of the tracer in the inner bay and the outer ocean are 1 and 0, respectively. As an external force, M_2 tide with amplitude of 1m and period of 12 hours is given at the open boundary. The

Environmental Coastal Regions 367

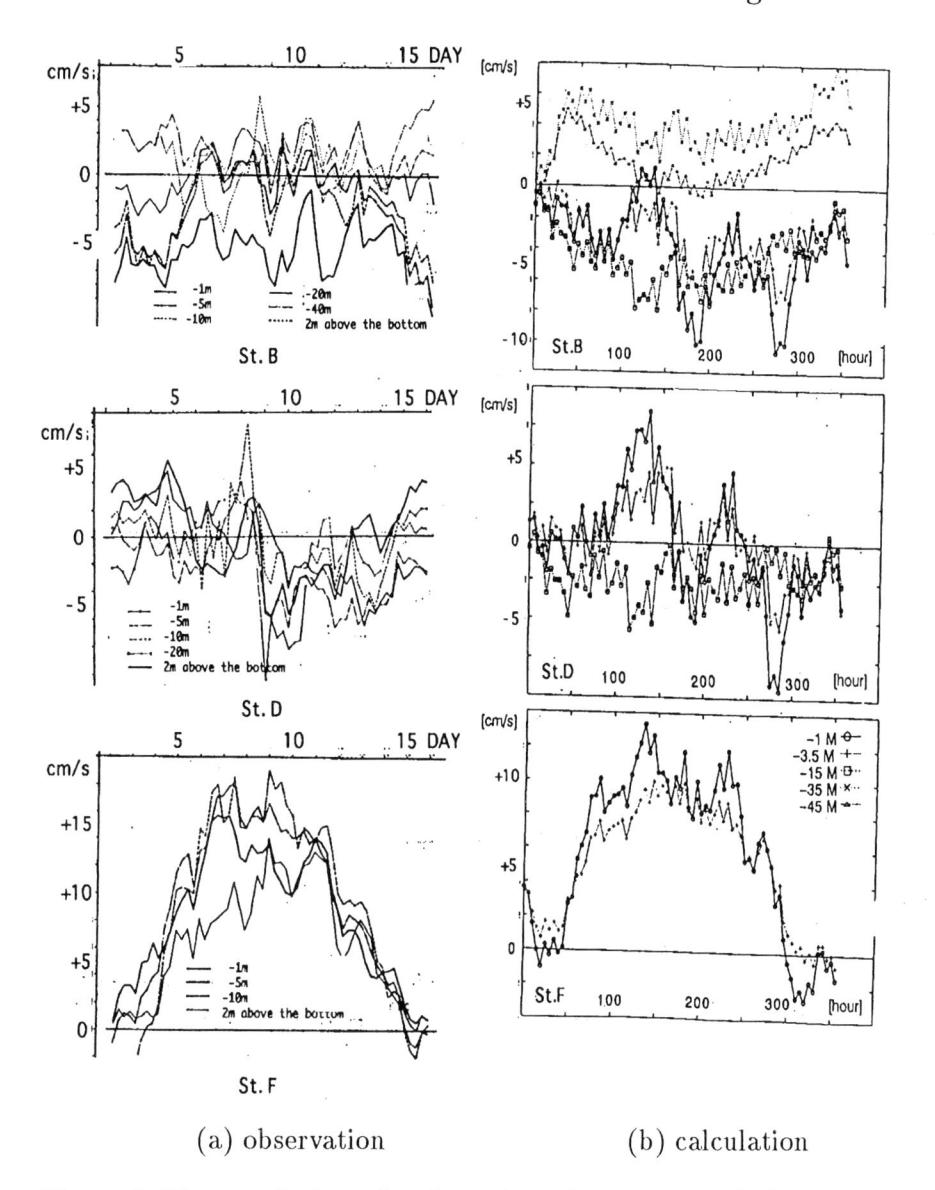

(a) observation   (b) calculation

Figure 2: Time variation of 25-hours' moving average of the current velocity near the side of Kannonzaki (St.B), about the center of the channel (St.D), and near the side of Futtsu (St.F)

368 Environmental Coastal Regions

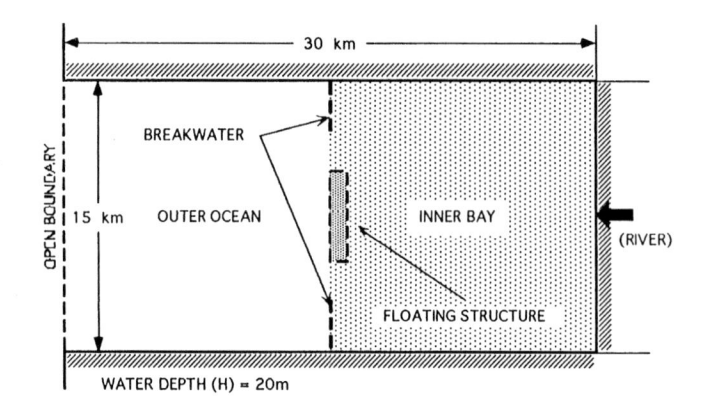

Figure 3: The rectangular basin used for numerical experiments of tidal exchange of sea water

calculations were carried out under four different conditions (CAL-1-4 in Table 1) corresponding to the hydraulic model experiments.

Figure 4 shows the decay curves of the amount of the tracer which remains in the inner bay region. Table 1 shows the averaged residence time calculated by the following formula [7].

$$T_r = \frac{1}{M_0} \int_0^\infty t \left[-\frac{dM(t)}{dt}\right] dt = \int_0^\infty r(t) dt \qquad (3)$$

Where $r(t)$ is remnant function which is defined as follows;

$$r(t) = \frac{M(t)}{M_0} \sim \exp(-\alpha t^\beta) \qquad (4)$$

in which M_0 is the amount of the substance at $t = 0$, $M(t)$ is that at $t = t$, and α and β are constants which are estimated from the decay curve by fitting. According to the figures, it is evident that the water exchange with density difference is much larger than the case of without one. The exchange ratio with the breakwater is larger than that without it when the density is homogeneous, while the exchange ratio with the breakwater is smaller than that without it when there is density difference. This tendency is recognized in the both cases of hydraulic experiments and numerical calculations.

The horizontal pattern of tidal residual current and distribution of the tracer at the surface layer in the cases without the density difference are shown in Figure 5. When there is no breakwater (CAL-1),

Environmental Coastal Regions 369

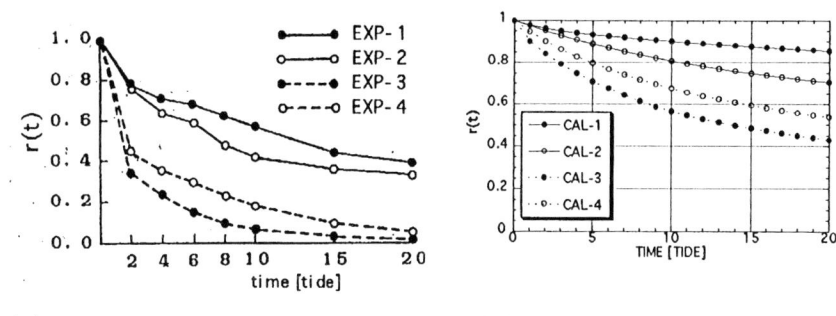

(a) hydraulic model experiments   (b) numerical simulations

Figure 4: Time variation of averaged concentration of the tracer in the inner bay

the residual current is very small. When the breakwater is installed (CAL-2), horizontal circulations are induced around the breakwater, which promote the exchange of sea water. Figure 6 is the pattern of residual current and the distribution of the tracer in the vertical section along the longitudinal axis of the basin. In the case without the density difference (CAL-1), the residual current is also very small. On the other hand, the vertical circulation by the density current appears in the case with density difference (CAL-3). With a breakwater (CAL-4), a vertical circulation is interfered by breakwater and the spread of the tracer is restrained. As shown above, the presence of breakwaters at the entrance of the bay promotes the water exchange by horizontal circulation, but suppresses the exchange due to

Table 1: Averaged residence time by hydraulic model experiments (EXP) and by numerical simulations(CAL)

	CONDITION		τ_r (days)	
CASE	DENSITY	BREAKWATER	EXP	CAL
1	HOMOGENEOUS	–	15.3	199.7
2	HOMOGENEOUS	○	13.2	40.8
3	DIFFERENT	–	1.3	15.9
4	DIFFERENT	○	2.7	21.5

density-induced vertical circulation.

3 Numerical Experiments of the Influence by VLFS

3.1 The conditions of calculations

In order to investigate the influence of VLFS installed in a bay on the exchange of sea water, some numerical experiments were carried out. The calculation domain for the experiments is the same rectangular basin used in the previous section. A floating structure with the horizontal dimension of 5km × 1km, which is supposed to be a floating airport, is installed at the position shown in Figure 3.

At first, the experiments of tidal exchange were carried out under the conditions with and without density difference for several different drafts of VLFS. The procedure of experiments was same as that of previous section.

Next, the spread of freshwater from the river was investigated. Initially, the bay was filled with the water which has same salinity of outer ocean water. From the river located at the interior part of the bay, fresh water continuously flows into the bay. At the same time, tidal oscillation was imposed at the open boundary. The experiments were carried out for the cases without artificial island and with a floating or reclaimed island located at the center of the basin.

3.2 Results

The averaged residence time estimated from the experiments of tidal exchange is shown in Figure 7. In the figure, the horizontal axis denotes the draft of VLFS (d) relative to the water depth of the basin ($H=20$m). Without difference of density, the residence time is smaller as the draft of VLFS becomes larger. On the other hand, under the condition with density difference, the larger the draft of VLFS is, the smaller the exchange ratio of sea water becomes.

Figure 8 shows the time variation of salinity about the center of the basin in the experiments for spread of freshwater from the river. The salinity in the bay remains higher when the artificial islands exist, because the islands obstructed the spread of fresh water. It is also recognized that the floating structure less influences the spread of fresh water than the reclaimed island.

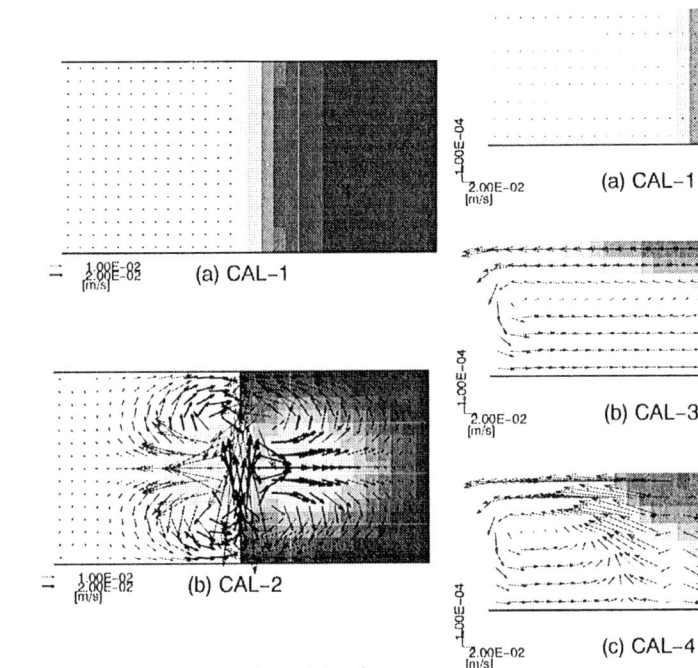

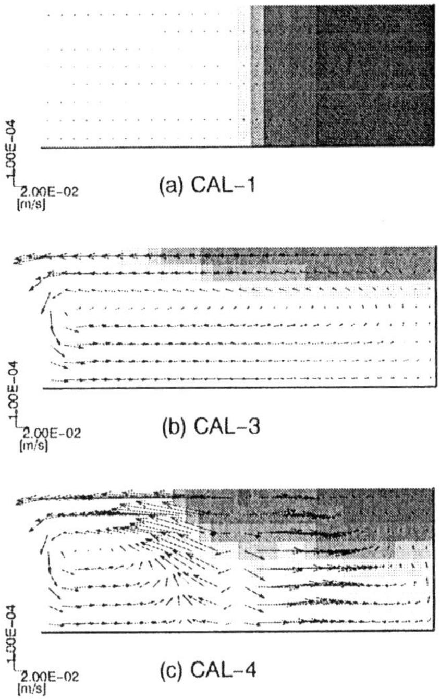

Environmental Coastal Regions 371

Figure 5: Horizontal residual current and concentration distribution of the tracer by the calculations without difference of density

Figure 6: Tidal residual current and concentration distribution of the tracer in the vertical section along the longitudinal axis of the basin

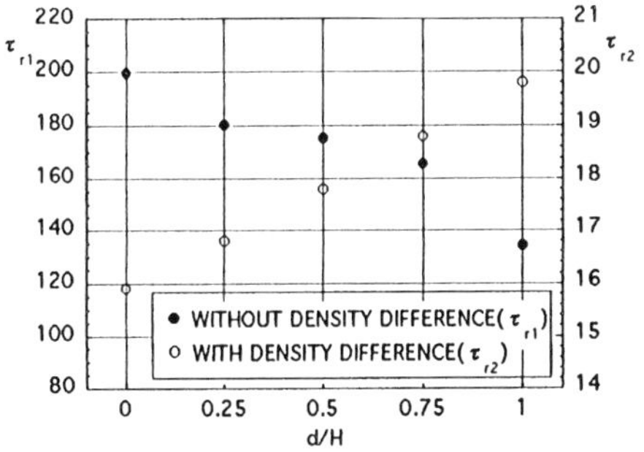

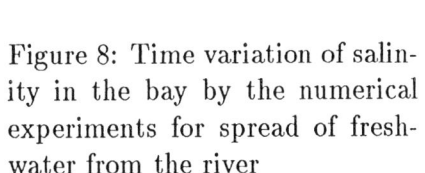

Figure 7: Averaged residence time by the numerical experiments of tidal exchange under the influence of VLFS with various drafts

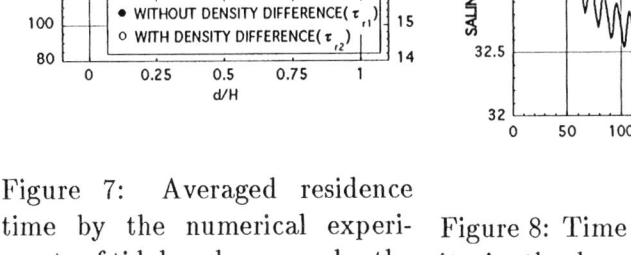

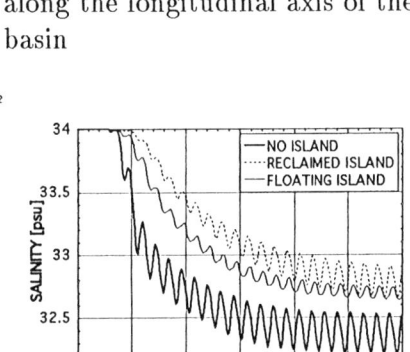

Figure 8: Time variation of salinity in the bay by the numerical experiments for spread of freshwater from the river

4 Concluding Remarks

In this study, the influences of large-scale ocean structures on exchange of sea water were investigated by numerical experiments. From the results of comparison with the field observation and the hydraulic model experiments, it can be said that the exchange of sea water including the effects of ocean structure is evaluated by the numerical model used in the present study. The results of experiments showed that the tidal exchange of sea water is promoted by the circulation caused by the structure, while it is suppressed because of the reduction of section area of the bay when a large-scale ocean structure is installed. The degree of each effect depends on the condition of current and density as well as the dimensions of the structure.

References

[1] Tabeta, S. & Inoue, Y., Interaction of Ocean Current and a Huge Floating Structure in Restricted Sea, *Proc. of Int. Conf. on Technology for Marine Environment Preservation,*, pp505-512, 1995.

[2] Fujino, M., et. al., Measurement of Marine Environment around Mega-Float Model Moored in Tokyo Bay, *Proc. of 6th Int. Conf. on Offshore Mechanics and Arctic Engg.*, Vol.VI, pp47-54, 1997.

[3] Matsumoto, T., et. al., *Proc. of Coastal Engineering, JSCE*, **21**, pp. 291–296, 1974 (in Japanese).

[4] Tabeta, S. & Fujino, M., Comparison between Simulation Results and Field Data about Currents and Density in Tokyo Bay, *Journal of Marine Science and Technology*, **1**, pp. 94-104, 1996.

[5] Parker, D.S., Morris, D.P., & Nelson, A.W., Tidal exchange at Golden Gate, *Proc. Amer. Soc. Civil Engg.*,**2**, pp. 305–323, 1972.

[6] Murakami, K., Hydraulic model experiments on water exchange for enclosed inner bay, *Proc. of Int. Symp. on Scale Modelling,* pp. 301-307, 1988.

[7] Takeoka, T., Fundamental concepts of exchange and transport time scale in coastal area, *Contineltal Shelf Research*, **3**, pp. 311-326, 1984.

A study of anoxic water structures generated in deep bay enclosed by tsunami breakwaters

Tsuruya, Hiroichi and Hibino, Tadashi
Marine Environment Division, Port and Harbour Research Institute, Ministry of Transport
3-1-1, Nagase, Yokosuka 239-0826, Japan
Email: tsuruya@cc.phri.go.jp

Abstract

Ohfunato Bay locates in the North-east part of Japan, and the water quality and physical structures inside the bay are strongly affected by climate such as air pressure arrangements around Japan and typhoons and ocean currents. From the present study, it is revealed that the water mass inside the bay is strongly related to the tidal level along Japanese coast facing to Pacific Ocean over the scale of 3,000km from Nemuro (Hokkaido) to Naha (Okinawa). Low salinity water which is a part of Oyashio ocean current intrudes in Ohfunato bay from April to October in the average year. Also, we found that low salinity water is carried by typhoon from ocean into Ohfunato bay.

1 Introduction

In the study to investigate and predict the water quality in an estuary, it is important to know the origin of water in an estuary. In this paper, the important roles of climate and weather to carry ocean water into an estuary are described.

A tsunami protection breakwater has been constructed at the mouth of Ohfunato Bay, Japan. The composite type breakwater, concrete caissons on rubble mound, reduced the bay opening area about 7% of the original one. Because of the existence of the rubble mound, the lower bay water cannot directly exchange with the water outside the bay.

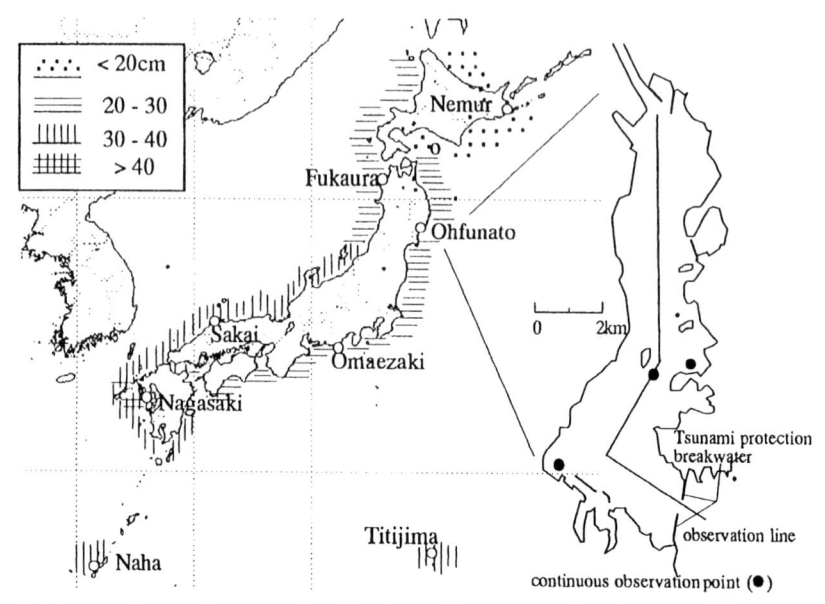

Fig. 1 Annual range distribution of mean sea level around Japan (Murakami and Yamada [1]) and plan of Ohnfunato Bay

In summer season, stratification enhances the generation of anoxic water in the lower layer. In order to investigate the water exchange ability through the tsunami protection breakwater, water quality (salinity, water temperature, etc.) in the bay will be the suitable tracers to distinguish ocean water from bay water.

2 Field observations

Field observations in Ohfunato bay were performed for three years from 1995 to 1997. Vertical structures of water temperature, salinity, dissolved oxygen (DO), etc. were measured at three points and along a line as shown in Fig.1. Figure 1 shows locations of the port where sea level variations and atmospheric pressures were observed. The figure also describes the ranges of yearly change of sea levels along Japanese coasts averaged with a month. Annual mean sea level variation is high in the south-western part of Japan and low in the northern part.

3 Seasonal ocean water inflow

Figure 2 shows the time variations of monthly averaged salinity at

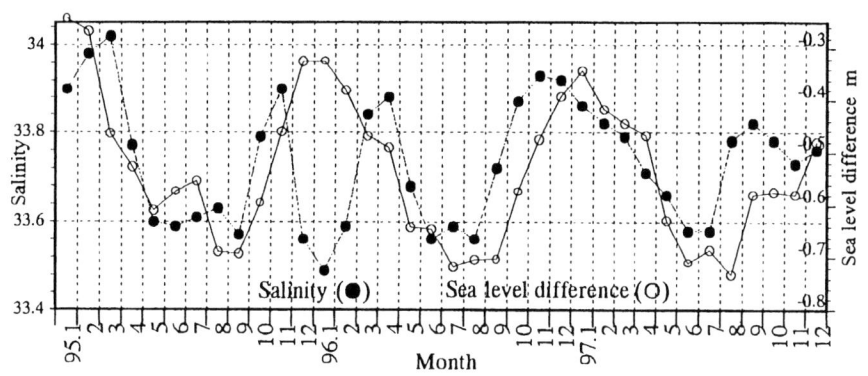

Fig. 2 Salinity variation at center of Ohfunato Bay and sea level difference between Nemuro and Naha (Nemuro - Naha)

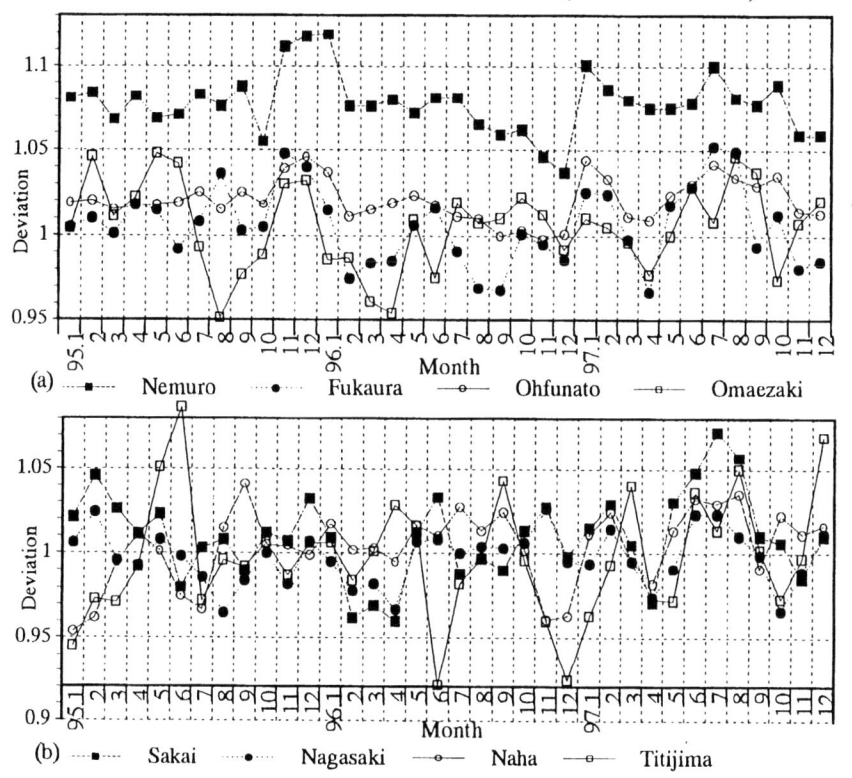

Fig. 3 Deviation of sea level average (11 years from 1987 to 1997)

Ohfunato Bay and sea level difference between Nemuro and Naha. Figure 3 shows the time variations of monthly averaged sea level (deviation of the sea level from the average at each observation point (11 years mean)) at the observation stations as shown in Fig.1.

376   Environmental Coastal Regions

It can be recognized from Fig.2 that low salinity water intrudes in Ohfunato Bay every summer season and it is closely related to the sea level difference between Nemuro and Naha. In three years when the observation was made, decrease in salinity in winter season was observed only from 1995 to 1996, although regular decrease in salinity was observed only in summer seasons. As shown in Fig.3, sea levels at north and east Japan (observation stations in Fig.3(a)) were from 3 to 10 % higher than that of 11 years average from 1987 to 1997. In recent 10 years, it has been relatively warm winter but that only from 1995 to 1996 was considered to be normal, because the pressure distribution was PNA pattern which is typical in winter.

It can be thought that as the pressure distribution for winter season dominates, sea water with low salinity and temperature from Bering Sea intrudes in Ohfunato Bay. 「Hibino and Tsuruya [2], [3]」 have shown that there is a strong relationship between air pressure distribution and sea level, and it was revealed that the sea level along the Japanese coast is controlled by the air pressure distribution for the space scale over 1000km.

4   Water structure characteristics

Figure 4 shows the time variations of water temperature (a), salinity (b) and DO (c) at the water depth of 10m and 27m at nearly the center of the bay (about 3.5km from the bay mouth) from the 1st of January to the 31th of December; 1997. After air temperature exceeded water temperature, thermocline began to form, and salinity decreased. At the same time DO decreased especially near the bottom.

Figure 5 shows the time variations of daily averaged atmospheric pressure and sea level, wind, daily rainfall amount and insolation time at Ohfunato Bay. At 190day, the atmospheric pressure and sea level begin to separate, and salinity increase (Fig.4). This is because of the change of pressure distribution to the summer type (Hibino and Tsuruya [3]). Figure 6 shows the time variations of daily averaged atmospheric pressure and sea level and daily rainfall amount, salinity (depth of 10m and 27m), DO (27m), water temperature (2,6,12,14,16,18,20,22,,26m) at Ohfunato from the 1st of September to the 9th of November, 1997. In this figure, rainfall amount at Titijima is also described. From the water temperature profile, it is found that the variation of water temperature at the deep layer is different from that of the upper layer. Comparing with salinity and water temperature variation from 10day (9/10) to 32day

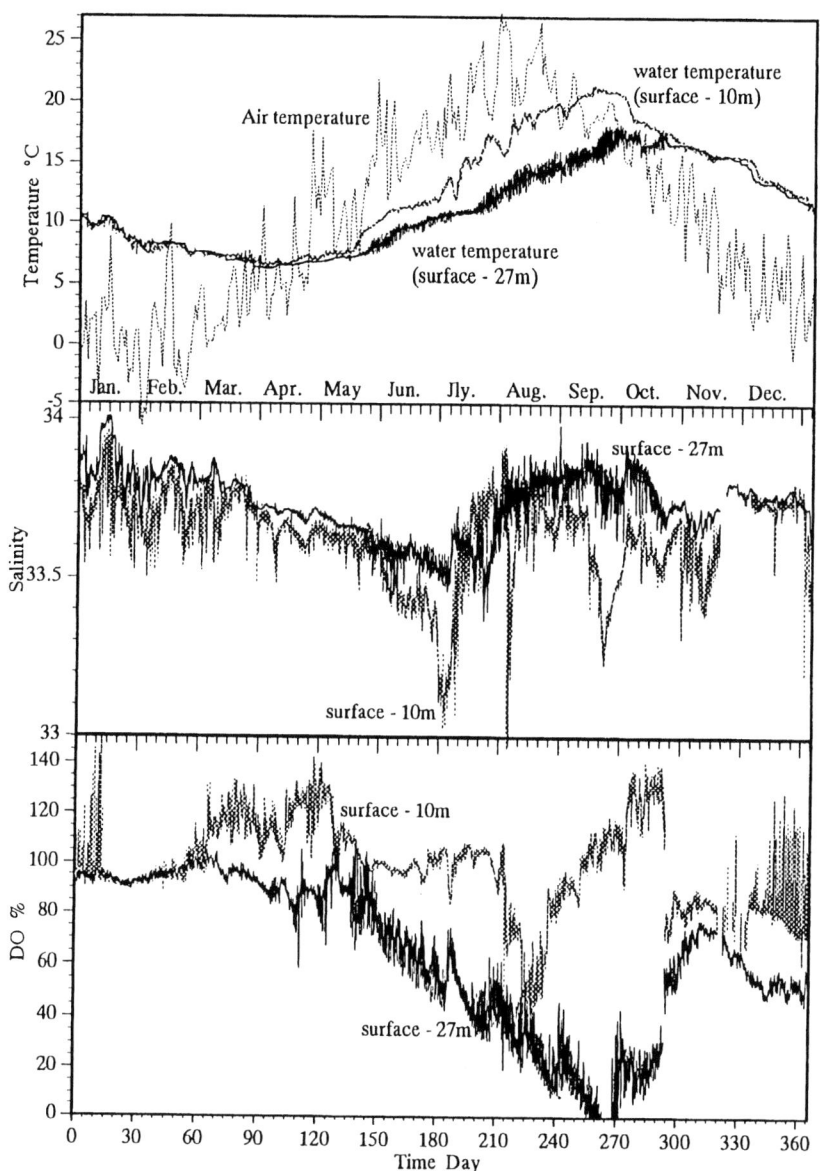

Fig. 4 Water temperature, salinity, DO variation at Ohfunato Bay in 1997

(10/2), it is found that before 20day (9/20) low salinity and warm water intruded and then high salinity and cold water intruded in the bay. From the rainfall at Titijima and Ohfunato, it can be recognized that low salinity and warm water appeared in the bay before 20day (9/20) came from the ocean.

378   Environmental Coastal Regions

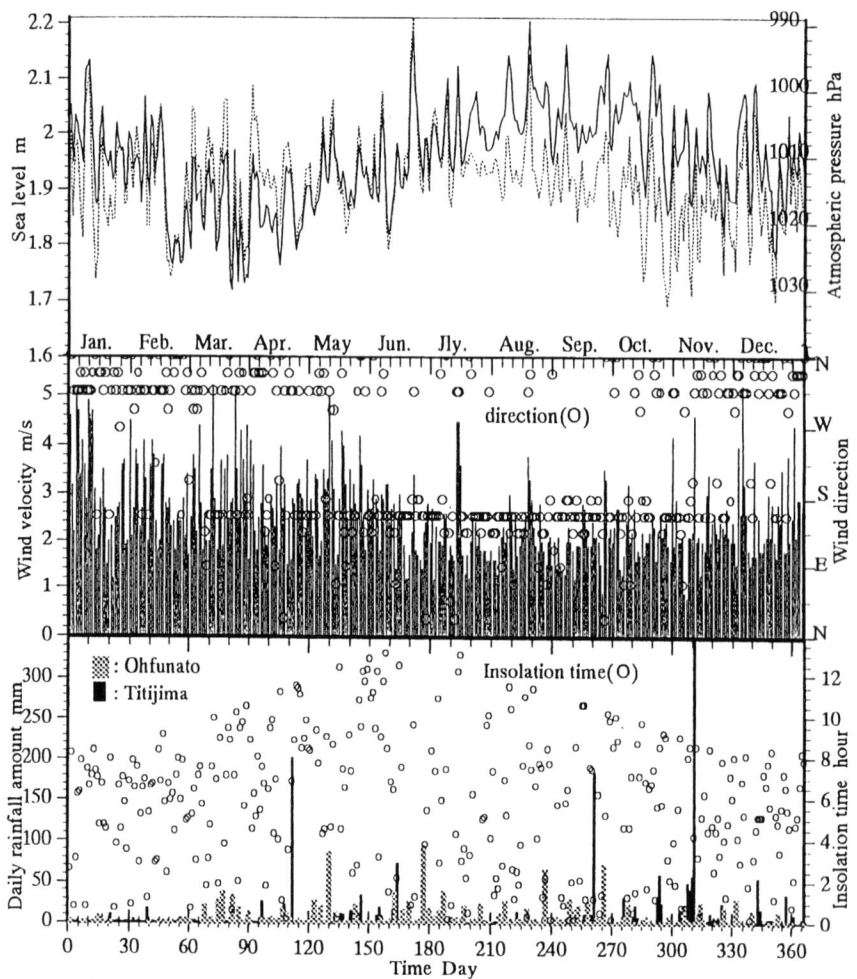

Fig. 5   Sea level, pressure, wind, rainfall amount, insolation time at Ohfunato Bay in 1997

In this period two big typhoon went across Japan. Figure 7 shows the courses of TY 9719 and TY 9720. The values in this figure show Day / Pressure. TY 9719 went ashore Kyushu prefecture, but TY 9720 grazed on the coast area of the Pacific ocean side.

About 0.4 decrease in salinity at September 19 corresponded to the rainfall at Titijima not at Ohfunato. When the center of TY 9720 passed near Ohfunato, the rainfall amount was 182.5mm/day at Titijima (in the Pacific ocean), but no rainfall at Ohfunato at September 19.

In addition, 348mm/day of rainfall at Titijima and decrease in salinity occurred at the same time at November 8.

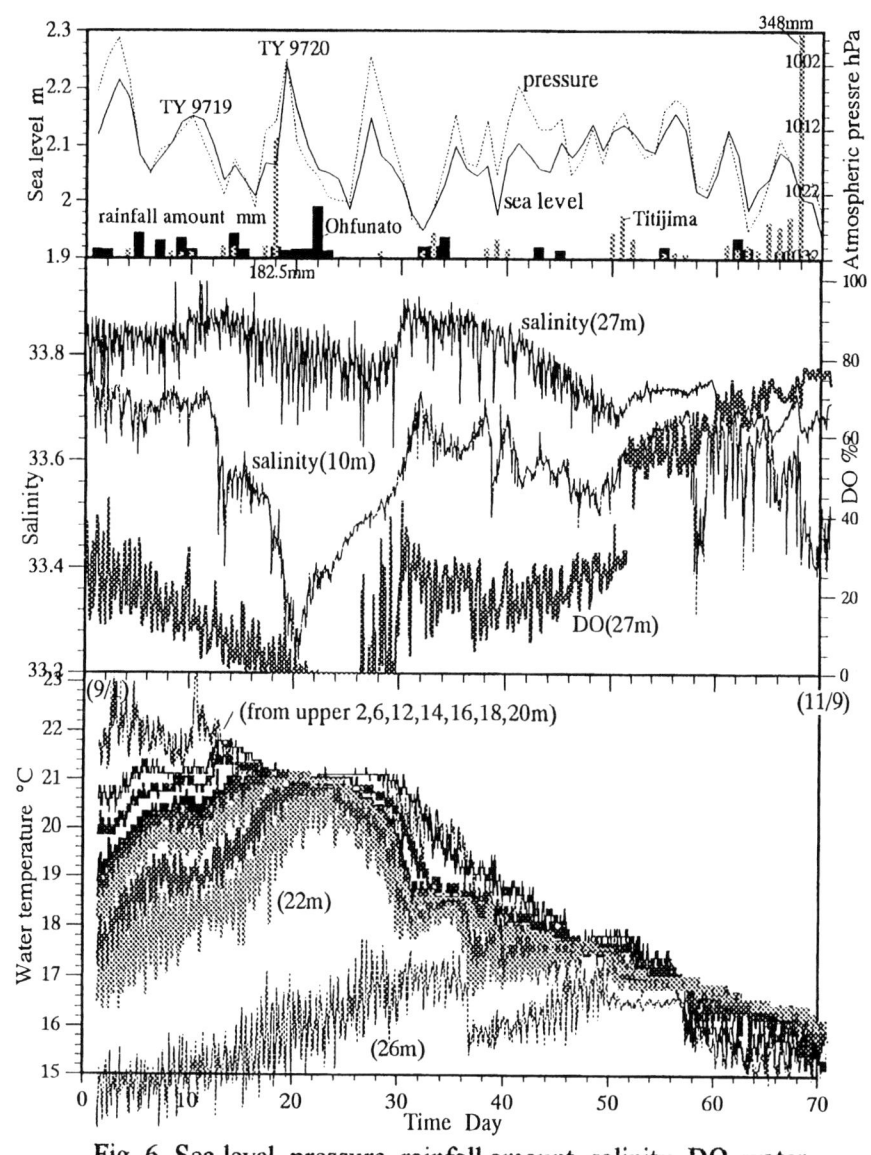

Fig. 6 Sea level, pressure, rainfall amount, salinity, DO, water temperature variation at Ohfunato Bay (1997.9.1 - 11.9)

It can be considered that decrease and increase in salinity inside Ohfunato Bay are not only by inflow of river water but also by intrusions of outer ocean water. These facts suggest that outer ocean currents play an important role in the formation and control of water quality inside a semi-enclosed bay.

380 Environmental Coastal Regions

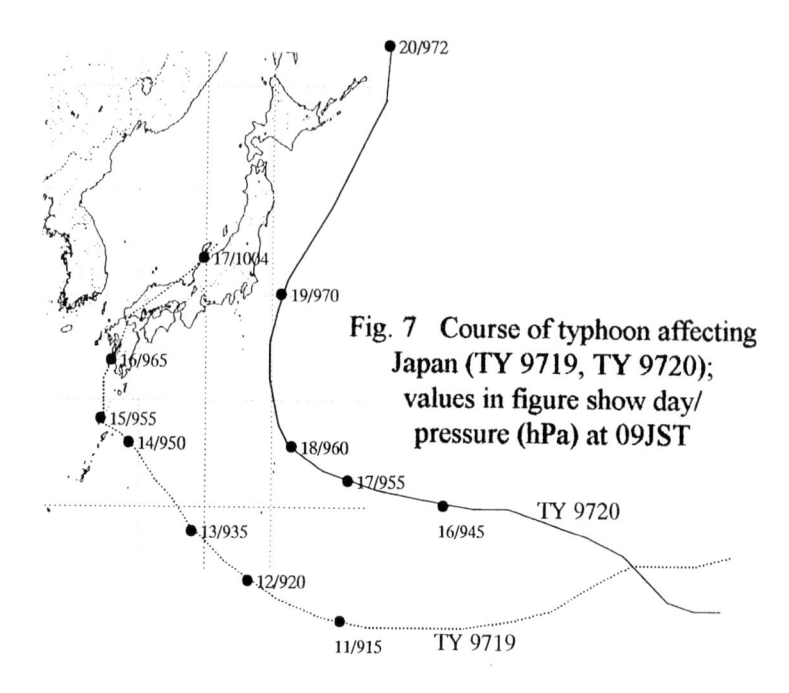

Fig. 7 Course of typhoon affecting Japan (TY 9719, TY 9720); values in figure show day/ pressure (hPa) at 09JST

5 Discharge of anoxic water

Figure 8 shows the temperature distributions around tsunami protection breakwater in Ohfunato Bay (see Fig.1). Measurements in the three figures were made during ebb tide at (a).9/2, (b) 9/30, and (c) 11./17 in 1997. Vertical temperature distributions at the locations of 5.7km and 6.4km from the river mouth are expressed on both sides of each figure. The solid lines correspond to the ebb tide and dotted lines for flood tide. On the left hand side of Fig.8(c), the vertical distributions of water temperatures at the time of three observations for ebb tides are described.

From Fig.8 it can be recognized that the amount of effluent of lower layer water inside the bay (anoxic water) is different for the observation period. At September 2, the lower layer water up to the depth of 20m was carried outside, but at September 30 water to the depth of 35m was effectively carried outside.

Further, it is found that water depths where densities (almost depend on water temperature) inside and outside the bay coincide are 11m at September 2 and 18m at September 30. Therefore, low layer water

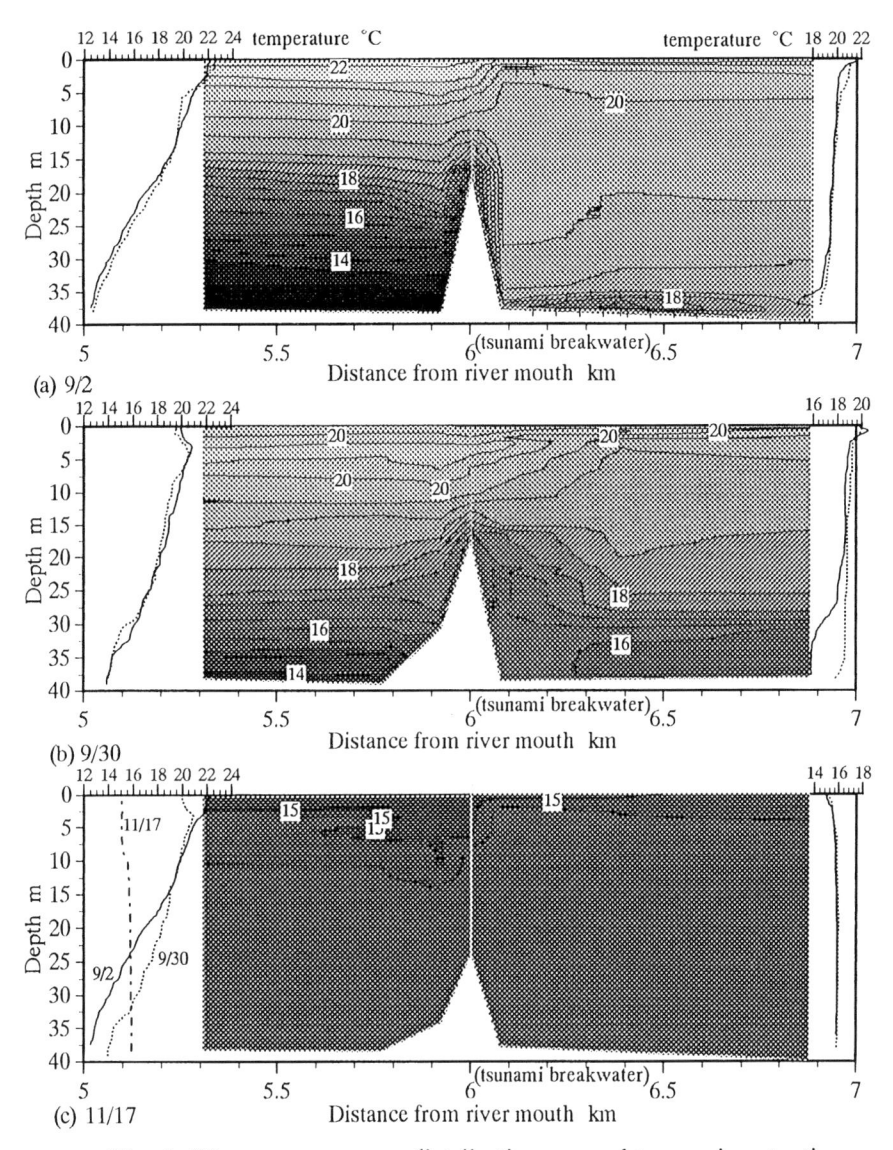

Fig. 8 Water temperature distribution around tsunami protection
breakwater duration of ebb
(Figures on both sides show water temperature profiles at 5.7km and
6.4km; solid line: duration of ebb tide; dotted line: flood tide. the left hand
side lines in (c) show water temperature profiles in three periods.)

inside the bay about 3℃ to 4℃ lower in temperature than ocean water was carried outside over the tsunami protection breakwater.

6 Conclusions

(1) Variation of climate and weather are one of the factors causing ocean water inflow into a bay.
(2) In Ohfunato Bay, it can be considered that intrusion of outer ocean water with low salinity and temperature enhances the stratification inside the bay and that with high salinity assists in breaking the stratification.
(3) It can be considered that decrease and increase in salinity inside the bay are not only by inflow of river water but also by intrusions of ocean water. These facts suggest that outer ocean currents play an important role for the water quality control inside a semi-enclosed bay.
(4) Water exchange type breakwater will be effective to break stratification and enhance the water outflow of low temperature and low oxygen bottom water.

Acknowledgments

The authors are greatly indebted to the Iwate Fisheries Technology Center and the Meteorogical Agency for the offer of the observation data.

References

[1] Kazuo MURAKAMI and Kuniaki YAMADA : Long-term variation of sea level and its factor, Proceedings of coastal engineering, JSCE VOL.39 (2), pp.1026-1030, 1992, (in Japanese).
[2] Hiroichi TSURUYA and Tadashi HIBINO : Water Circulation System in Enclosed Bay and its Utilization for Water Quality Management, The 1st Joint Meeting of the CEST Panel of the UJNR, pp.14.1-20, 1998.
[3] Tadashi HIBINO and Hiroichi TSURUYA : Connection between Atmospheric Pressure Distribution and Sea level, Bulletin on Coastal Oceanography, CORC, The Oceanographical Society of Japan, in press, 1998, (in Japanese).
[4] Yoshihiko SEKINE : Anomalous Southward Intrusion of the Oyashio East of Japan, Long-term variability of pelagic fish populations and their environment, ed. by T. Kawasaki et al., Pergamon Press, pp.61-75, 1989.

Section 9: Groundwater Studies

Preliminary signals of groundwater contamination of stressed coastal aquifers: the case of the Eastern Mediterranean groundwater basins

A. Melloul & M. Collin
Hydrological Service, P.O. Box 6381, IL-91063 Jerusalem, Israel
Email: avimel@hju.uil

Abstract

Coastal areas are in general characterised by high on-going levels of population growth and domestic water demand. Major sources of pollution are urbanisation and overpumpage of aquifers. The result is an increase in salinity and pollution from sea water intrusion and other high-salinity sources, and from anthro-pomorphic percolation from ground surface to the water table. The purpose of this study is to delineate guidelines to identify preliminary signals of groundwater salinisation in areas under high land-use contamination pressure, and under impact of stress-management.

The study is based on analyses of groundwater chlorographs and nitrographs of wells located along the Israel and the Gaza Strip Coastal aquifers, in inland areas with high anthropogenic activities, as well as in areas where groundwater under stress-management is influenced by sea water intrusion. In these areas, Cl^- graphs indicate that after some years of monotonous behaviour corresponding to a steady-state flow regime, a fluctuation rate of 3 to 5 years sets in previous to the onset of final assault of groundwater salinisation. This intermediate stage can be considered an early-warning signal, occurring most clearly when the contaminant source is as massive as sea water. In the other cases, the amplitude of the fluctuation stage also depends upon properties and magnitude of contaminants, the flow regime, and aquifer matrix characteristics. Such observations enable the undertaking of early operational activities needed to rectify adverse groundwater quality trends, as well as planning and choosing adequate aquifer remedial measures. Such graphs can therefore be utilised as a tool for groundwater quality monitoring and control.

1 Introduction

Long term overpumpage of Coastal aquifers leads to a stressed situation which appears as decline in groundwater levels, along with alteration of the natural circulation direction of groundwater flow. This reduces the normal washing of salts into the sea, and increases inland aquifer salt and pollutant concentrations (Fig.1). While contamination effects increase significantly in the aquifer, self-restoration capacity diminishes, e.g.USEPA[12], Travis & Doty[11]. Therefore, the state of stress of groundwater resources should be a concern for persons and organisations tackling water issues around the world with the objective of sustainable aquifer management, e.g. Ableson[1], Davis[2]. For this purpose, the aquifer's eco-hydrogeological characteristics must be understood, including various salt and pollutant trends, in order to formulate a framework within which effective decisions can be taken.

The central objective of this study is therefore to determine guidelines which can isolate preliminary aquifer signals preceding the initial assault of salinisation. Such guidelines are proposed here, involving observation of chloride and nitrate curves of groundwater under stress-management for the coastal phreatic aquifers of Israel and Gaza Strip .

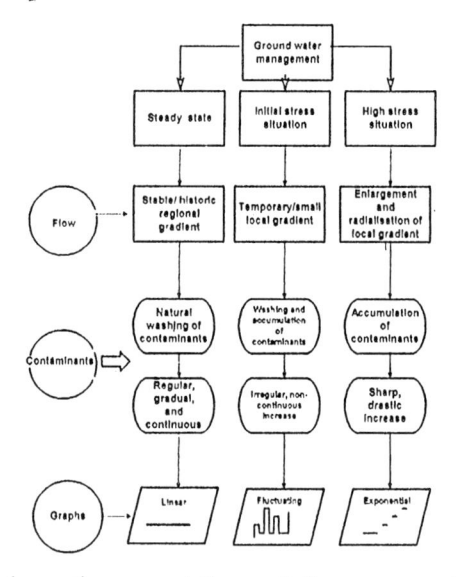

Figure 1. Schematic presentation of the process of groundwater contamination in an aquifer

2 Methodology

This study investigates the pattern of groundwater contaminant changes in coastal aquifer areas under stress-management, involving such concepts as:
1. Choice of parameters which are routinely reported and can represent salinisation and /or pollution processes in groundwater.
2. Utilisation of data which include a period of time preceding the onset of significant adverse effects from anthropogenic activity upon groundwater quality.
3. Sampling from pumping wells located in areas where there is a high probability of sea-water intrusion, as well as inland areas (more than 2000m from the seashore), where anthropogenic activities and groundwater management contribute relative stress.
4. Identification of contamination impact upon groundwater: a.) by use of vertical profiles of groundwater chemical parameters to assess land-usage effects; and b.) by analysing groundwater quality curves of specific wells to detect impact of groundwater stress-management involving such major sources of salinization as sea water and brines upon groundwater aquifer quality.
5. Utilisation of laboratory findings to gain a better understanding of groundwater quality and aquifer change-rates

The parameters used in this study are chlorides and nitrates. Chlorides, as conservative, non-reactive constituents within aquifer matrix material, can be used as good tracers to follow the processes of groundwater salinisation, especially from sea water and brines e.g. White[13], Konikow & Rodriguez[6]. Nitrates, although not good tracer parameters, may be utilised to identify anthropogenic pollution in wells.

3 Hydrogeology of the study area

The study area is located along about 120 km of the Mediterranean coast, and between 5 to 15 km inland (Fig. 2), involving a highly populated area, with intensive levels of industrial, agricultural, and other land-use activities. The Israel and the Gaza Strip Coastal aquifers are involved. These are composed of sand, sandstone, and silt, interbedded with clay lenses, containing several water producing zones e.g.Tolmach[10].

The aquifers are recharged by: rainfall, water percolating from agricultural, industrial, and domestic sources, and along Israel's coast, by artificial recharge. These aquifers have been over-exploited, most specifically the Gaza Strip aquifer. The hydrological situation is characterised there by a declines in groundwater levels of more than 10 centimetres per year. Close to the seashore, this results in an increase in groundwater salinity due to sea water and other salt water intrusion. Inland, this results in a significant increase in groundwater contaminants e.g. Melloul & Gilad[8], Melloul & Collin[7], Hydrological Service Situation Report[5].

These Coastal aquifers are used here to characterise a general pattern to be expected upon onset of groundwater quality deterioration within such phreatic coastal aquifer conditions. All data involved in this work have been collected by the Hydrological Service of Israel (IHS) for periods between 1952 to 1992.

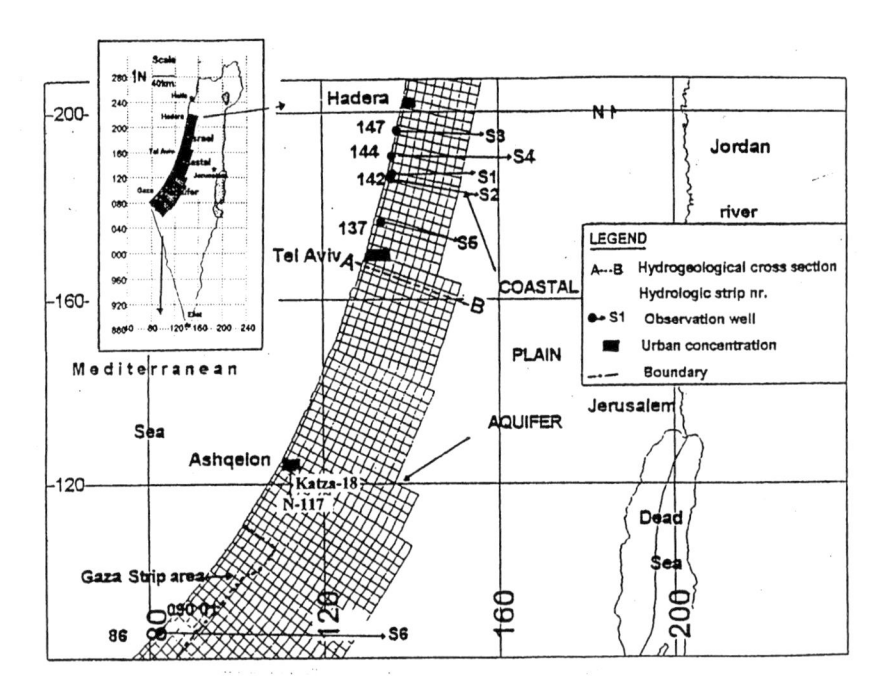

Figure 2. Location map of the Coastal aquifer of Israel and Gaza Strip Authority

4 Observation of chlorographs and nitrographs in groundwater along the eastern Mediterranean coastal ground water basins

4.1 Depth profile in areas under stress from land-use and environment

Figures 3 and 4 respectively represent vertical profiles of groundwater chloride and nitrate levels, based upon averaged values of around 50 wells located in northern and central portions of the Israel Coastal aquifer, at a distance of 2 to 7 km from sea (Fig. 2).

Figures 3 and 4 indicate that between 1967 and 1985, shallow layers, at depths to 40 meters below sea level (b.s.l.) present larger fluctuations, highest levels of groundwater salinity and pollution and rates of salinisation and pollution. Deeper layers (depths greater than 80 meter b.s.l.) yield groundwater quality fluctuations with smoother curves. For all layers, nitrate fluctuation changes with depth are more significant than those for Cl⁻. Thus, for these aquifers, sources of pollution are mainly from the ground surface, where the phreatic aquifer is overlayed by high levels of heavy-contaminant land-use activity. These data also testify to the fact that the closer groundwater is to contaminant sources, the larger the fluctuations.

4.2 Observation of chlorographs and nitrographs in specific wells located near sea shore

Groundwater quality deterioration behaviour is more significantly illustrated by wells located in proximity to the sea shore. Figures 5 and 6 present respectively nitrographs and chlorographs of wells S1, S2, S4, S5, S6, S10, S11, and S12. These wells are located along the Coastal aquifer of Israel and the Gaza Strip(Well S6), up to 1500 m from the sea shore, and in areas where sea-water intrusion can be expected.

Figure 5 represents nitrate data behaviour between 1968 and 1994 along the seashore of Israel's Coastal aquifer. There, nitrates are at low levels and show no significant increase. However, when considering the Gaza Strip's Coastal aquifer, values of NO_3 in some wells exceed 100mg/l e.g. Melloul & Collin[7].

Figure 6 represents chlorographs, mostly evidencing three different portions or stages. Stage a entails initial chloride concentration data characterised by the lowest levels and smallest fluctuations. Stage b is characterised by more significant fluctuations, lasting three to five years. In this stage, Cl^- values are still at low levels, and fluctuate from 50 to 100 % around an approximate mean value. Stage c commences only when Cl^- values begin to increase rapidly at rates higher than 20 mg/l Cl^- per year. The figure indicates that despite varied distances between wells (Fig. 2), chlorographs have similar behaviour. This phenomena also appears in Suleiman[9] as regards groundwater salinity in the Suani wellfield of the Coastal Plain aquifer along the eastern portion of the Mediterranean sea at Tripoli, Libya.

From Figure 6 one notes that for certain wells, the fluctuation stage is not distinct. This is due to significant differences in Cl^- concentrations between the initial and final portion of these graphs (Cl^- values varying from 50 to around 2500 mg/L). Additionally, one notes that in some cases, as in the Gaza Strip well S6, the fluctuation stage of the chlorograph indicates relatively higher Cl^- levels. This can be explained by anthropogenic stress and polluted conditions in this aquifer e.g. Melloul & Collin[7]. Laboratory findings of Goldenberg et al.[3,4] explain that the magnitude of such fluctuations may be related to lithologic matrix changes, aquifer heterogeneity, flow characteristics, and contaminant properties.

5 Discussion

This study shows by means of observation graphs the response of an aquifer to a variety of eco-hydrological stress scenarios. From initial stages of abstraction, when management is under a steady state regime Cl^- and NO3 hydrograph curves respond to contaminant input as continuous and monotonous lines (Fig. 1).

With low levels of contaminants but with increased groundwater management and environment stress, groundwater quality changes are related to the significance of contamination sources and to local groundwater gradients produced (Fig. 1). It does appear that the closer the phreatic aquifer water table is to an intensive land-use activity area, the stronger the fluctuations may be, their values exceeding ambient fresh groundwater background values.

In inland areas, where contamination comes from the ground surface and there is a non-steady-state flow regime, Cl⁻ as well as NO_3 graphs exhibit fluctuations (Fig. 1), with a continuous increasing trend of contamination of around 1 to 10 mg/l per year.

Near the seashore, as illustrated by Fig. 6, the contamination process more clearly involves three stages. Stage a is characterised by a steady-state regime in which groundwater chemical concentration levels and trends are relatively stable and low. In stage b Cl⁻ graph curves show significant fluctuations which can last between 3 to 5 years before the onset of a sharp increase in groundwater salinity. This stage can be augmented because of aquifer heterogeneity, flow domain, contaminant properties, as well as anthropogenic contamination sources, as seen in the case of Gaza well S6. Fluctuation stage characteristics can be a guide to identify the type and order of magnitude of contamination sources influencing groundwater in these areas, before the water arrives at stage c, where salinity concentration increases sharply until groundwater well production is threatened. This last stage appears more significantly in the presence of a major contaminant source, such as sea water intrusion into the aquifer. Stages b and c are, in fact, related to the contamination process of groundwater, stage b being the intermediate stage at which is desireable to remedy the situation as soon as possible before further alteration of groundwater takes place. Utilisation of nitrates has been utilised in this case mostly to demonstrate impact of manure and domestic pollution upon groundwater, as well as to indicate the degree of anthropogenic load, particularly along the seashore.

6 Conclusions

Analysis of chlorographs and nitrographs of groundwater from areas under different stressed eco-hydrological scernarios leads to the conclusion that curve characteristics in the fluctuation portion of a graph can indicate type and magnitude of contaminants affecting groundwater. It is important to focus on the fluctuation stage in order to take appropriate early measures to rectify adverse groundwater quality trends and promote sustainability management of the water resource. Graphic curve observations may be utilised

392 Environmental Coastal Regions

as a tool to detect preliminary signals of groundwater quality degradation, and gain fuller understanding of conceptual models of aquifer contamination.

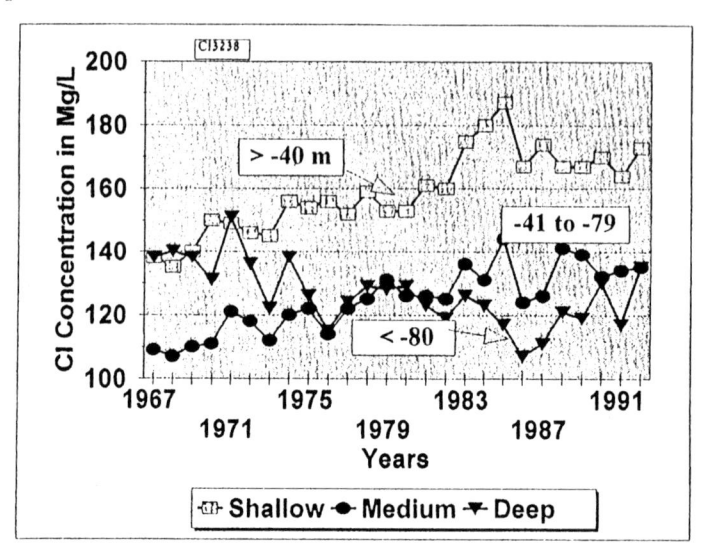

Figure 3. Vertical depth profiles of chlorides in the Israel Coastal aquifer.

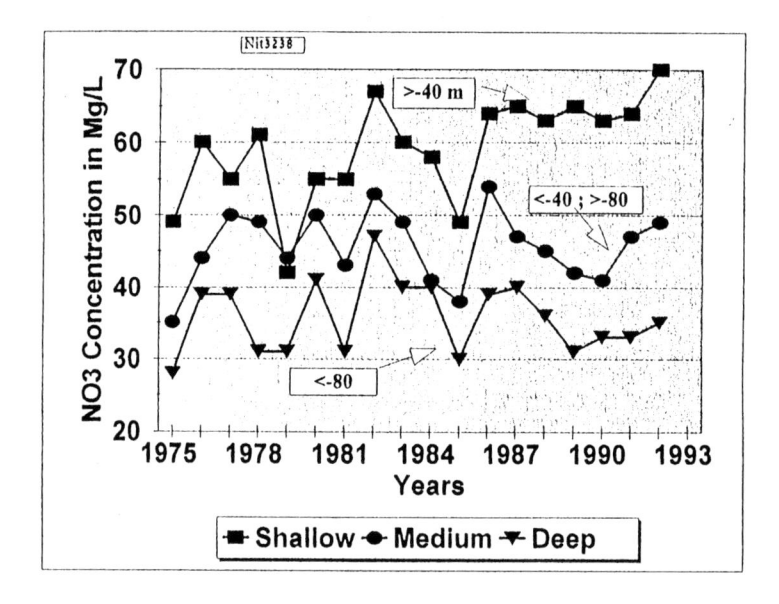

Figure 4. Vertical depth profiles of nitrates in the Israel Coastal aquifer.

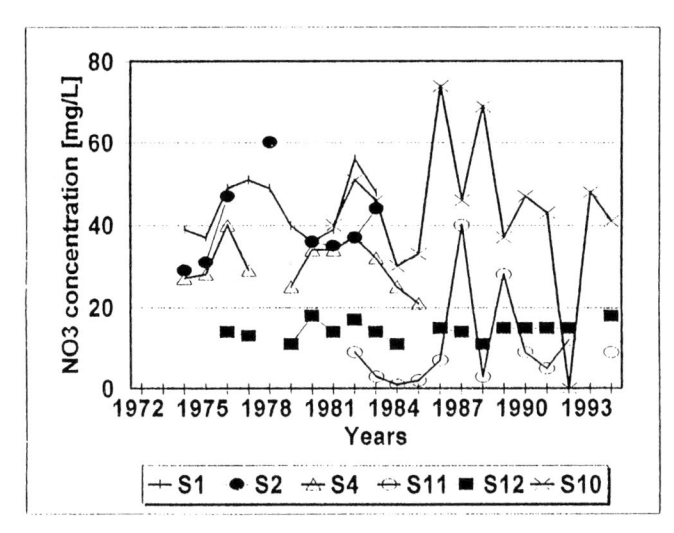

Figure 5. Nitrographs of groundwater wells near the sea shore of Israel's Coastal aquifer.

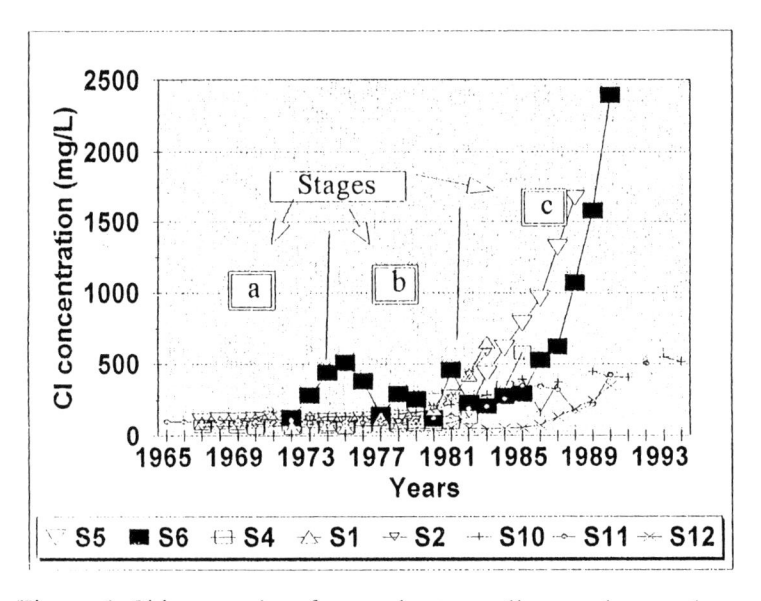

Figure 6. Chlorographs of groundwater wells near the sea shore along the Israel and Gaza Strip Coastal aquifer.

References

[1] Ableson P.H., *Sustainable future for planet earth,* Science 253, 117pp, 1991

[2] Davis C., *The Need for Ecological Cooperation in Europe* Int. Journal of the Unity of the Science, 4(2), pp 201-204, 1991

[3] Goldenberg L.C., Mandel S. and Magaritz M., *Fluctuating non-homogeneous changes of hydraulic conductivity in porous media,* Q J Eng Geology, 19: pp 183-190, 1986

[4] Goldenberg L.C., Hutcheon I., Waldlaw N., and Melloul A., *Rearrangement of fine particles in porous media causing reduction of pemeability and formation of preferred pathways of flow:experimental findings and a conceptual model,* Transport in Por. Med., 13: pp 221-237, 1993

[5] Hydrological Service Situation Report, *Development of groundwater resources in Israel in 1993,* Report 93-1093, p 171, 1994

[6] Konikow L.F. and Rodriguez A.J., *Advection and diffusion in a variable-salinity confining layer,* Wat. Resour. Resear. 29 (8): pp 2747-2761, 1993

[7] Melloul A.J. and Collin M., *The hydrological malaise of the Gaza strip,* Israel Journal of Earth Sciences, 43(2), pp105-116, 1994

[8] Melloul A. and Gilad D., *Movement of sea-water intrusion in the coastal aquifer of Israel over the years 1983-1990,* Wat. and Irrig. Journal 318, pp 46-48, 1993 (in Hebrew)

[9] Suleiman S.E., *Deterioration of quality of groundwater from Suani Wellfield, Tripoli, Libya 1976-93,* Hydrogeol. Journal, 3(2), pp 58-64, 1995

[10] Tolmach Y., *Atlas of Gaza region hydrogeology,* ed. Melloul A., Hydrological Service report, Jerusalem, 35pp, 1991 (in Hebrew)

[11] Travis C.C. and Doty A., *Can contaminated aquifers at Superfund sites be remediated,* Environ. Sci. Technol. Vol. 24:10, pp 1464-1466, 1990 [12] US EPA, *Progress in groundwater protection and restoration,*Washington 440/6-90-001, pp40, 1990

[12] US EPA, *Progress in groundwater protection and restoration,* Washington 440/6-90-001, pp 40, 1990

[13] White K.E., *Tracer methods for the determination of groundwater residence-time distributions,* in Groundwater quality-measurement, prediction and protection, Medmenham, England, Water Research Centre, pp 246-273, 1977

Numerical modeling and management of saltwater seepage from coastal brackish canals in southeast Florida

M. Koch[a], G. Zhang[b]

[a] *Department of Geohydraulics and Engineering Hydrology, University of Kassel, Kurt-Wolters Straße 3, D-34109 Kassel*

[b] *Geophysical Fluid Dynamics Institute, Florida State University, Tallahassee, FL, 33206*

E-mail: kochm@hrz.uni-kassel.de

Abstract

The phenomenon of density-driven vertical salt water intrusion from brackish open sea-canals in southeast Florida has been simulated with a density-coupled groundwater flow and transport model. The underlying conceptual model includes seasonal changes of the groundwater level, tidal variations of the canal stage, rainfall recharge, and a low permeability canal bed. As expected, whether brackish canal water intrudes into the aquifer, depends on the adjacent groundwater table elevation: lowering the latter during a dry season may initiate the seepage process which then becomes essentially irreversible. A significant influence of the short-term tidal fluctuations on the long-term dispersion of the vertical saltwater plume in the aquifer is found. The model is applied to simulate how the large Hollywood well field affects salt water intrusion from the adjacent C-10 tidal canal, and to determine possible mitigating water management strategies to prevent further intrusion. The models show that, for attaining this objective, a minimum threshold water level must be maintained in the well field during the dry seasons. However, raising the water table cannot be achieved by artificial injection of reclaimed wastewater, but may be accomplished by placing a freshwater canal along the brackish C-10 tidal canal.

1 Introduction

Salt water intrusion has been of concern in southeast Florida for some time *(Henry, 1964; Segol and Pinder, 1976; Andersen et al., 1989; Zhang, 1995)*. In addition to classical, horizontal oceanic intrusion which in Florida, as well as in many other coastal aquifers, has become an alarming problem due to increasing groundwater pumping (saltwater upconing), the situation in southeast Florida is exacerbated by the presence of numerous open canals connected to the Atlantic Ocean *(Fig. 1)*. These canals serve as waterways for boat owners and/or flood control drainage. They often carry brackish, saline water from the ocean, especially during high tides. The salinity in the canal increases during the annual dry season when, as a consequence of a lowered groundwater table, the natural groundwater inflow into the canal is reduced and the hydrostatic balance between the saline canal and the fresh ambient groundwater is disturbed. Saltwater intrusion from the canal into the surficial aquifer may occur under such unfavorable conditions *(Chin, 1990)*.

This adverse situation is accentuated for brackish canals close to the wellfields in the region, such as the C-10 canal in the vicinity of the Hollywood wellfield *(Fig. 1)*. Brackish water has been detected in some of the the wells there *(Bearden, 1974)* limiting their use for the groundwater supply. The prediction of the future migration of the saltwater intrusion plume and the development of mitigating management strategies are of utmost importance for securing the groundwater supply in the area. Numerical models are the most efficient tools for this task, allowing the simulation of various scenarios.

In this paper the density-coupled flow and transport model SUTRA *(Voss, 1988)* will be used. In the first part of the paper the general characteristics of the canal intrusion process are investigated by means of a sensitivity analysis. In the second part the model is applied to test several management strategies on how to inhibit or to revert future saltwater intrusion.

The phenomenon of infiltration of saltwater into an aquifer is one representative of the general class of density-driven instability problems in miscible transport, the study of which has been of growing interest in recent years for practical purposes as well as for their peculiar theoretical aspects. Density effects on the migration of contaminant plumes have been observed by *Paschke and Hoopes (1984)* and numerically modeled by *Koch and Zhang (1992)*. Experimental investigations of variable density flow and mixing in porous media have been carried out by *Schincariol and Schwartz (1990)*. A numerical analysis of the physics underlying this so-called Rayleigh-Taylor instability problem was presented by *Koch (1992)* where it was found that the hydrodynamic dispersion was a major controlling factor of the instability.

2 Mathematical and numerical formulation

Flow in a porous media is governed by *mass-conservation (continuity)*

$$\frac{\partial (n\rho)}{\partial t} + \nabla \cdot (n\rho \boldsymbol{v}) = Q, \tag{1}$$

and *Darcy's law* for the flow (seepage)-velocity v [L/T]

$$\mathbf{v} = -\frac{k}{n\mu}(\nabla p + \rho g \mathbf{k}_z) \qquad (2)$$

which results in the *groundwater flow equation* for the pressure p [M/L/T²]

$$S_0\rho\frac{\partial p}{\partial t} + \left(n\frac{\partial \rho}{\partial c}\right)\frac{\partial c}{\partial t} = \nabla\cdot\left(\frac{k\rho}{\mu}(\nabla p + \rho g \mathbf{k}_z)\right) + Q \qquad (3)$$

(cf. Bear, 1979; Voss, 1984). Notations are: n porosity; ρ [M/L³] density; μ [M/L/T] dynamic viscosity; k [L²] permeability; g [L/T²] gravity acceleration; Q [M/L³/T] flow source/sink term; $\mathbf{k}_z$ vertical unit vector; $S = (1-n)\alpha + n\beta$ [LT²/M] specific storativity; with α, β [LT²/M], the compressibility of the aquifer matrix and of water, respectively.

The concentration c is described by the solute transport equation

$$n\rho\frac{\partial c}{\partial t} + n\rho \mathbf{v}\cdot\nabla c = \nabla\cdot(n\rho D \nabla c) + Q_s \qquad (4)$$

where Q_s [M/L^3/T], a solute source/sink term; $\mathbf{D} = D^* + \alpha_T |\mathbf{v}|\delta_{ij} + (\alpha_L - \alpha_T) v_i v_j / |\mathbf{v}|$ [L²/T] hydrodynamic dispersion tensor; D^* [L²/T] molecular diffusion; α_L, α_T [L], longitudinal and transversal dispersivity, respectively.

The flow eq. (3) and the transport eq. (4) are coupled through an equation of state, $\rho = \rho(c)$ which can be linearized as $\rho = \rho_r + \partial\rho/\partial c^*(c-c_r)$, with ρ_r the reference density for pure water ($c_r = 0$), and $\partial\rho/\partial c$ a constant.

The coupled flow and transport eqs. (3) and (4) are solved by the well-known SUTRA (Saturated-Unsaturated-TRAnsport) *(Voss, 1984)* 2D finite element model. SUTRA can either be used in the horizontal x-y domain or in a vertical x-z cross-section of a saturated-unsaturated zone. The latter set-up is employed in the present application. The discretization of the x-z domain is done by means of 4-node quadrilateral elements. A bilinear approximation for p and c is used in each element and the classical Galerkin formulation of eqs. (3) and (4) is established. For each timestep the resulting linear matrix systems for the nodal unknowns p or c are solved consecutively by an efficient iterative conjugate gradient technique, after the corresponding updates for the density $\rho(c)$ and the seepage velocity $\mathbf{v}$ (eq. 2) have been made. The integration in time is performed by means of a fully implicit backward Euler method which is unconditionally stable. Nevertheless, because of the strong coupling of ρ, p, and c, the timesteps Δt have to be taken sufficiently small to ensure convergence and stability, especially in those models where the semidiurnal tidal variations of the canal gage are considered.

3 Sensitivity study of the canal intrusion process

3.1 Objectives

To understand the physical dynamics of the canal saltwater intrusion and its sensitivity to various geohydraulic model parameters, numerous simulations were run, to answer the following questions: 1) for a given ambient surficial groundwater head h_a above the canal stage what is the critical canal concentration c_{cr} that starts intrusion? 2) for a given canal concentration c_o, what is the minimal ambient groundwater head h_{acr}, that prevents intrusion? 3) what are the separate and combined effects of seasonal and tidal variations?

3.2 Conceptual model; boundary and initial conditions; input parameters

The basis cross-sectional conceptual model consists of an aquifer section 2500ft wide and extending 220ft deep from the surface to the base of the Biscayne aquifer and which is incised in the middle by a tidal canal of depth H_w=10 ft and width of 80 ft. This model area is discretized by an 81x21 finite element grid which is horizontally refined underneath the canal (*Fig. 2*).

Boundary conditions are as follows: *bottom boundary:* no flow; *left and right boundaries:* periodically varying pressure that simulates seasonal changes of the ambient groundwater table (period T_s =12 months; i.e. $p_a = p_o * \sin(2\pi t / T_s)$); *top boundary:* specified only at the position of the canal as periodically varying pressure due to the semidiurnal tidal gage height variations. In the brackish canal a Dirichlet b.c., $c = c_0$ for the chloride concentration is also prescribed when solving the transport eq. (4). An average annual rainfall recharge of 12 inches for the region is included by assigning the upper boundary appropriate source nodal values. As for the initial condition, the concentration is set to $c_i = 0$ throughout the aquifer, simulating the intrusion of brackish water into a fresh groundwater aquifer.

Further aquifer parameters are set as follows (Zhang, 1995): *Porosity* n =0,25; *aquifer compressibility* α= 10^{-9} ms²/kg; *horizontal permeability* k_x = 1,9 x 10^{-11} m²; *vertical permeability* k_z = 1,9 x 10^{-11} m² ; *longitudinal dispersivity* α_L= 1m; *transverse dispersivity* α_T= 0,1 m.

3.3 Simulation results

Fig. 3 illustrates results of the simulation family P2 answering question 1) above. The ambient head h_a is fixed to 1m above canal stage (see *Fig. 2*) and is time-invariant, i.e. neither seasonal nor tidal variations are included here. Saltwater fronts are shown after 15 years for two canal chloride concentrations of c_0=13000 and c_0=15000 ppm, respectively. Whereas for the former case, no intrusion is observed, for c_0= 15000 ppm the plume has not only sunk to the bottom of the aquifer, but has also widely spread horizontally. Thus, for a head h_a=1m the critical concentration is c_{cr}~14000 ppm. Of course, increasing h_a will result in yet higher values of c_{cr} for intrusion to occur.

By reverting the above reasoning, question 2) will be answered in the simulation family P4. Here c_0 is fixed at 13000 ppm and, in order to mimic seasonal changes, the ambient groundwater table h_a is varied periodically with an amplitude of 0.5m around an average head h_{ao} which is altered within this simulation set. Note that tidal effects are still discarded here. *Fig. 4* shows that

Environmental Coastal Regions   399

whereas for h_{ao} = 1.1ft the plume has fully penetrated into the aquifer after 15 years, canal intrusion is inhibited by merely increasing h_{ao} to h_{ao} = 1.2 ft. The latter thus represents the critical average groundwater table height h_{a0cr} that must be maintained permanently to prevent intrusion. It should be noted that for the same model without seasonal changes, h_{a0cr} is only about 0.8ft (*Zhang, 1995*). The higher value found for the average head h_{a0cr} in the former simulation set P4 is required to counterbalance the accelerated intrusion during the dry-season interval with lower-than-average water table heights.

The additional effects of semidiurnal tidal variations of the canal gage are illustrated in the model set P10. Because of the very small timesteps Δt <3h required for the resolution of the tides, the integration was only carried out up to 10 years. Following observed tides in the New River (*Fig. 1*), a tidal amplitude of 1.5ft is used for the canal b.c., on top of the seasonal variations of model set P4. *Fig. 5* depicts two cases with different average groundwater heights h_{ao}. Whereas no canal intrusion is obtained after 10 years for h_{ao} = 1.8ft, the saline plume has fully seeped into the aquifer for h_{ao} = 1.2ft, so that the former value is the critical height h_{a0cr} required to permanently prevent canal intrusion.. Note that this value of h_{a0cr} is higher than the h_{a0cr} =1.2ft obtained previously for model set P4 which included only seasonal groundwater changes. The short-periodic canal tides have therefore a non-negligible effect on the long-term canal intrusion. Visual differences are also observed in the plume contours of the two simulation sets P10 (*Fig. 4*) and P4 (*Fig.5*) which are smoother in the former (with tides) than in the latter case (no tides), providing evidence for a larger effective hydrodynamic dispersion generated by the high-frequency tidal velocity fluctuations.

4 Application to the C-10/Hollywood field area

4.1 Effects of the Hollywood groundwater pumping

The foregoing results demonstrate that canal intrusion will occur only if the groundwater level falls temporarily, e.g. during a dry season, below a critical threshold. Clearly, heavy groundwater pumping close to a canal will magnify such a situation. This appears to be the case for the C-10 canal/ Hollywood wellfield (HW) area *(Fig. 1)*, where Chloride (Cl⁻) concentrations of ca 10000 ppm in the canal and ca 500 ppm in the aquifer have been detected (*Bearden, 1974*). Geophysical resistivity surveys have delineated a foot-shaped saline plume between the C-10 and the HW (see *Zhang, 1995*, for details).

To investigate the effects of the HW on the future migration of this intrusion, several models similar to the ones before were run, but which mimic an 8400 ft wide aquifer cross-section starting west of the HW and extending east of the C-10 (see *Fig. 1*) and that include pumping of the HW. Because its wells extend essentially in the NS- direction, parallel to the C-10, the HW can be modelled as a line source, and the 2D model approach is applicable.

Fig. 6 shows the intrusion fronts after 20 years for a model with the present-day pumping rate Q_p of the HW (27.6mgd) and another one with twice that value (Q_f =2*Q_p), representing anticipated future groundwater demands. Both seasonal and tidal variations are included and a canal Cl⁻-concentration of 6000ppm is assumed. The model is calibrated for the aquifer permeability and the dispersivity, using head- and Cl⁻-measurements in the two monitor wells G-1240 and G-1548 located between the HW and the C-10 canal.

The notable feature of *Fig. 6* is the insensitivity of the $0.1c/c_0$ isoline (≥600ppm which are above the 250ppm Florida Drinking Water Standard) to the pumping rates Q_p and Q_f. Contrary to expectation, the toe of that isoline for the double Q_f pumping has not intruded further than that for the present-day Q_p. The time histories of the intrusion fronts show (*Zhang, 1995*) that for both models these have reached approximately steady-state after 20 years.

4.2 Management of canal intrusion using hydraulic barriers

Although the previous results indicate that increasing the pumping rate in the HW should not trigger a further advance of the present-day C-10 brackish intrusion front, the question arises if and what kind of management strategies can be taken to stop or even revert intrusion. The models discussed so far suggest that to attain this objective, a sufficiently high groundwater table has to be created, especially during dry seasons. Two variants of such a "hydraulic barrier", whose effectiveness and practicality will be tested in this section are: (1) aquifer injection of reclaimed wastewater, proposed by Floridan agencies, which would create a permanent groundwater mound; (2) construction of a freshwater canal whose gage height h_c would be maintained at an all-season sufficiently high level, with the effect of also raising the groundwater table.

For the numerical test of the injection barrier model (1), an injection well with a recharge rate equal to the today pumping of the HW (27.6mgd) was placed as a source node 120m deep in the aquifer, halfway between the HW and the C-10 canal. Since all other model parameters are identical to those used in the previous section, the effects of such an injection well on the intrusion fronts can be directly seen by comparing *Fig. 7, Top* with *Fig. 6, Top*. Obviously, even after 20 years of simulation the 0.1 c/c_0-isoline for the injection model in *Fig. 7, Top* has barely retreated from its former position in *Fig. 6, Top*. Other models with higher injection rates were run, but could not improve this negative outcome much neither. Therefore, one can conclude that this injection variant of the hydraulic saltwater barrier is not a viable option, given also the huge quantities of reclaimed wastewater that would be needed.

The freshwater canal option (2) is modeled by imposing a Dirichlet bc. $h = h_c$, for a top boundary node at the hypothetical canal position, midway between the HW and the C-10 (width of 100 m). The average canal stage height h_c was chosen as $h_c = 1.6$ ft which is sandwiched between the seasonal changes of the groundwater table, ensuring the natural canal recharge from the aquifer during the wet season, but raising the groundwater table during the critical dry season. The effectiveness of such a 'hypothetical' freshwater canal can be clearly seen from *Fig. 7, Bottom*. Compared with the reference model of *Fig. 6, Top*, after 20 years, the saltwater intrusion plume has been pushed back towards the brackish canal and has maximum concentrations $c/c_0<0.1$. A budget analysis shows that the amount of freshwater infiltrating from the hypothetical canal into the aquifer during the dry season is about 9 times the today HW pumping rate. Although a portion of this fresh canal water is naturally flowing back out of the aquifer during the wet seasons, additional surficial water will be needed to make this kind of hydraulic barrier feasible.

References

Andersen, P.F., J.W. Mercer and H.O. White, Numerical modeling of saltwater intrusion in Hallandale, Florida, *Groundwater*, 26, 203-210, 1989.
Bear, J., *Hydraulics of groundwater*, Mc Graw Hill, New York, 1979.
Bearden, H.W., Groundwater resources of the Hollywood area, Florida, *USGS Water Resources Investigations Report*, 74-015, 1974.
Chin, D.A., A method to estimate canal leakage to the Biscayne aquifer, Dade county, Florida, *USGS Water Resour. Invest. Rep.*, 90-4135, 1990.
Henry, H.R., Effects of dispersion on salt encroachment in coastal aquifers, *US. Geol. Survey Water Supply Paper*, 1613-C, 1964.
Koch, M., Numerical simulation of finger instabilities in density and viscosity dependent miscible solute transport, In: *Proc. IX Int. Conf. Comput. Meth. Water Resour.*, Russel, T.F. et al. (ed.), pp. 155--162, Computational Mechanics Publications, Southampton, UK, 1992.
Koch, M., and G. Zhang, Numerical simulation of the effects of variable density in a contaminant plume, *Groundwater*, 5, 731-742, 1992.
Paschke, N.W., and J. Hoopes, Buoyant contaminant plumes in groundwater, *Water Resour. Res.*, 20, 1183--1192, 1984.
Schincariol, R.A., and F.W. Schwartz, An experimental investigation of variable density flow and mixing in homogeneous and heterogeneous media, *Water Resour. Res.*, 26, 2317--2329, 1990.
Segol, G. and G.F. Pinder, Transient simulation of saltwater intrusion in southeast Florida, *Water Resour. Res.*, 12, 65--70, 1976.
Voss, C.I., SUTRA, A finite element simulation model for saturated unsaturated fluid-density-dependent ground energy transport or chemically reactive single species solute transport, *U.S. Geol. Survey Water Resour. Investig. Report* 84--4369, 1984.
Zhang, G., Numerical models of saltwater intrusion from brackish canals in southeast Florida, *Ph.D. Thesis*, FSU, Tallahassee, Florida, 1995.

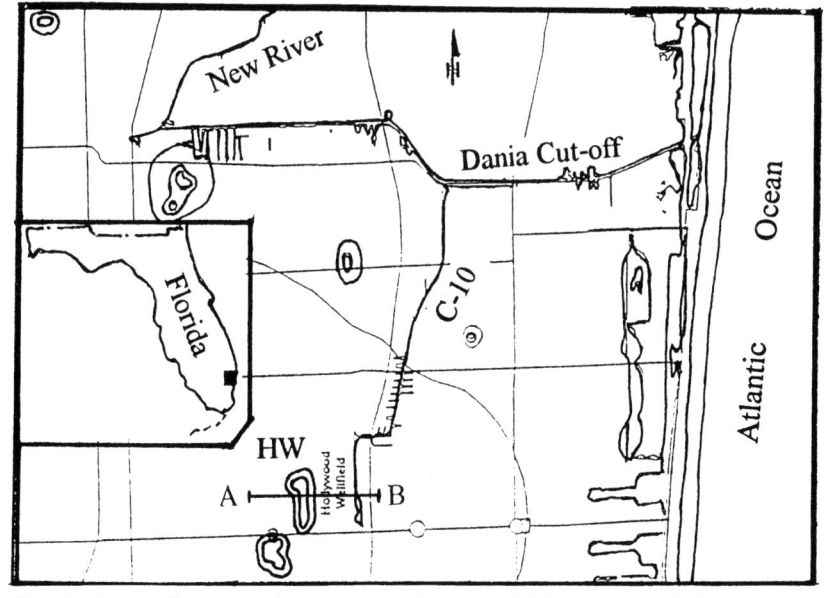

Fig. 1 Map of Broward county, with main tidal canals and the HW. The line AB marks the aquifer cross-section modelled in Section 4.

402   Environmental Coastal Regions

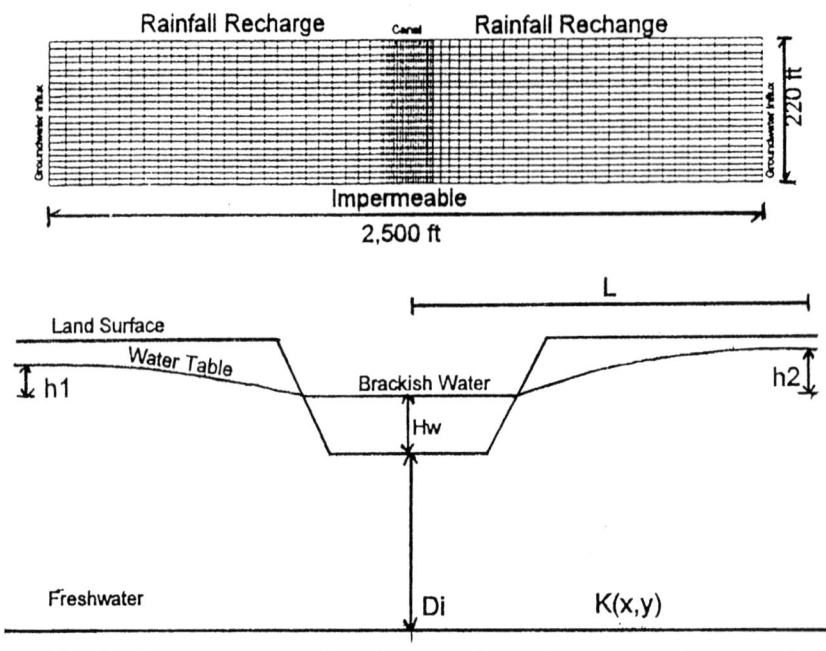

Fig. 2 Conceptual model of the canal intrusion process (top panel) and the finite element grid used (bottom panel). (Not to scale.)

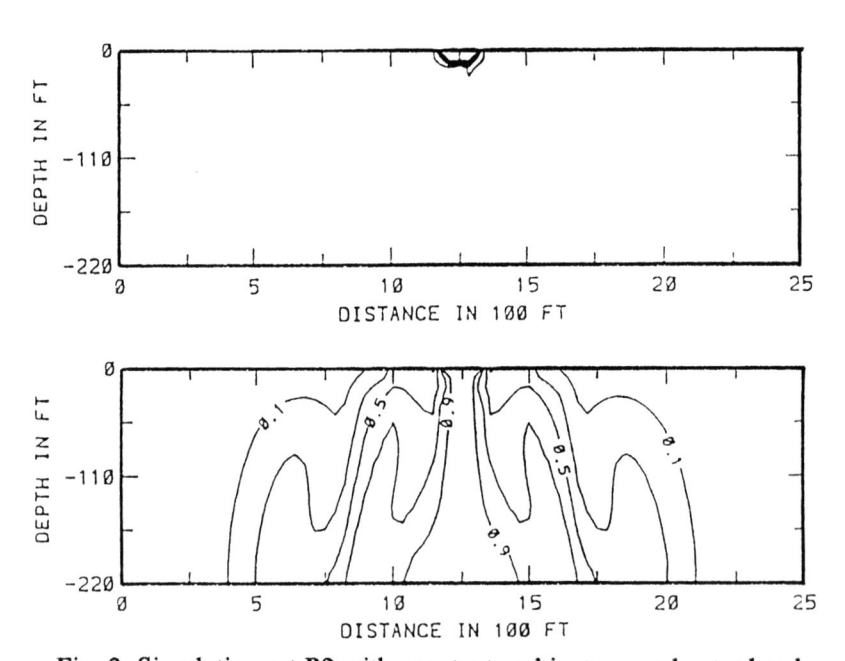

Fig. 3 Simulation set P2 with constant ambient groundwater head $h_a=1$m. Normalized (c/c_0) saltwater intrusion isolines after 15 years for Cl$^-$- concentration of $c_0=13000$ (top) and $c_0=15000$ ppm (bottom).

Environmental Coastal Regions 403

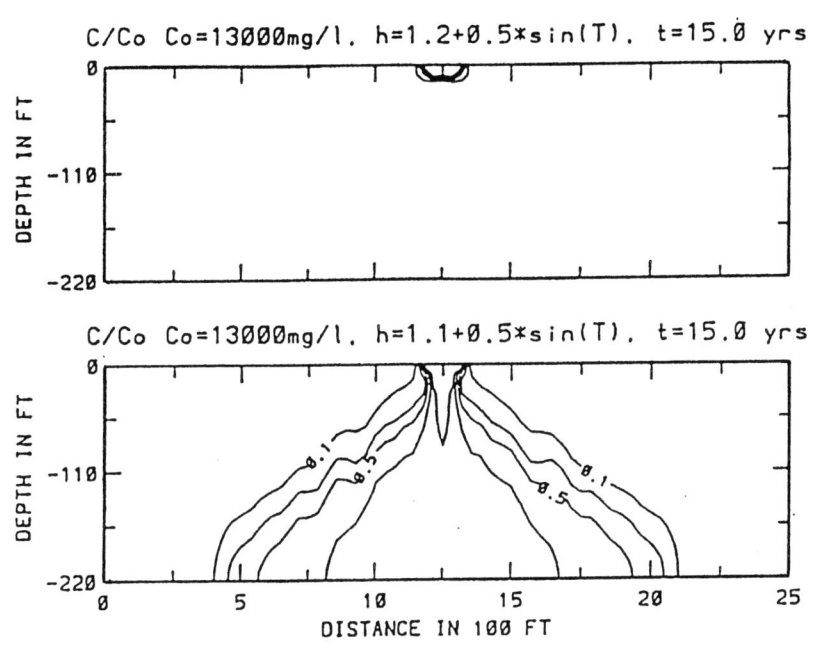

Fig. 4 Simulation set P4 including seasonal variations. Saltwater intrusion fronts after 15 years for two models with ambient groundwater head h_{a0} of 1.2m (top) and 1.1m (bottom) respectively.

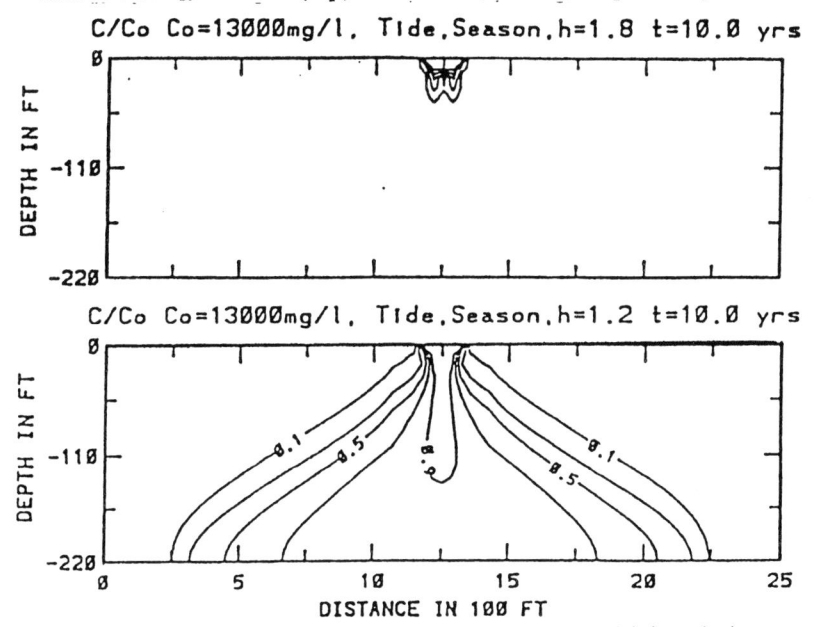

Fig. 5 Simulation set P10 including seasonal and tidal variations. Intrusion fronts after 10 years for two models with ambient groundwater heads ha0 of 1.8m (top) and 1.2m (bottom) respectively.

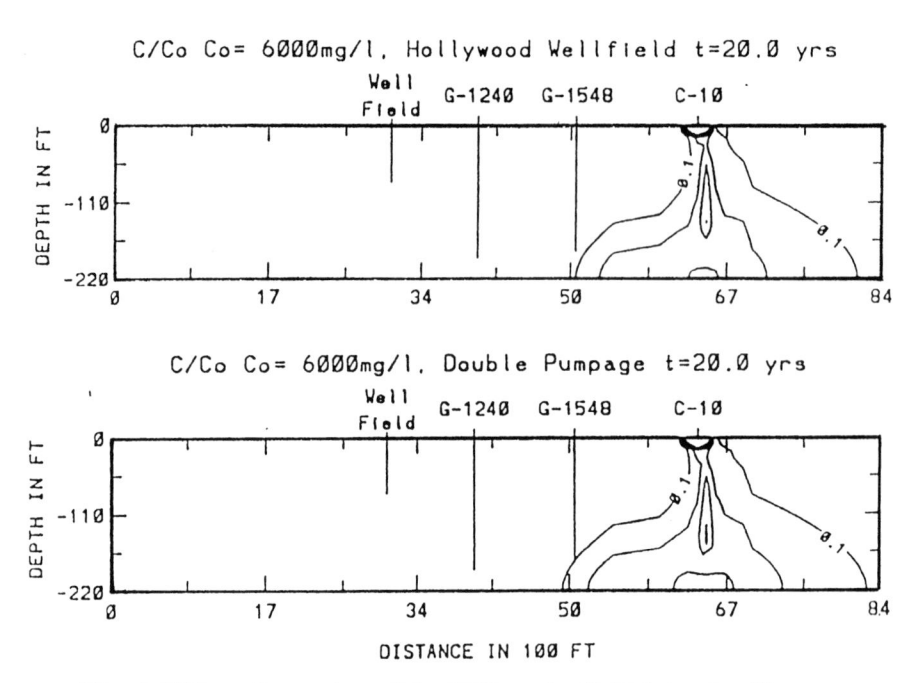

Fig. 6 Effect of pumping of the HW on the C-10 intrusion. Top: present-day pumping rate; bottom: twice the present-day pumping rate.

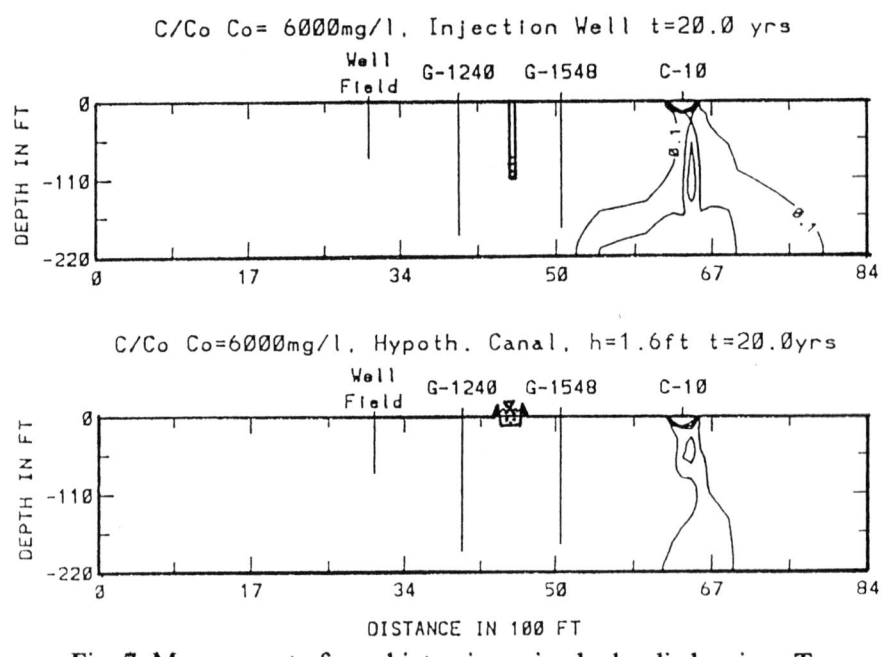

Fig. 7 Management of canal intrusion using hydraulic barriers. Top: Effect of an injection well. Bottom: Effect of a freshwater canal.

Cyclic movement of water particles in a coastal aquifer

H. Sun[a], M. Koch[b]

[a]*Department of Geological Sciences, Rider University, Lawrenceville, NJ 08648, USA*

[b]*Department of Geohydraulics and Engineering Hydrology, University of Kassel, Kurt-Wolters Straße 3, 34109 Kassel, Germany*

Abstract

The movement of water particles in a coastal aquifer is controlled not only by the regional flow, but also by the rising and falling of the tidal levels in the coastal water. This combined influence results in periodic variations of the hydraulic gradient with ensuing cyclic movements of water particles in a coastal aquifer. Based on a two-dimensional analytical series solution for the piezometric head in response to tidal loading, these cyclic movements are theoretically investigated. The approach is then applied to the New Jersey coastal aquifer near Atlantic City. Amplitudes and phases of 25 tidal constituents are calculated using standard harmonic analysis. After appropriate filtering of the observed piezometric heads, a calibration for the storativity/transmissivity (S/T) ratios is carried out. The latter are found to decrease with increasing hydraulic pressure which appears to be consistent with the idea of a reduced aquifer compressibility with higher pressure. Eventually, water particle trajectories in the aquifer are calculated using Darcy's law. The trajectories of a ground water particle demonstrate a cyclic flow path caused by the tidal wave, creating a unique 'flushing' effect in the coastal aquifer. This 'flushing' phenomenon should be taken into consideration when the transport or removal of a solute or contaminant within a coastal aquifer is modeled.

1 Introduction

Piezometric heads in a coastal aquifer rise during high tide and fall during low tide, causing periodic variations of the hydraulic gradient, with a subsequent change also of the direction and magnitude of the flow. On top of these tidal influences the movement of water particles is driven by the regional flow, with the result that zigzag flow paths arise in a coastal aquifer. This cyclic movement of water particles in a coastal aquifer is similar to the progressive movement of water particles in open water except that it is confined in the pore space of rocks, and therefore its magnitude limited by the hydraulic properties of the aquifer, e.g. transmissivity and storativity.

While there have been a few studies of one-dimensional (1D) piezometric head's response to tidal loading (e.g. *Greg, 1966; Carr and Van Der Kamp, 1969; Hsieh et al.,1989; Nielsen, 1990; Serfes, 1992; Yim and Mohsen, 1992*), no comprehensive investigations of 2D water particle movements have been carried out to-date. This study is an extension of earlier work of *Sun (1997)* on the 2D analytical solution for the groundwater response to tidal loading, but with emphasis on the calibration of the transmissivity/storativity (S/T)-ratio and on the analysis of water particle trajectories in coastal aquifers. The results of this investigation are important for further understanding coastal hydrology and for modeling contaminant transport in a coastal aquifer.

2 Mathematical theory

2.1 Analytical solution for piezometric head's response to tidal loading

The hydraulic head h [L] in a horizontal (x-y domain) confined, nonleaky, homogeneous aquifer is governed by the 2D *groundwater flow equation*

$$\frac{\partial^2 h}{\partial x^2} + \frac{\partial^2 h}{\partial y^2} = \frac{S}{T} \frac{\partial h}{\partial t} \qquad (1)$$

(cf. *Bear, 1972*), where t [T] is the time; S, storativity; T [L^2/T] = kB, transmissivity; k, hydraulic conductivity [L/T]; B [L] aquifer thickness. Eq. (1) is solved in a domain that extends in x-direction from the coast x=0, inland to x =∞, and whose y-axis is aligned along the coast or an estuary.

The initial condition is $h(x,y,t=0)=0$ and the two boundary conditions are (1) the coastal tidal bc.

$$h_{tide}(x,y,t) = H_0 + \sum_{k=1}^{n} A_k e^{-p_k x - m_k y} \cos(\omega_k t - b_k y - q_k x + \varphi_k) \qquad (2)$$

and (2) the far-inland bc. $h(\infty,y,t)=0$. Eq (2) describes a damped tidal wave consisting of a superposition of n tidal constituents traveling along-shore in y-direction into an estuary *(Dean and Dalrymple, 1984)*. Notations are: A_k, amplitude; m_k, damping factor; $\omega_k = 2\pi/T_k$, angular tidal velocity; T_k, tidal period; b_k, wavenumber; and φ_k, phase of the kth tidal constituent. ω_k and b_k are related through $b_k \approx \omega_k / c_0$, where c_0 is the shallow-water phase speed.

With the coastal bc. (2), the tidal solution of the groundwater flow eq. (1) can be written as a linear superposition of all tidal constituents

$$h_{tide}(x,y,t) = H_0 + \sum_{k=1}^{n} A_k e^{-p_k x - m_k y} \cos(\omega_k t - b_k y - q_k x + \varphi_k) \quad (3)$$

Insertion of this expression into eq. (2) results in terms for *(Sun, 1997)*

$$p_k = \left[\sqrt{(b_k^2 - m_k^2)^2 + (\omega_k S/T - 2b_k m_k)^2} + b_k^2 - m_k^2\right]^{1/2} / \sqrt{2} \quad (4)$$

and

$$q_k = (\omega_k S/T - 2b_k m_k) / 2p_k . \quad (5)$$

With regional flow included, using the superposition principle *(Bear, 1972)*, the total 2-D potentiometric head in a coastal aquifer is then written as

$$h(x,y,t) = h_{reg}(x,y,t) + h_{tide}(x,y,t) \quad (6)$$

2.1 Calculation of water particle trajectories

Assuming a constant regional hydraulic gradient $\nabla h_{reg} = (f_x, f_y)^T$, the application of Darcy's law $v = -k \nabla h$ to eq. (6), together with eq. (3), results in the following expression for the x-component $u(t)$ of the flow velocity v

$$u = -kf_x + k\sum_{k=1}^{n} A_k p_k e^{-p_k x - m_k y} \cos(\omega_k t - b_k y - q_k x + \varphi_k)$$
$$- k\sum_{k=1}^{n} A_k q_k e^{-p_k x - m_k y} \sin(\omega_k t - b_k y - q_k x + \varphi_k) \quad (7)$$

and a corresponding one for the y-component $v(t)$. Knowing these local velocities u and v at a particular location (x_0, y_0) and time t_o, the x_1-coordinate of the new position (x_1, y_1) of the particle at time t_1 can be computed through

$$x_1 = x_0 + \int_{t_o}^{t_1} u(t) dt$$

$$= x_o - kf_x t \Big|_{t_o}^{t_1} + k\sum_{k=1}^n \left(\frac{A_k p_k}{\omega_k} e^{-p_k x - m_k y} \sin(\omega_k t - b_k y - q_k x + \varphi_k) \Big|_{t_o}^{t_1} \right.$$
$$\left. + k\sum_{k=1}^n \frac{A_k q_k}{\omega_k} e^{-p_k x - m_k y} \cos(\omega_k t - b_k y - q_k x + \varphi_k) \Big|_{t_o}^{t_1} \right) \quad (8)$$

and a corresponding one for the the y_1-coordinate. Repeated application of eq. (8) between times t_n and t_{n+1} ($n=1,...,nmax$) provides the loci of the trajectory of a water particle in the aquifer between x_1 and x_{nmax}.

3 Application to the New Jersey coastal aquifer

The above theory is applied to a section of the New Jersey coastal aquifer near Atlantic City, NJ. Hourly piezometric head data from the USGS monitor wells *Galen*, 590m inland from the coastline, and *Firehouse*, 350m inland (*Fig. 1*)---that both tap the confined 800-foot sand of the Kirkwood Formation--- were obtained for July 1994 (*USGS, 1994*), together with tidal levels (NOAA). The analysis of this data in accordance with the theory of Section 2 is carried out in three steps.

3.1 Harmonic tidal analysis

Tidal amplitudes and phases of up to 25 total tidal constituents for the use of eq. (2) are calculated from the NOAA tidal data series by means of a standard harmonic analysis (*Boon and Kiley, 1978*) Results for the six major components are shown in *Table 1*. The wavenumber b_k is calculated from $b_k = \omega_k/c_0$, where c_0, the phase speed is given by $c_0 = (gd)^{1/2}$, with d the average water depth (=2m for the NJ coastal section in the study area), and g, the gravity acceleration.

Table 1: Results for the first six components of the tidal series expansion

No	Tide	Velocity ω_k [degree/h]	Amplitude A_k [cm]	Phase φ_k [degree]
1	M2	28.98	57.82	148.83
2	S2	30.00	10.80	148.83
3	N2	28.44	12.91	148.83
4	K1	15.04	12.22	74.42
5	O1	13.94	7.78	74.42
6	M4	57.97	1.10	297.67

3.2 Calibration for the storativity/transmissivity ratio

Knowing the tidal parameters as in Table 1, the series expansion (3) for the

q_k's are calculated from eq. (4) and (5), respectively, assuming an initial S/T ratio which is then refined during the calibration procedure. Moreover, since tidal friction in the open water along the shore can be neglected, the damping coefficient m_k in eq. (3) has been set to zero.

During the calibration of the S/T-ratio, the predicted heads are compared with the measured ones. However, before doing so, the raw data has to be filtered to remove all non-tidal fluctuations caused e.g. by precipitation and storm surges. This "detrending" is performed by means of standard 24- 25-hour moving- average filters *(Fig. 2)* *(Godin, 1972; Hsieh et al.,1987)*.

Results of a few calibration runs are shown in *Fig.3*. For the *Galen well* one notices that the *S/T*-ratio has a stronger effect on the phase than on the amplitude (compare *Figs.3a* and *3b*). In accordance with eq. (3) the amplitude of the predicted head is more controlled by the inland-distance x from the coastline than by the *S/T*-ratio. Moreover, it can be seen that the match between the predicted and the measured head data is not equally good over the one-month observation period. For example, with a preselected storativity value of $S=0.002$, the values for the transmissivity in the two *S/T*-calibrations for the *Galen well* are $T=93$ m²/day in *Fig. 3a* and $T=98$ m²/day in *Fig. 3b*, respectively. With the former value, the predicted head matches the measured data very well in late July *(Fig.3a)*, but is approximately one and a half hour out of phase in early July. With the latter value, on the other hand, the situation is just the opposite, with a good match now in early July *(Fig. 3b)*. Re-examining the original head data *(Fig. 2a)*, it can be seen that in the beginning of July 1994 the piezometric head is almost 2m higher than in late July. Since for a confined aquifer the head h is directly correlated with the hydraulic pressure p through $p = \rho g h$ [M/LT²] (ρ = density of water), one may conjecture that the *S/T*-ratio itself is not constant, but varies with the pressure. Theoretically, the storativity S is more likely to be sensitive to pressure than the transmissivity T, because S for a confined aquifer is calculated through the well-known formula $S = \rho g B(\alpha + n\beta)$, where α [LT²/M] and β [LT²/M] are the compressibilities of the aquifer matrix and the water, respectively, and n is the porosity *(Bear, 1972)*. Since it is conceivable that α decreases with higher pressure, a reduced value for S is consistent with the lower *S/T-ratio* needed to get the good match to the high measured hydraulic heads in early July.

Whereas the *Galen well*, because of its location, is affected by the open Atlantic Ocean tide, the water levels in the *Firehouse well*---which is located close to the inland estuary *(Fig. 1)*---react more to the tide there. An approximate six-hour phase difference exists between the tides at these two locations. Note the very good fit of both the amplitude and the phase of the predicted data with the filtered measured data for the *Firehouse well* *(Fig.4c)* which might be a consequence of the stronger observed signals at this well, but shows clearly the superiority of the present 2D-calibrations over the 1D approach of earlier studies mentioned in the introduction.

3.3 Calculation of water particle trajectories

Once the S/T-ratio has been calibrated, Darcy-flow velocities of the water particles are calculated using eq. (7). It should be noted that large depression cones exist near the study site, as a consequence of heavy pumping that has occurred recently in the wake of construction of several new casinos in Atlantic City *(Smith and Sun, 1996)*. The groundwater flow along the coast is, therefore, slightly directed toward Atlantic City. *Fig. 4* shows stick plots of the predicted Darcy-flow velocities over a period of 3 days. The tidal effects can be clearly seen from the varying magnitudes and the changing directions of the velocity vectors. However, because of the superposition of regional- and tidally induced flow (eq. 6), the average total Darcy-flow velocity does not change in a coastal aquifer, despite a longer path traveled by water particles near the shore.

Particle trajectories calculated with eq. (8) are shown in *Fig. 5* for the *Galen well* calibration and in *Fig. 6* for the *Firehouse well* calibration. Water particles are moving back and forth as the hydraulic gradient changes every 12 and 24 hours. This zigzag movement of the groundwater front is stronger at a location closer to the coast (*Figs. 5a* and *6a*) than farther inland (*Figs. 5b,c* and *6b*) where the damping effect comes into play. This is also the reason for the overall stronger undulations of the water particles for the *Firehouse well* (*Fig. 6*) which is closer to the shore than the *Galen well* (*Fig. 1*). In any case, a unique 'flushing' effect is created in the coastal aquifer that would carry away and disperse dissolved solute (contaminant) particles in the groundwater more effectively than regional flow alone. This 'flushing' effect should be taken into consideration when the hydraulic removal of a contaminant along the coastal aquifer is modeled.

References

Bear, J., *Dynamics of fluids in porous media,* Elsevier, New York, NY, 1972.
Boon, J. D. and K.P. Kiley, *Harmonic analysis and tidal prediction by the method of least squares,* Special Report No.186, Virginia Institute of Marine Science, Glocester Point, Virginia, 1978.
Carr, P. A. and G.S. Van Der Kamp, Determining aquifer characteristics by the tidal method, *Water Resour. Res.,* 5, 1023-1031, 1969.
Dean, R.G. and R.A. Dalrymple, *Water wave mechanics for engineers and scientists,* Prentice Hall, Inc., Englewood Cliffs, NJ, 1984.

Godin, G., *The analysis of tides.* University of Toronto Press, Toronto, 1972.
Gregg, D. O., An analysis of ground water fluctuation caused by ocean tides in Glynn County, Georgia. *Ground Water*, .4, 211-232, 1966.
Hsieh, P.A., J.D. Bredehoeft and J.M. Farr, Determination of aquifer transmissivity from earth tide analysis, *Water Resour. Res.*,.23, 1824-1832, 1987.
Nielsen, P., Tidal dynamics of the water table in Beaches, *Water Resour. Res.*, 26,. 2127-2134, 1990.
Parker, B. B., *Tidal Hydrodynamics*, John Wiley & Sons, New York, NY, 1991.
Serfes, M. E., Determining the mean hydraulic gradient of ground water affected by tidal fluctuations, *Ground Water*, 29, 549-555, 1992.
Smith, J. and H. Sun, Potential problems of groundwater resources depletion and water quality deterioration in the state of New Jersey in 1997. *American Geophysical Union, Transactions (EOS)*, 77, 17, 1996.
Sun, H., A two-dimensional analytical solution of groundwater response to tidal loading in an estuary, *Water Resour. Res.*, 32, 1997.
USGS, Water Resources Data New Jersey, Water Year 1994, U.S. *Geological Survey Water Report NJ-94-2*, 1994.
Yim, C.S. and M.F..N. Mohsen, Simulation of tidal effects on contaminant transport in porous media, *Ground Water*, 30, 78-86, 1992.

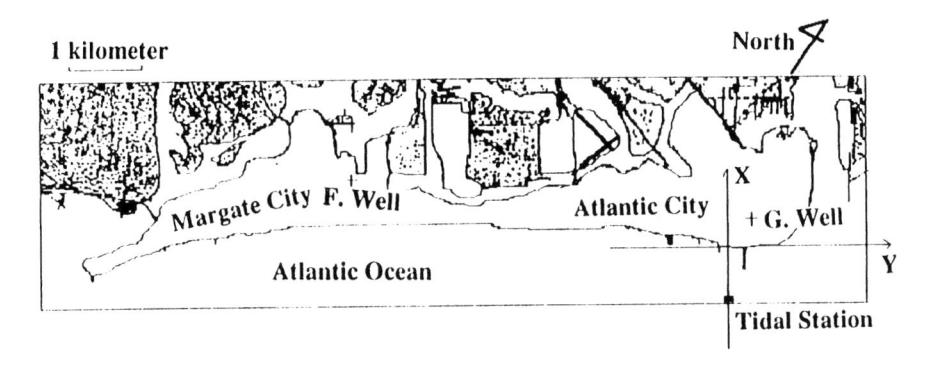

Fig. 1 Map of Atlantic City area with locations of the wells and the tidal station. G.Well = Galen well, F. Well = Firehouse well.

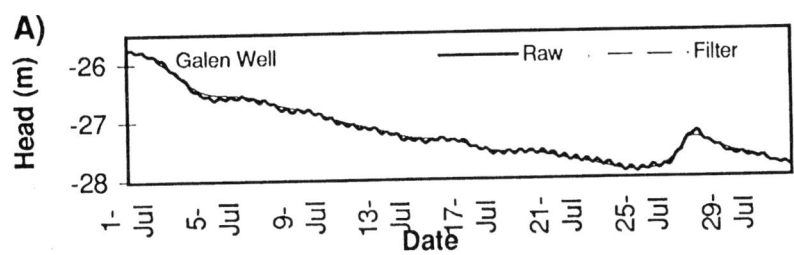

Fig. 2

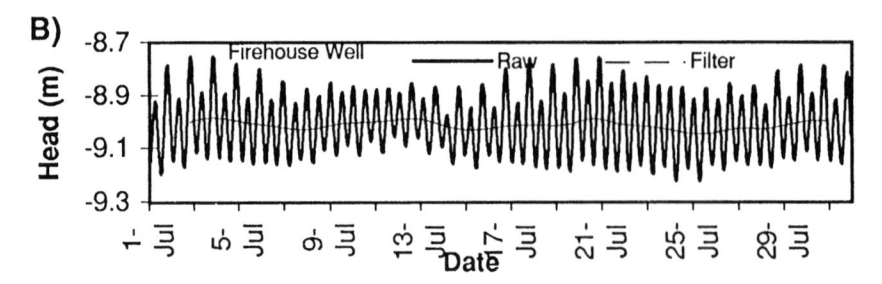

Fig. 2 Diagrams of the raw piezometric data (thick line) and of the filtered data (thin line). A) Galen well; B) Firehouse well.

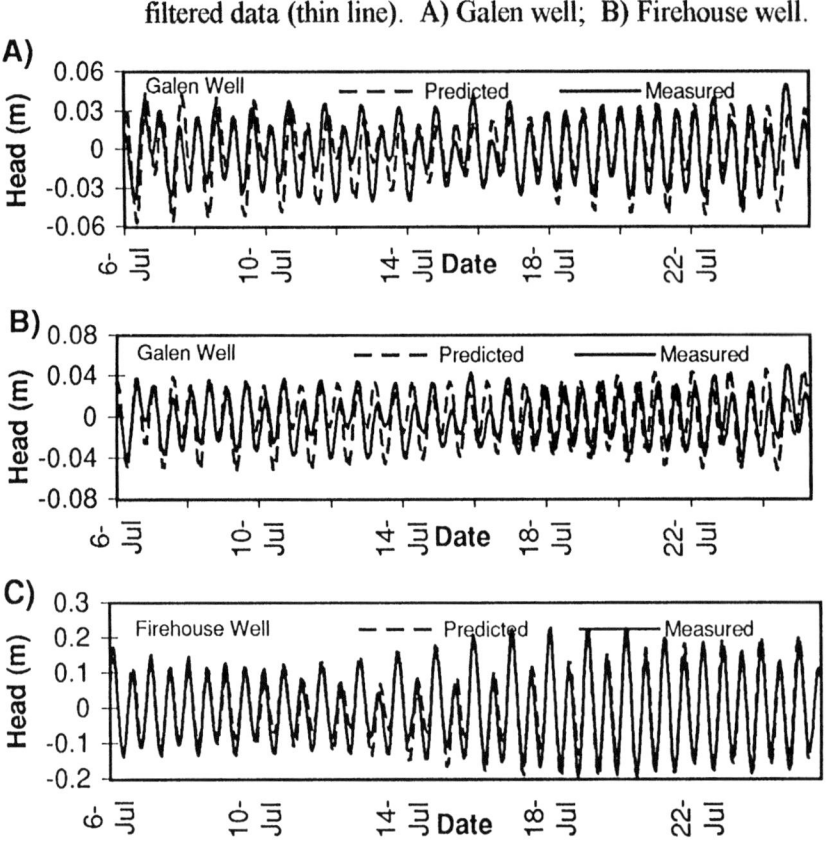

Fig. 3 Results of S/T-ratio calibration. With a constant S=0.002, T is varied. A) Galen well, S/T=2.15*10-5 day/m^2 (T=93m2/day); B) Galen well, S/T=2.04*10-5 day/m^2 (T=98m^2/day); C) Firehouse well, S/T=2.14*10-5 day/m^2 (T=93.5m^2/day).

Environmental Coastal Regions   413

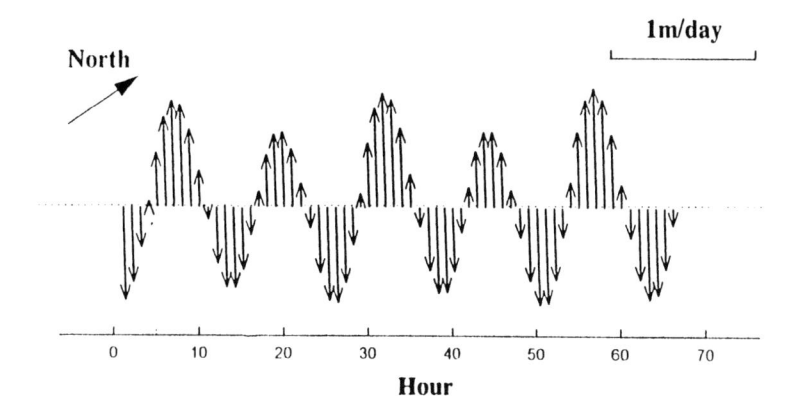

Fig. 4   Stick plots of the calculated hourly flow velocity from July 6 to July 8, 1994 with T=93 m^2/day at location x=5, y=5 meters near Firehouse well.

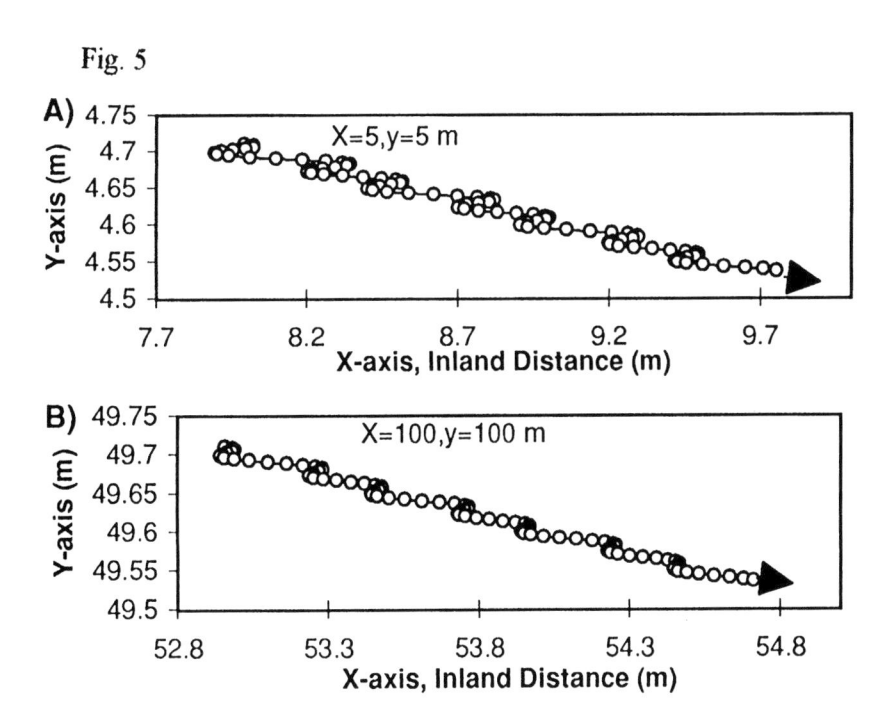

Fig. 5

A)

B)

414 Environmental Coastal Regions

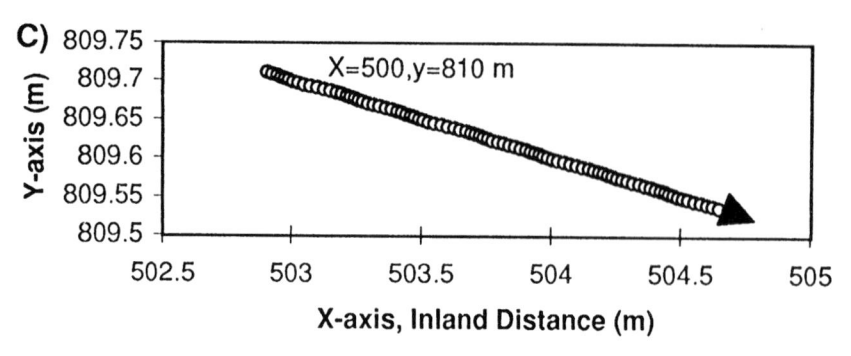

Fig. 5 Trajectory of water particle for the flow at Galen Well from July 6 to July 8, 1994. A) at location x=5, y=5 meters; B) at location x=100, y=100 meters; C) at location x=500, y=810 meters.

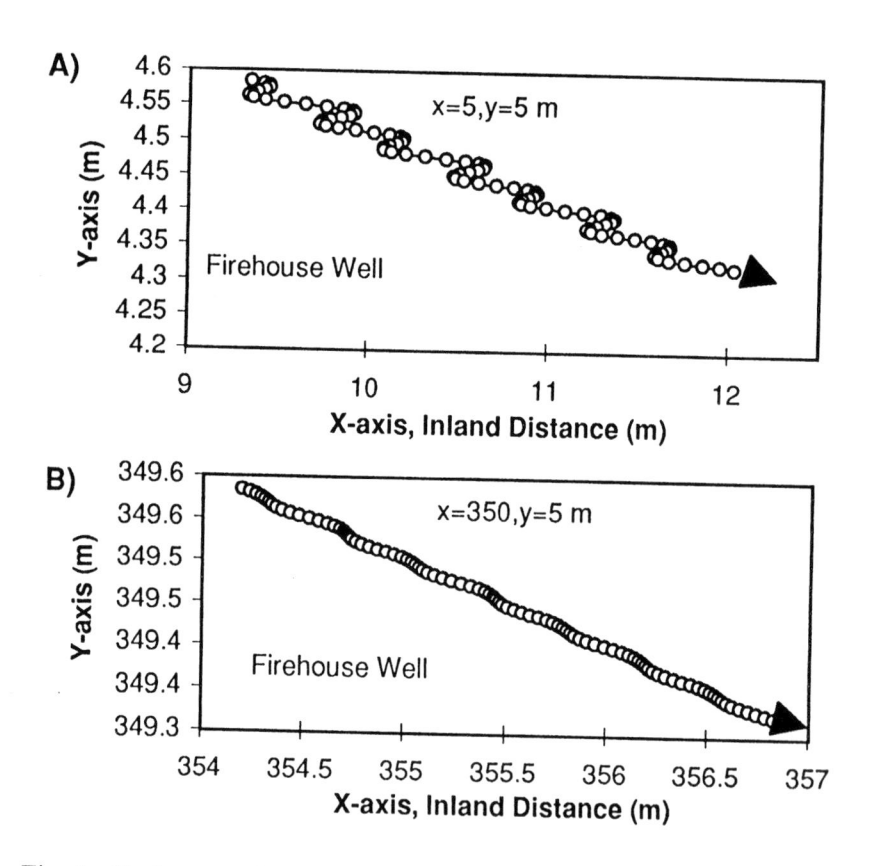

Fig. 6 Trajectory of water particle for the flow at Firehouse well from July 6 to July 8, 1994. A) at location x=5, y=5 meters; B) at location x=350, y=5 meters.

Section 10:
Siltation and Dredging

Sediment resuspension in hurricane conditions

Katherine A. Larm and Billy L. Edge
Ocean Engineering Program, Department of Civil Engineering, Texas A&M University, College Station, TX 77843-3136, USA; EMail: b-edge@tamu.edu
Norman Scheffner
U. S. Army Corps of Engineers, Waterways Experiment Station, Vicksburg, Mississippi, USA

Abstract

Sediments are suspended and mobilized by wave-induced fluid motion as well as tidal and riverine currents in the coastal zone. For a site in Lavaca Bay, Texas, the wind and hydrodynamic climates are such that sediments are deposited due to every day conditions. The depositional environment has created a lens of clean sediments that overlies a layer of contaminated sediments. The Texas Gulf coast is also prone to hurricanes, which may supply sufficient energy to the system to resuspend sediments. A numerical study modeling historical hurricanes has been performed, using ADCIRC to model the hydrodynamics, SWAN to model the wave environment, and the sediment relationships of Krone to estimate the resulting sediment erosion. Eleven hurricanes are considered, and the erosion estimates are provided.

1 Introduction

Lavaca Bay, located off Matagorda Bay on the Texas coast, is a low energy environment (Fig. 1). Of particular interest is the area of Cox Bay, which is typically a depositional environment. With its location adjacent to the Gulf of Mexico, infrequent hurricanes result in high energy events with storm surges and increased wave heights. These events contribute to the resuspension of bottom sediments and the erosion of banks. For the purposes of this paper, the resuspension of bottom sediments due to hurricane conditions will be examined.

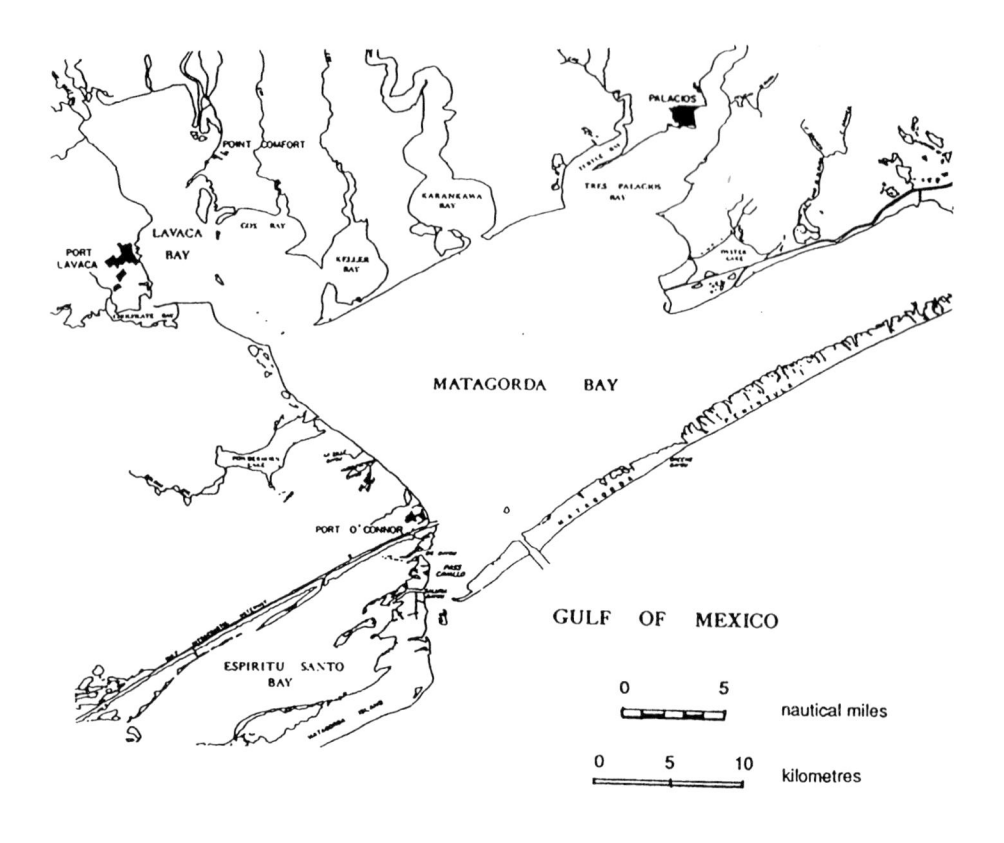

Fig. 1. Location of Lavaca Bay, Texas

Historical activities in Lavaca Bay resulted in the localized presence of contaminated sediments. Since their initial deposit, the contaminated sediments are mostly covered with a layer of clean sediments. Since the source of contamination has been contained, the covered contaminated sediments may present a concern if they are resuspended and distributed throughout Lavaca Bay.

Hurricanes are events with ample energy to erode the clean layer of sediments and perhaps expose the contaminated sediments. Using the parameters of previous storms that have traversed the area, both storm surge and wind wave models are used to simulate the Bay climate and estimate the probable sediment resuspension. Erosion results for a fixed point in Cox Bay will be presented for the storm set considered.

2 Historical Hurricane Set

A set of eleven storms that affected the Lavaca Bay area is listed in Table 1. Parameters to define each storm, such as the location of the center and the pressure of the eye, are given in the National Hurricane Center Hurricane Database (HURDAT) for the North Atlantic basin. The category of the storm is relative to the Safir-Simpson scale at the point when it neared the Matagorda Bay area. This information was obtained from the Purdue Weather Processor website (http://wxp.atms.purdue.edu).

Table 1. Lavaca Bay Area Storms

Storm #	Year	Category
5	1886	2
218	1916	3
295	1929	1
324	1933	1
405	1941	1
445	1945	4
602	1961	5
690	1970	3
703	1971	1
704	1971	TS
783	1980	4

The Unified Planetary Boundary Layer (PBL) model of Cardone et al. (1992) is used to simulate the cyclonic wind fields that would result under hurricane conditions. The PBL model solves the horizontal equations of motion averaged over the boundary layer.

3 Hydrodynamic Model

The hydrodynamics of the Bay were estimated by the numerical model ADCIRC (ADvanced CIRCulation) (Luettich et al. 1994). ADCIRC is a two-dimensional, depth-integrated model that is forced by the winds from the PBL model, the pressure fields, runoff, and tidal conditions. ADCIRC solves the generalized wave continuity equation, yielding the free surface displacement (including storm surge) and the depth-averaged velocities (Luettich et al. 1994).

The ADCIRC model uses an irregular, finite element grid. For this study, a larger domain which includes the North Atlantic Ocean, the Caribbean Sea, and the Gulf of Mexico was used (Fig. 2). The water surface elevations from this large domain were used to force the model using a smaller domain consisting of Matagorda and Lavaca Bays (Fig. 3). The larger Eastcoast grid is comprised of approximately 31,400 nodes and 58,400 elements. The smaller Matagorda grid contains approximately 4,200 nodes and 7,700 elements.

4 Wave Model

The wave model SWAN (Simulating WAves in the Nearshore) was selected to estimate the wave climate due to each storm. SWAN is a finite difference spectral model that solves the action balance equation (Ris et al. 1996). The finite difference grid used in SWAN simulations is shown in Fig. 4. Model runs were conducted in 2nd generation mode; this mode allows for the dissipation of wave energy due to currents, wave breaking, bottom friction, and nonlinear wave interactions. SWAN includes the refraction of waves due to changes in bathymetry and current interaction. Diffraction is not accounted for with the model (Ris et al. 1996).

Wind fields from the PBL model provide the source of energy for SWAN runs; the water elevations and currents from the ADCIRC runs are also included to simulate the storm conditions. SWAN was run in a series of static simulations at 0.5 hour intervals over a period of 24 hours surrounding the point of maximum surge in Cox Bay. SWAN results include the significant wave height (energy-based), mean wave direction, mean wavelength, and mean wave period.

5 Sediment Resuspension

The sediments of Cox Bay are characterized as primarily fine silts with some clays. smaller fraction consists of fine to medium sands. To

Environmental Coastal Regions 421

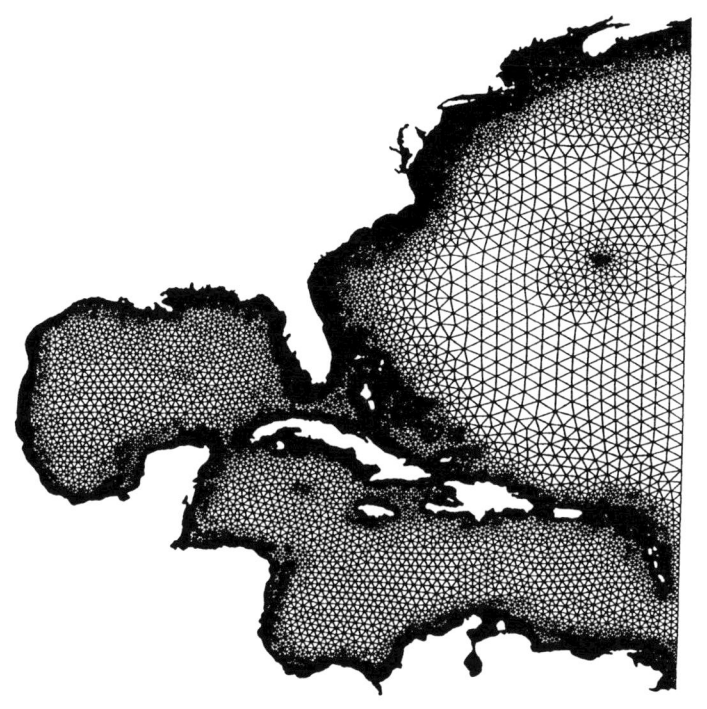

Fig. 2. Finite Element Grid of Gulf of Mexico and Atlantic Ocean

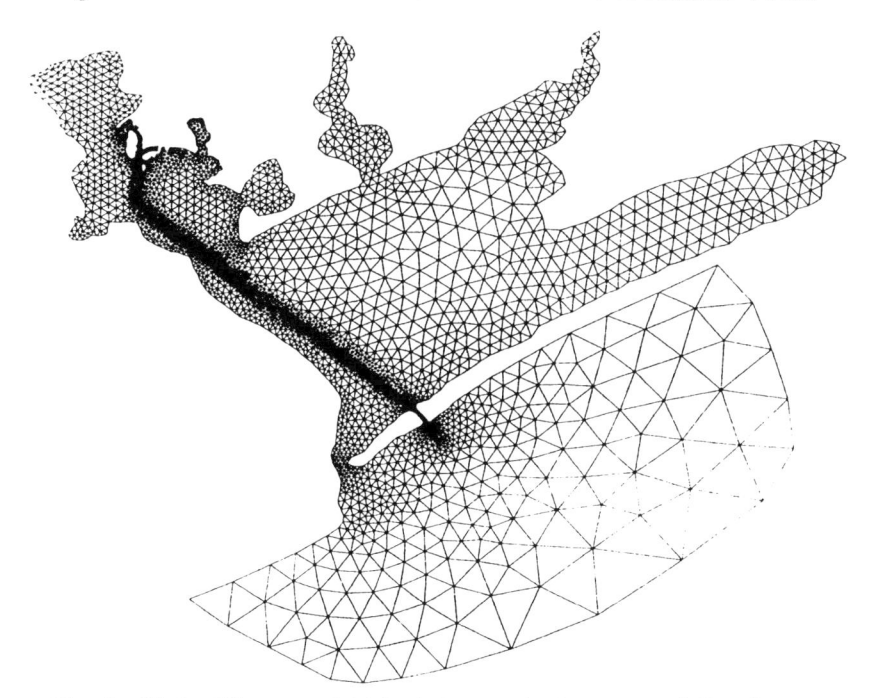

Fig. 3. Finite Element Grid for Matagorda, Lavaca and Cox Bays

estimate sediment resuspension, the cohesive relationships of Krone (1962) were used.

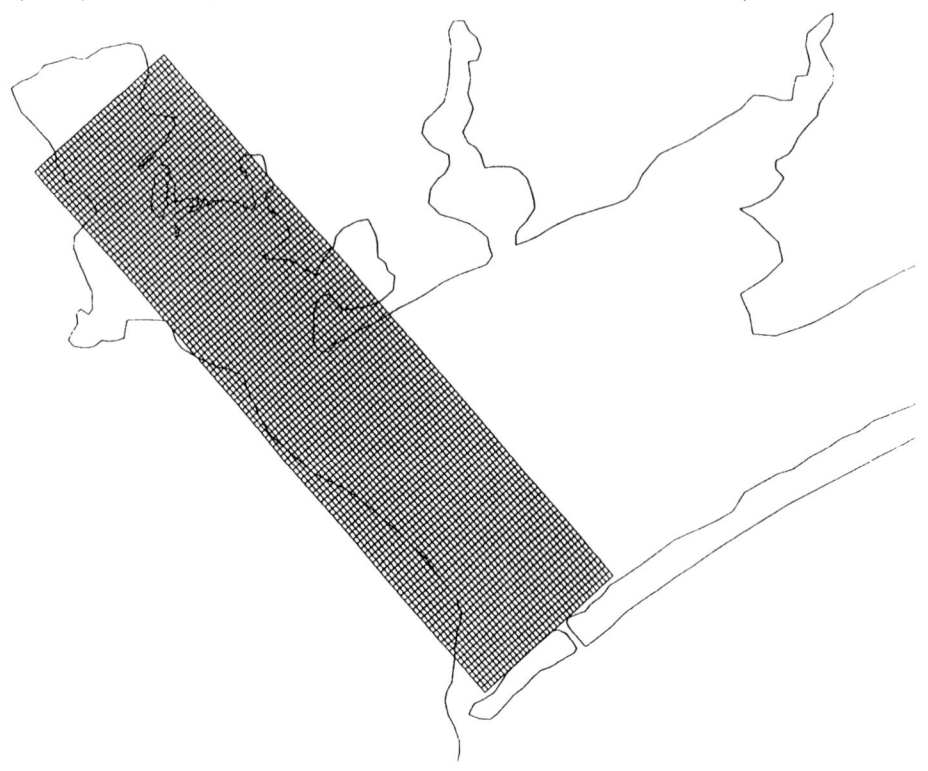

Fig. 4. Finite Difference Grid Used for SWAN Wave Model

5.1 Sediment Deposition

Krone attributed sediment transport processes to sediment deposition, bed failure, erosion, and suspension. Krone found that the deposition of sediments depended on the suspended sediment concentration, flocculation rates, and the probability that the sediment would stick to the bed. This probability of deposition is given as:

$$P_d = 1 - \frac{\tau}{\tau_{cd}} \qquad (1)$$

where τ is the bottom shear stress, and τ_{cd} is the critical shear for deposition. The bottom shear is related to the shear velocity (Krone 1962):

$$\tau = \rho u_*^2; \quad u_* = \sqrt{g}\,\frac{V}{C_z} \tag{2}$$

where V is the average velocity, and C_z is the Chezy coefficient.

Deposition of suspended sediments occurs below a critical shear stress. The deposition rate becomes:

$$\frac{dc}{dt} = -\frac{P_d w_s c}{h} \tag{3}$$

where w_s is the particle settling velocity, c is the suspended concentration, and h is the depth through which the particles settle.

5.2 Sediment Erosion

As the shear increases past a critical value, failure of the bed results in an almost immediate suspension of sediment (Krone 1962). Erosion occurs after the shear is greater than the critical shear for erosion:

$$\frac{dc}{dt} = \frac{P_e e_r}{h} \tag{4}$$

where e_r is the erosion rate constant, and P_e is the probability the sediments will be eroded:

$$P_e = \frac{\tau}{\tau_e} - 1 \tag{5}$$

where τ_e is the critical shear stress for erosion, or bed failure. In the absence of shear strength data for Cox Bay sediments, values for τ_{cd}, τ_e, and e_r were obtained from Krone's sediment transport study (1962).

5.3 Wave Effects

The concepts of Bijker (1967) and Swart (1976) are used to incorporate surface waves and currents in the sediment suspension process. The effect of waves is included as an increase in the depth-averaged velocity, V_{wc}, based on the velocity in the presence of currents only, V.

$$V_{wc} = V\left[1 + \frac{1}{2}\left(\frac{\xi u_o}{V}\right)^2\right]^{\frac{1}{2}} \qquad (6)$$

where ξ is a function of the hydraulic bed roughness and orbital excursion at the bed, and u_o is the orbital velocity at the bed based on linear wave theory.

5.4 Erosion Results

A rectangular grid in Cox Bay with dimensions 3.5 km by 3.4 km was selected to model the sediment suspension that results from the storms; the location of the sediment grid with respect to Cox Bay is shown in Fig. 5. The discretization of the sediment grid is consistent with the SWAN computational grid: 250 m in the x-direction, 340 m in the y-direction. Estimates for the total erosion resulting from a 24 hour period (surrounding the point of maximum surge) are given for a specific fixed location A (Fig. 5) in Table 2. Mean values for the water depth, current, and significant wave height (averaged over the 24 hr period) at point A are also provided in the Table.

Table 2. Total Erosion at Point A in Cox Bay

Storm Number	Erosion (cm)	Depth (m)	V (m/s)	Hs (m)	Storm Category
5	3.2	2.47	0.09	0.508	2
218	3.4	2.78	0.14	0.670	3
295	2.4	2.35	0.10	0.496	1
324	0	2.34	0.05	0.301	1
405	4.1	1.84	0.16	0.339	1
445	0.6	2.62	0.11	0.481	4
602	10.2	3.85	0.21	1.318	5
690	2.5	3.04	0.13	0.707	3
703	0	2.19	0.04	0.246	1
704	3.0	2.96	0.16	0.739	TS
783	0.7	3.02	0.10	0.527	4

Since the depth, current, and wave height values are averaged over the 24 hour duration, it is not readily discernible which may contribute more to the total erosion. It appears that the wave height lends a marked contribution to the erosion of sediments when the water depth is somewhat shallow. As the water depth increases, both the wave height and current are important factors of sediment erosion.

Deposition of suspended sediments is included in the model; however, the transport of sediments to adjoining cells is not. Also, the erosion depths are not coupled with the hydrodynamic or wave models.

The potential and extent of total erosion depends on the severity of the storm. Estimating the likelihood that the clean layer of sediments will be eroded, potentially exposing and/or eroding the contaminated sediments will help to delineate areas that may require special consideration or protection from certain storm events. Linking the total erosion with the category of a storm will assist in identifying the degree of risk of exposure that may occur.

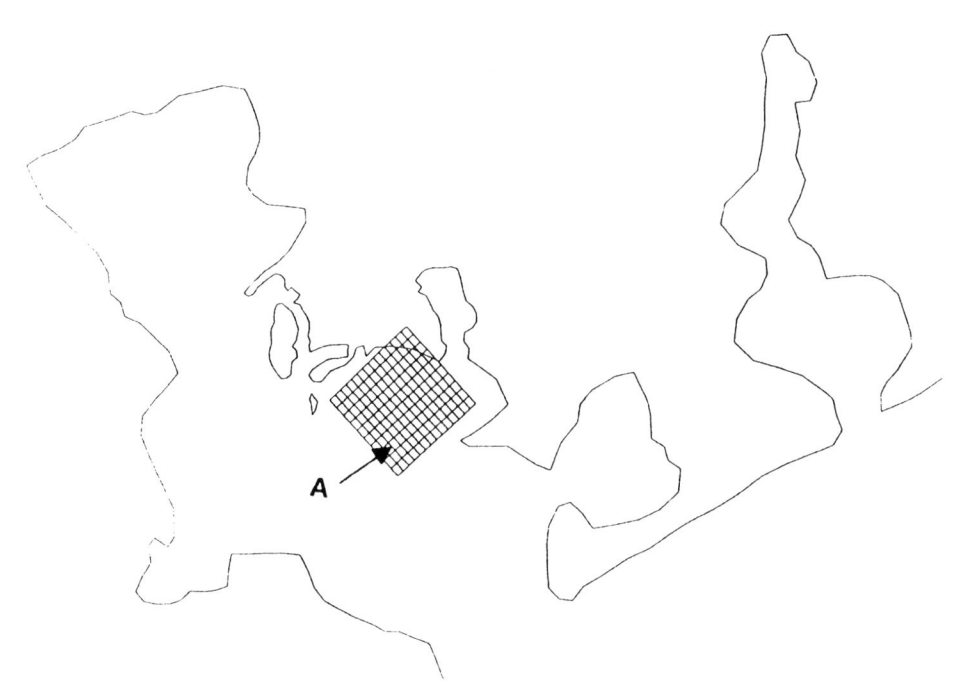

Fig. 5. Location of Point "A" in Cox Bay

References

[1] Bijker, E.W., Some Considerations about Scales for Coastal Models with Movable Bed, Publ. No. 50, Delft Hydraulics Laboratory, Delft, The Netherlands, 1967.

[2] Cardone, V.J., Greenwood, C.V., & Greenwood, J.A., Unified Program for the Specification of Hurricane Boundary Layer Winds over Surfaces of Specified Roughness, Contract Rep. CERC-92-1, U.S. Army Eng. Waterways Experiment Station, Vicksburg, Miss, 1992.

[3] Krone, R.B., Flume Studies of the Transport of Sediment in Estuarial Shoaling Processes, Final Report, Hydraulic Engineering Laboratory and Sanitary Engineering Research Laboratory, Univ. of Calif. Berkeley, 1962.

[4] Ris, R.C., Booij, N., Holthuijsen, L.H., & Padilla-Hernandez, R., SWAN Cycle 2 User Manual, Delft University of Tech., the Netherlands, 1996.

[5] Swart, D.H., Predictive equations regarding coastal transports, Proc. of the 15th Coast. Engrg. Conf., ASCE, New York, pp. 1113-1132, 1976.

[6] Westerink, J.J., Blain, C.A., Luettich, R.A., & Scheffner, N.W., ADCIRC: An Advanced Three-Dimensional Circulation Model for Shelves, Coasts, and Estuaries, Tech. Rep. DRP-92-6, U.S. Army Eng. Waterways Experiment Station, Vicksburg, Miss., 1994.

Modelling of the sediment transport at the Belgian coast

Dries Van den Eynde
Management Unit of the North Sea Mathematical Models, Gulledelle 100, B-1200 Brussels, Belgium
EMail: D.VandenEynde@mumm.ac.be

Abstract

A vertically integrated and a three dimensional sediment transport model are used to simulate the dispersion of dredged material at the Belgian coast. Both models are Lagrangian models, based on the Second Moment Method. Different sediment types can be taken into account. The bottom stress is calculated under the influence of prevailing currents and waves. The results of radioactive tracers, executed by HAECON NV, are used for the validation of the models. For the time being, the most satisfying results are obtained with the two-dimensional sediment transport model. Therefore, only this model and its results will be presented. By a careful interpretation of the results of the experimental and model results, a better insight can be gained on the transport of the dredged material at the Belgian coast.

1 Introduction

Between 1980 and 1989 each year around 33 million m³ dry material was dredged to maintain and to deepen the fair channels and the harbours in the Belgian coastal waters (BMM & AWZ[2]). Most of this material is dumped back into sea (Malherbe[9]). The selection of dumping sites with a high efficiency is essential. Firstly the dumped material should return as little as possible to the place where it was dredged. Further the chemical and biological effects, caused by these dumping activities, should stay as localised as possible. Since the efficiency of the current dumping sites varies between 25 % and 0 % (BMM & AWZ[2]) an investigation to select dumping sites with a higher efficiency was necessary.

Different numerical models are developed to study the transport and dispersion of the dredging material at the Belgian coast. Since the dispersion of cohesive material is a complex phenomenon, of which not all essential relations are yet well understood, the models can provide qualitative results only. The validation of these models therefore is of great importance. The models can be used in many different simulations and in sensitivity studies to evaluate the importance of different factors on the sediment dispersion. By careful interpretation of the model and experimental results, a better insight can be gained on the transport of the sediments in the area of interest.

In the present paper, some results of the sediment transport models are presented. In a first section, the numerical models are shortly introduced. In a second section, the validation of the models will be presented. Some conclusions are formulated in the last section.

2 Description of the numerical models

2.1 Introduction

Since the water at the Belgian coast is well-mixed during the entire year a two-dimensional vertical integrated model was developed in a first stage. Later a full three-dimensional (layer-)model was developed to account for the vertical structure of the water velocities and the suspended matter. The results of the latter model obtained so far are less satisfying then that provided by the former. Therefore, only the two-dimensional model and its results will be presented and discussed further.

In a first section, a short description is given of the hydrodynamic model. The wave model is discussed in section 2.3, while the sediment transport model is presented in section 2.4. A more complete description of the models can be found in Van den Eynde[14].

2.2 The hydrodynamic model

The two-dimensional hydrodynamic model MU-BCZ calculates the depth-integrated current velocity and the mean water level over the model grid under the influence of the tides and the meteorological effects. The bottom stress is computed using a quadratic friction law. The equations are solved using a fully explicit finite difference method on a staggered Arakawa-C grid.

The model is implemented on a grid, covering the Belgian coastal area and the Flemish Banks, with a resolution of 25" x 40" (approximately 750 x 750 m²). The model bathymetry is presented in Figure 1. At the open sea boundaries, the model is coupled with a larger hydrodynamic model of the North Sea and the Channel. At the outflow of the Scheldt river, the model is coupled with a one-dimensional model for the Scheldt estuary.

2.3 The wave model

For the calculation of the wave environment in the Belgian coastal area, the MU-WAVE model (Van den Eynde[12]) is used. The core of the model is formed by the HYPAS wave model (Günther & Rosenthal[4]), which is a second generation wave model, which combines the independent calculation of swell energy for each frequency and direction through a ray technique, with a parametrical wind sea model, using the JONSWAP parameters and the mean wind sea direction as prognostic variables. For the current application, the model is implemented on two nested grids. In the Southern North Sea a grid resolution of 5 x 5 km² is used.

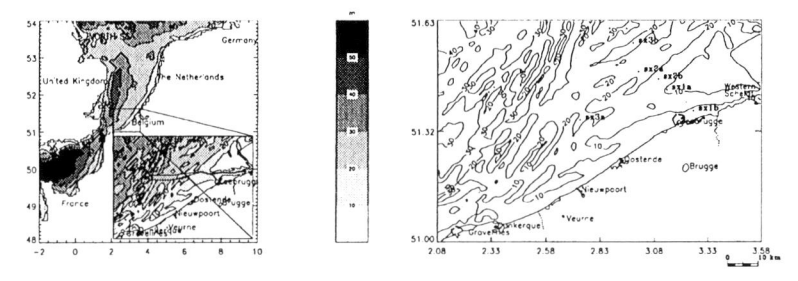

Figure 1: (a) Bathymetry of the Belgian coastal zone and the Flemish Banks; (b) Initial position of radioactive tracers (experiments executed by HAECON NV).

2.4 Sediment transport model

The sediment transport model MU-STM is based on the Second Moment Method (de Kok[3]), which introduces less numerical diffusion than classical Eulerian methods. Using this Lagrangian method, the dumped material directly can be followed through the model grid and the return to the fair channels and the sea harbours can be calculated easily.

The model can account for different sediment classes and calculates for each of them the advection and diffusion of the material in suspension under the influence of the tidal currents and of the currents generated by the waves (Stokes drift). The bottom stress is calculated under the combined effect of currents and waves (Bijker[1]). The erosion and the deposition are modelled according to Partheniades[10] and Krone[8].

3 Application to Belgian coastal zone

3.1 Introduction

Six radioactive tracer experiments, executed in the Belgian coastal zone between 1992 and 1993 by HAECON NV[5,6,7], are used for the validation of the model (Van den Eynde[13]). In a first section, the tracer experiments will be shortly described. Further, the model simulations will be discussed. The most important results are formulated.

3.2 Tracer experiments

Six radioactive tracer experiments were executed by HAECON NV, to investigate the recirculation of the dumped material to the Belgian coast. The position of the six sites are given in Figure 1. Note that five sites are within a range of 21 km of the coastline, while the last site (sx3b) is about 30 km from the coastline.

After the dumping of the radioactive mud, bottom samples were taken at different stations and over a period of several months to trace the dispersion of the dumped material. The main findings of these experiments can be summarised as follows.

The material dumped at the five stations within 21 km of the coast, clearly recirculates very rapidly to the coast. In some cases, material is found back in the harbours after two or three days already. The material spreads out over the entire Belgian coast and even migrates into the Western Scheldt. During storm periods, the material can disappear for a shorter or longer period, but after time, material is found back at the coast. In strong contrast with this, the material dumped at site sx3b does not recirculate to the coast. During the entire sampling period, no tracer is detected on the Belgian continental shelf.

Although it is clear that these tracer experiments are a very valuable source of data for the validation of the model, due to several reasons, the results can only give some first indications. First of all, during the sampling of the bottom, no attention was made to the tidal cycle. Measurements and model results clearly indicate that the mud comes into suspension during periods of high currents and settles down during slack water periods. Material thus can be present in the water column at a certain station while no radioactivity will be found in the bottom samples.

Further most of the bottom samples were taken near the coast and consequently not much information on the dispersion of the material in the open sea is available. Furthermore, many of the samples were taken in the harbours themselves, where local effects, e.g. of harbour dams, can play an important role, which can not be represented in the numerical models. At last, from the calculation of activity balances, which are probably questionable, it appears that in some cases only a very small amount of the material is found back during the detection campaigns. This implies that the complete dispersion of the material can not be derived from these results. The results from the tracer experiments therefore must be interpreted with the necessary precautions.

3.3 Model simulations

For each of the tracer experiments the dumping of 5500 ton of mud is simulated. The critical stress for erosion and for deposition are both set to 0.5 Pa, the erosion constant is taken as 0.12 g/m²s and the fall velocity has a value of 0.01 m/s. These parameters were selected after a literature study and some sensitivity tests. A period of 14 days after the dumping is simulated. For four tracer experiments a longer simulation was executed.

3.4 Results

3.4.1 Tracer experiments sx1a and sx1b (22/4/1992)
During the first fourteen days after the dumping of the radioactive material, the meteorological conditions were relatively quiet. As well in the results of the tracer experiments as in the model results the sediments returned to the Belgian coast quite rapidly, although the recirculation and the spreading over the entire Belgian coast are in the model results not as fast as observed in the experiments. In the model results, the material is dispersed over a large area. The lack of bottom samples taken in the open sea prevents to confirm this by the experiments.

Note that, as shown in Figure 2, the moment in the tidal cycle has a large effect on the material that can be found at the bottom. Once again, the experimental results must be interpreted with the necessary precautions.

3.4.2 Tracer experiments sx2a and sx2b (18/1/1993)
The first week after the dumping of the radioactive material at sites sx2a and sx2b, the weather conditions were relatively rough. Wind speeds up to almost 20 m/s and wave heights up to 2.5 m were observed

432 Environmental Coastal Regions

in front of Zeebrugge on 24-25/1/1993. Both wind and waves were directed to the north to northeast.

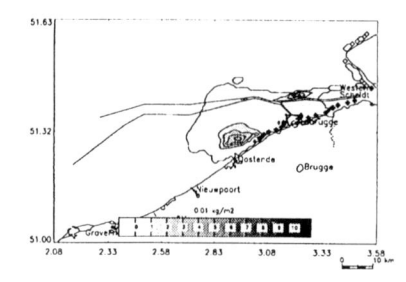

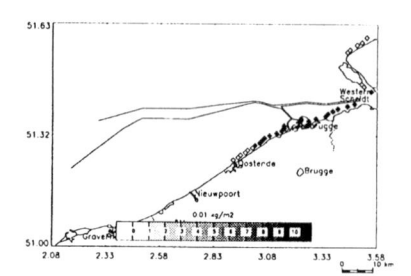

Figure 2: Results of MU-STM for the dispersion of mud for tracer experiment sx1b: mud at the bottom at (a) 6/5/1992 12h24 and at (b) 6/5/1992 14h24 (14 days after the dumping).

In the few bottom samples, taken during this first week, some radioactivity could be detected in the sea harbours. After the storm, material has been detected at different coastal stations only after a much later date (24 of more days after the beginning of the experiment). During different campaigns material has been detected at the open sea stations.

In the model results the dumped material mainly is transported in the direction of the wind and wave fields. A considerable amount of the mud (especially for the material dumped at site sx2b) leaves the model area towards the Netherlands and the Eastern Scheldt. A certain fraction enters the Western Scheldt. As showed in Van den Eynde[14] the residual currents and transports indeed are strongly influenced by the meteorological conditions. Therefore, it is reasonable to assume that the mud will be transported mainly in the direction of the strong wind fields and that only a minor part of the mud will be transported, due to the dispersion, towards the coast. This also agrees with the fact that material is only detected at a much later date and that material is found in open sea stations. The material that shows up again at the Belgian coast a later date could be migrated into the Western Scheldt and washed out later, a process that can not be represented in the model.

Remark at last that, although the site sx2a is further located from the coastline then the site sx2b, more material is detected in the bottom samples that originated from the former station. This is also the case in the model results.

As an example, in Figure 3 the results of the tracer experiments and the model results are presented 10 days after the dumping of the radioactive material at site sx2a. In the tracer experiments no material is

Environmental Coastal Regions 433

detected in the bottom samples, except in Heist en Blankenberge, in station Breskens in the Western Scheldt and in the station in open sea Wandelaar. Due to the rough weather conditions also in the model results no material can be found at the bottom. The core of the material in suspension, however, is located near station Wandelaar.

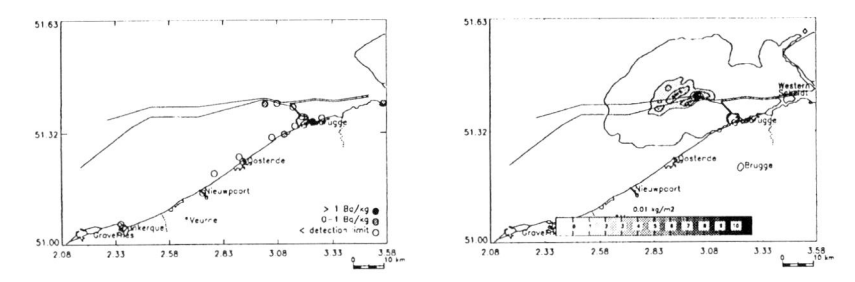

Figure 3: Results for the simulations for tracer experiment sx2a: (a) measured radioactivity in the bottom samples on 28/1/1993 (10 days after the dumping); (b) distribution of mud (in suspension and at the bottom) at 28/1/1993 14h00, calculated with the MU-STM model.

3.4.3 Tracer experiments sx3a en sx3b (29/9/1993)

The first fourteen days after the dumping at the sites sx3a and sx3b, calm weather conditions occurred. During these tracer experiments, the radioactive material, which was dumped at both sites, behave very differently. While the material, dumped at site sx3a, recirculates to the coast rapidly, the material, dumped at site sx3b, is not detected anymore at the Belgian coast. Only in the Western Scheldt, some low values of radioactivity were sometimes detected. These results are well reproduced in the model. This is illustrated in Figure 4 where for both simulations, the mud in suspension and at the bottom is presented at the end of the simulation. While for the material dumped at site sx3a 5258 ton is still located on the model grid, almost all material dumped at site sx3b disappeared. Only 8 ton is still on the model grid located. Remark that a part of this material is located near the Western Scheldt, which agrees with the measurements.

3.4.4 Long term simulations

For the tracer experiments sx1a and sx1b and for experiments sx3a en sx3b, the simulations were continued over a longer period (69 and 80 days). During the simulations, the mud patch is displaced under the influence of the prevailing meteorological conditions. When the meteoro-

434 Environmental Coastal Regions

logical conditions remain quiet for a sufficiently long period, however, the material returns back to the coast and a tendency for the formation of a turbidity maximum located between Ostend and Zeebrugge can be noted (see Figure 5). The presence of such a turbidity maximum in the area is well-known and is e.g. described by Malherbe[9]. Also in satellite images, such a turbidity maximum is often observed. An example of an AVHRR satellite images with a clear turbidity maximum in the area is given in Figure 5 (from Ruddick et al.[11]).

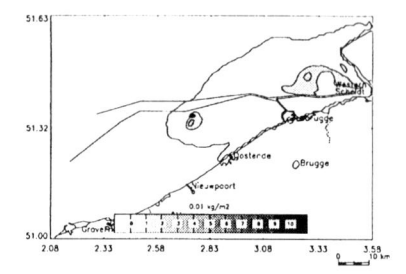

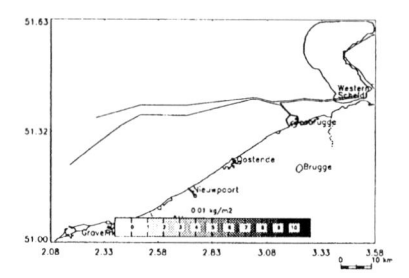

Figure 4: Distribution of mud (in suspension and at the bottom) calculated with the MU-STM model (a) at 13/10/1993 10h00 for the material dumped at site sx3a and (b) at 13/10/1993 13h36 for the material dumped at site sx3b.

4 Conclusions and further work

In the paper, a vertically integrated sediment transport model, developed to simulate the dispersion of dredged material at the Belgian coast, is presented. First the hydrodynamic and wave model are discussed, whereafter some specifications of the sediment transport model are given.

The model is used to calculate the dispersion of radioactive traced mud. The main features of the behaviour of the mud are well reproduced by the model. In five experiments, the material can recirculate and spread out over the entire Belgian coast quite rapidly, although the recirculation in the model is not as fast as that observed. The exact reason for this very rapid return to the coast is not yet well identified. The model further indicates that the material is spread out over a wider area and that it will be displaced under the influence of the prevailing meteorological conditions. It will be eventually transported to the northeast. This seems to be confirmed by some of the experimental data. The model results may therefore provide added value for the interpretation of the results of the tracer experiments.

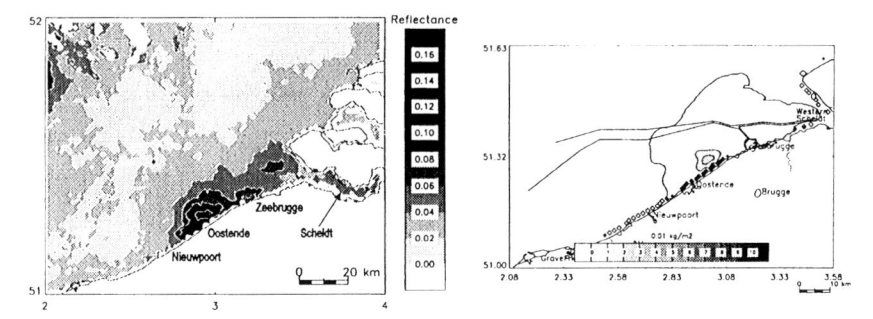

Figure 5: (a) Sub-surface irradiance reflectance for AVHRR channel 1 from NOAA-14 taken on 11/7/1997 at 13h24 (from Ruddick et al.[11]); (b) distribution of mud (in suspension and at the bottom) at 28/1/1993 14h00 calculated with the MU-STM model.

In the last experiment, no recirculation of material to the Belgian coast was observed, either in the measurements or in the model results. Also these results thus agree very well with observations

It has to be mentioned that so far, the results of the two-dimensional model were superior to those of a three-dimensional model. Clearly some further work has to be executed for the calibration and the fine tuning of the sediment parameters and the parametrisation of the vertical diffusion coefficient in the three-dimensional model, to improve its results.

Acknowledgements

José Ozer is thanked for useful discussion of the model results and for his constructive criticism on this paper. Kevin Ruddick is thanked for providing the AVHRR satellite image. The CAMME team is acknowledged of computing support. This study was supported by the Flemish government under the "STM-II" and "VESTRAM" contracts.

References

[1] Bijker, E.W., The increase of bed shear in a current due to wave motion, in: *Proc. 10th Conf. Coastal Eng.*, Tokyo, 746-765, 1966.

[2] BMM & AWZ, Ecologische impact van baggerspecielossingen voor de Belgische Kust. *Eindrapport*, Brussel, 101 pp., 1993.

[3] de Kok, J.M., Numerical Modelling of transport of processes in coastal waters, *Ph.D. Thesis*, Univ. Utrecht, 158 pp., 1994.

[4] Günther, H. & W. Rosenthal, The Hybrid Parametrical (HYPA) Wave Model. In: *Ocean Wave Modelling*, the SWAMP Group, Plenum Press, New York, 211-214, 1985.

[5] HAECON NV, Recirculatie-proef Sx1 aan stortplaatsen ZB-Oost en S2, Synthese der meetresultaten, *Rapport MSB0721/93/00232*, 43 pp., 1993.

[6] HAECON NV, Recirculatieproef Sx2 aan de noordflank Akkaertbank en ebschaar S1 & S2, Synthese der meetresultaten, *Rapport MSB0721/93/00287*, 42 pp., 1993.

[7] HAECON NV, Recirculatieproef Sx3 aan de noordflank Thorntonbank en de Negenvaam, Synthese der meetresultaten, *Rapport MSB0721/95/00418*, 43 pp., 1995.

[8] Krone, R.B., Flume studies of the transport of sediment in estuarial shoaling processes, *Final Report*, Hydr. Eng. Lab. and Sanitary Eng. Research Lab., Univ. California, Berkeley, 1962.

[9] Malherbe, B., Case study of dumping in open seas, in: *Proc. Int. Seminar on the Environmental Aspects of Dredging Activities*, Nantes, France, Ch. Alzieu & B. Gallenne (eds.), 227-261, 1989.

[10] Partheniades, E., A study of erosion and deposition of cohesive soils in salt water, *Ph.D. Thesis*, Univ. California, Berkeley, 1962.

[11] Ruddick, K., F. Ovidio, D. Van den Eynde & A. Vasilkov, The redistribution and dynamics of suspended particulate matter in Belgian coastal waters derived from AVHRR imagery, accepted for publication in: *Proc. 9th Conf. on Satellite Meteorology. and Oceanography*, Paris, 1998.

[12] Van den Eynde, D., MU-WAVE: an operational wave forecasting system for the Belgian coast, in: *Proc. 3rd Int. Workshop on Wave Hindcasting and Forecasting*, May 19-22, 1992, Montréal, Canada, 313-324, 1992.

[13] Van den Eynde, D., Validatie van de sedimenttransportmodellen MU-STM, MU-STMF en MU-STM3, *Rapport STM2/1/DVDE/ 199708/NL/TR/5*, Beheerseenheid van het Mathematisch Model van de Noordzee, Brussel, 117 pp., 1996.

[14] Van den Eynde, D., Project 'Sedimenttransportmodel - Fase II', Eindrapport, *Rapport STM2/1/DVDE/199709/NL/ER/1*, Beheerseenheid Mathematisch Model Noordzee, Brussel, 14 pp., 1997.

INDEX OF AUTHORS

Agirre, E. 155
Albizu, M. 155
Amin, M. 109
Appendini, C.M. 99
Arcilla, A.S. 331
Ardeshir, A. 67

Bach, H.K. 33, 45, 251
Baldasano, J.M. 135
Barbosa, P.C. 143
Benedetti, C. 197
Benitez, C. 231
Bergamasco, A. 341
Bergamasco, A.A. 341
Berge, E. 175
Blanc, S. 231
Borrego, C. 265

Caamaño, J. 155
Cinquepalmi, F. 197
Collin, M. 385
Covelli, S. 11

Davenport, I.J. 109

Ebecken, N.F.F. 165
Elías, A. 155
Edge, B.L. 417
Eloheimo, K. 123

Faganeli, J. 11
Fathi, M. 67
Fischer, D.W. 99
Flather, R.A. 109
Foster, T. 251
Franic, Z. 241, 351

Gurney, C. 109

Haiduk, A. 21
Hannonen, T. 123
Härmä, P. 123
Heemink, A.W. 311
Hendee, J.C. 57
Hibino, T. 373
Horvat, M. 11, 289

Ibarra, G. 155

Jensen, A. 45
Jensen, K. 33, 45
Juárez, R. 231

Kirkkala, T. 123
Koch, M. 395, 405
Kock Rasmussen, E. 251
Koponen, S. 123
Kuspilić, N. 301
Kutser, T. 123

Lamont, E. 3
Larm, K.A. 417
Lascalea, G. 231
Leal, L. 209
Lee, J.L. 277
Lin, H.X. 311
Lizárrago-Arciniega, R. 99
Logar, M. 11
Lopes, M. 265

MacDonald, N.J. 321
Madariaga, I. 155
Mandić, V. 11
Marovic, G. 241, 351
Mason, D.C. 109
Maurício, A.M. 77, 87
Melloul, A. 385

Mestres, M. 331
Milou, M. 231
Mosto, P. 231

Nicholson, J. 321
Nunes, R.A. 143

O'Connor, B.A. 321

Pacheco, A.M.G. 77, 87
Paciornik, S. 143
Pan, S. 321
Pastakia, C.M.R. 33
Payne, T.E. 219
Pecly, J. 209
Planinc, R. 11
Pullianen, J. 123
Pyhälahti, T. 123

Rajar, R. 11, 289
Ramalingam, P. 185
Robinson, G.J. 109
Rodríguez, A. 331
Roldão, J. 209

Schiuma, D. 197
Sencar, J. 241
Sierra, J.P. 331
Širca, A. 11, 289
Smith, J.A. 109

Soriano, C. 135
Stalter, R. 3
Stijnen, J.W. 311
Sun, H. 405
Szymczak, R. 219

Tabares, R.H. 143
Tabeta, S. 363
Tagliapietra, D. 197
Taher shamsi, A. 67
Tandon, P.N. 185
Tsuruya, H. 373
Tsyro, S.G. 175

Umgiesser, G. 341
Uria, J. 155

Valentini, E. 209
van den
   Boogaard, H.F.P. 311
Van den Eynde, D. 427
Vuković, Ž. 301

Waite, T.D. 219

Žagar, D. 11, 289
Zampato, L. 341
Zhang, G. 395
Zitelli, A. 197, 341

Computational Mechanics Publications

Measuring and Modelling Investigation of Environmental Processes

Edited by: R. SAN JOSE, Technical University of Madrid, Spain.

Computerisation has allowed a greater analysis and simulation of environmental processes, an area of increasing interest for those with responsibility for predicting environmental effects, and with a need to better understand their fundamental processes. This book relates the state of the art practice among researchers around the world which it is hoped will contribute to this need and assist other researchers in their work.

ISBN: 1853125660   1998
apx 300pp   apx £95.00/$155.00

Environmental Science

These proceedings of the ASE 98 Conference is aimed at mathematicians, physicists, chemists, environmental scientists and engineers who are involved in the solution of environmental problems. The proceedings are planned to cover measurement and data analysis, pollutants and environmental chemistry, mathematical modelling, environmental management and control, and removal and remediation technology.

ISBN: 1853126039   1998
apx 500pp   apx £155.00/$248.00

Measurements and Modelling in Environmental Pollution

Edited by: C.A. BREBBIA, Wessex Institute of Technology, UK and R. SAN JOSE, Technical University of Madrid, Spain.

This book contains the proceedings of the First International Conference on Measurements and Modelling in Environmental Pollution. It focuses on all aspects of measuring and modelling of different environmental problems relating to pollution.

ISBN: 1853124613   1997
560pp   £149.00/$238.00

Environmental Engineering and Management

The importance of sustainable development requires familiarisation with current environmental issues, so that sound engineering action can be taken to remedy deteriorating environmental situations. These proceedings aim to highlight the trend from pollution prevention to pollution control and cover such issues as broad strategy, environmental management tools, modelling, prevention methods, education and training.

ISBN: 1853126020   1998
apx 500pp   apx £155.00/$248.00

Computational Mechanics Publications

HydroTrack

Developed by: R. Garcia-Martinez and J. J. Rodriguez, Universidad Central de Venezuela.

Computational Mechanics Publications introduces HydroTrack, a sophisticated hydrodynamic and pollutant transport software developed to help professionals and researchers solve water pollution problems. The system consists of two closely linked programs: the hydrodynamic kernel and the pollutant transport model. One of the key advantages of the HydroTrack software lies in its user-friendly interface which simplifies the data input and allows the simulation of complex pollutant transport scenarios in just a few minutes. Full graphics and comprehensive report writing facilities are also available. HydroTrack runs under the Windows operating system on IBM-PC and compatibles.
ISBN: 1853124869   1997
£990.00/$1490.00

A demonstration disk of the HydroTrack software can be downloaded from the CMP web site at: http://www.cmp.co.uk or obtained on diskette on request from CMP

HydroTrack Starter Pack

This includes all the functionality of the full version but has limitations on the size of the problems that can be solved (up to 250 computational cells and 500 particles). It is a professional system as well as an useful educational tool to obtain quick results of practical hydrodynamic and pollutant transport problems. Users of HydroTrack Starter Pack may later upgrade to the HydroTrack full version taking advantage of a special offer.
ISBN: 1853126314   1998
£250.00/$375.00

Geographical Information Systems in the next Millennium

Geographical Information Systems are rapidly penetrating a wide range of applications and these proceedings are planned to cover their use in environmental conservation, economic planning, resource utilisation, cartography, urban planning, risk assessment, pollution control, transport management systems and many others.
ISBN: 1853125954   1998
apx 300pp   apx £95.00/$155.00

Computational Mechanics Publications

OilTrack

Developed by: R. GARCIA-MARTINEZ and J.J. RODRIGUEZ, Universidad Central de Venezuela, Venezuela.

OilTrack is a simple to use and powerful oil spill trajectory simulation software. The system includes a hydrodynamic model to calculate wind or tidal generated water velocities and elevations in any complex geometry site. The oil trajectory model uses state-of-the-art three dimensional particle tracking techniques to determine oil spreading and dispersion depending on oil properties, ocean diffusion and Fay's spreading regimes. OilTrack calculates oil trajectories in the 3D water space allowing simulation of instantaneous or continuous spills. One of the key advantages of the OilTrack model lies in its user interface that simplifies the data which is thoroughly validated to minimise common input errors. Full graphics such as water velocity vector plots and oil trajectories may be generated and also included in most popular word processors to facilitate report preparation. OilTrack runs under the MS-Windows operating system on IBM-PC and compatibilities.

ISBN: 1853125598   1998
CD Rom   apx £990.00/$1490.00

*Look for more information about CMP on the internet:
http://www.cmp.co.uk*

OilTrack Starter Pack

This includes all the functionality of the full version but has limitations on the size of the problems that can be solved (up to 300 computational cells and 500 particles). It is a professional system as well as an useful educational tool to obtain quick results of practical hydrodynamic and oil spill trajectory calculations.

Users of OilTrack Starter Pack may later upgrade to the OilTrack full version taking advantage of a special offer.

ISBN: 1853126306   1998
CD Rom   apx £250.00/$375.00

Oil and Hydrocarbon Spills, Analysis and Control

These proceedings of Oil Spill 98 deal with advanced theoretical and practical aspects of oil spills in land and water environments. Scheduled topics include modelling of trajectory and fate of spills, contingency planning, operational procedures for storage, handling and transportation, biological impact, treating agents, risk and financial assessment, case studies and others.

ISBN: 1853125261   1998
apx 450pp   apx £138.00/$220.00

All prices correct at time of going to press. All books are available from your bookseller or in case of difficulty direct from the Publisher.